Wisconsin State Parks, Forests, and Recreation Areas— A Ranger's Guide

Human History, Geology, Campgrounds, Day-Use Areas, Beaches, Wildlife, Nature Programs, Hiking and Biking Trails, Accessibility, and other Recreation Opportunities at Wisconsin's 80 State Parks and Forests

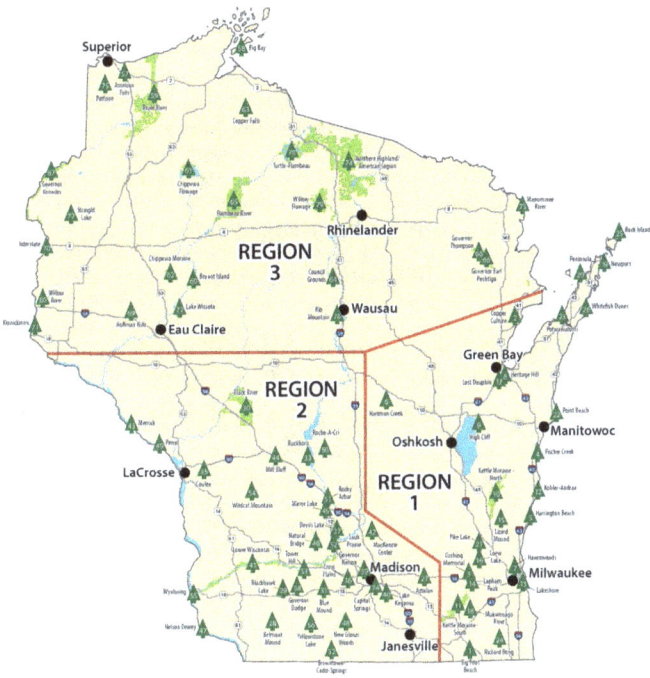

James Buchholz

Henschel HAUS
www.henschelHAUSbooks.com

Milwaukee, Wisconsin

All photographs provided by the author.
Map created by James Buchholz, Jr.

ISBN: 978159598-994-9
LCCN: 2024939833

Published by HenschelHAUS Publishing, Inc.
www.henschelHAUSbooks.com
Milwaukee, Wisconsin

Cover photo:
Mirror Lake State Park, Echo Rock hiking trail

Printed in the United States of America

This book is dedicated to the men and women who work to protect, preserve, maintain, interpret, and enhance Wisconsin's state parks, forests, and recreation areas for all to enjoy.

Superior

Big Bay 36

Amnicon Falls
Pattison

2

Brule River

Copper Falls

51

Turtle-Flambeau 78

Northern Highland/ 74
American Legion

45

Governor
Knowles

Straight
Lake

Chippewa
Flowage

65 Flambeau River

Willow
Flowage 79

Rhinelander

Governor
Thompson

Governor Earl 8
Peshtigo

Menominee 72
River

Interstate 8

63

Willow
River 80

Kinnickinnic 71

Chippewa Moraine

Brunet Island 66

Lake Wissota

53

REGION
3

Council
Grounds

61

Rib
Mountain 26

Wausau

46

45

Copper
Culture

41

Rock Island 45

Peninsula

Newport

Whitefish Dunes

Potawatomi

Green Bay

Lost Dauphin

Heritage Hill

Point Beach

Manitowoc

Fischer Creek

Perrot 49

LaCrosse

Coulee

Wildcat Mountain

14

Hoffman Hills 69

Eau Claire

10

Merrick

Black River 30

REGION
2

Roche-A-Cri

Buckhorn

Mill Bluff 44

Mirror Lake

Rocky
Arbor

Hartman Creek

High Cliff

45

Oshkosh

Kettle Moraine
North

Kohler-Andrae

Harrington Beach

Devils Lake

Natural
Bridge

Tower
Hill

Sauk
Prairie

Cross
Plains

MacKenzie
Center

Pike Lake

Cushing
Memorial

Lapham
Peak

Lizard
Mound

Loew
Lake

REGION
1

Kettle Moraine-
South

Lower Wisconsin

61

14

Wyalusing

18

Blackhawk
Lake

Governor
Dodge

Blue
Mound

Governor
Nelson

Capital
Springs

Madison

Lake
Kegonsa

Aztalan

12

Havenwoods

Milwaukee

Lakeshore

Nelson Dewey 17

61

Belmont
Mound

Yellowstone
Lake

New Glarus
Woods

Browntown-
Cadiz-Springs

Janesville

Mukwonago
River

Richard Bong

Big Foot
Beach

TABLE OF CONTENTS

Region Two—South Central and Southwestern Wisconsin: Mississippi, Kickapoo, Black and Lower Wisconsin rivers, Castle Rock Flowage, Central Sands Region, Blue Mounds, New Glarus, Baraboo Hills, Wisconsin Dells and Madison Area. (p. 155)

CHAPTERS

Region Three—Northern Forests, Parks and Scenic Water Flowage Areas: Northern Highlands/American Legion Forest and Lake Superior area, Wausau, Marinette, Chippewa Moraine, Apostle Islands, and Brule, Flambeau, Brule, Peshtigo, Willow, St. Croix, Menomonee, and the Upper Wisconsin rivers. (p 329)

CHAPTERS

INTRODUCTION

Wisconsin's scenic state parks, forests, and recreation areas are often described as the crown jewels of the Midwest. Each year, more than 20 million people travel from throughout the country to visit one or more of these treasured sites. What draws so many people to these state-owned properties? For many, it's the desire to experience the natural world and enjoy the great outdoors. State parks and forests have been the favorite destination for outdoor recreation for generations of people thanks to Wisconsin's 150-year-legacy of acquiring the most beautiful natural areas of the state for everyone to enjoy.

So what natural wonders and scenic vistas await your discovery at Wisconsin's state parks and forests? To some, it's viewing the orange glow of a sunset over the Mississippi River from the bluffs at Wyalusing State Park or catching a glimpse of the evening sun as it dips into the waters of Green Bay from Peninsula State Park. Equally spectacular is watching the morning sun emerge from the shimmering waters of Lake Michigan at Kohler-Andrae State Park or as it ascends above Lake Superior at Big Bay State Park. Many would argue the deep blue waters of Devil's Lake viewed from the park's towering 500-foot-high bluffs can't be surpassed, except perhaps at Pattison State Park, where Wisconsin's tallest waterfall thunders over a 164-foot-tall rocky precipice. These are just a few of the many scenic wonders to be experienced within Wisconsin's state parks and forests.

Devil's Lake

1

As you will discover, the outstanding scenery and unique landforms of each state park and forest were shaped by natural geologic events of the past including rock-crushing glaciers, tectonic plate upheavals, earthquakes, volcanic eruptions, and ancient ocean deposits. All have withstood thousands or even millions of years of wind and water erosion as well. Human activities, both prehistoric and modern, have also had profound impact on many of these sites. The combination of these natural and man-made influences has ultimately led to the preservation and protection of Wisconsin's most scenic areas.

HOW TO USE THIS BOOK

This guidebook will assist readers in discovering all the wonderful sights, sounds, amenities, and best-kept secrets to be found at all 80 of Wisconsin's state parks, forests, recreational areas, and scenic water flowages. Each property has been assigned to one of three distinct regions of Wisconsin as shown on the property locator map at the beginning of this book and in the individual regions.

- ◆ *Region One* includes parks and forests within eastern and southeastern Wisconsin, including the Door County Peninsula, the Kettle Moraine Forests, and the Greater Milwaukee area.

- ◆ *Region Two* highlights properties located in the south central and southwestern part of the state including the Central Sands, Lower Mississippi and Wisconsin rivers, and the Madison area.

- ◆ *Region Three* features Wisconsin's northern forests, parks, and scenic water flowage areas.

The book can be used as a guide while visiting a particular state park, forest, or recreation area or as a reference for those planning a trip to one of these sites. Each chapter begins with a brief snapshot of the property and its geologic and human history. This is followed by more detailed information of the site's recreational amenities including hiking and biking trails, campgrounds, swimming beaches, picnic areas, interpretive programs, wildlife viewing, boating and fishing opportunities, and winter activities. Accessible resources available for mobility-impaired visitors are also included. "Ranger Picks and Tips" are offered to assist visitors in selecting the best hiking routes, nature trails, overlooks and other attractions while visiting each park or forest property. "Ranger Trivia" narratives reveal little-known facts or anonymities specific to that property or a nearby community. Offsite attractions such as local parks, museums, interpretive centers, natural areas, or nearby state bike trails are highlighted at the end of each chapter.

PARK AND FOREST MAPS AND LOCAL INFORMATION SOURCES

State park, forest, and recreation maps can be obtained from brochure racks outside most visitor entrance stations. Property maps can also be viewed online at www.wiparks.net, which also has winter use, cross-country ski and hunting/trapping maps for each site. A listing of upcoming interpretive programs and special events or warnings such as closed areas or construction alerts are also posted online for each property.

Campsite reservations are required for most state park and forest campgrounds. Reservations for any vacant (unreserved) campsite can be made upon arrival at each property. Campsites can also be reserved up to eleven months in advance of the arrival date. Campers can reserve a campsite by phone at (888) 947-2757 or online at www.wisconsin.goingtocamp.com. Information regarding private or county camping areas, motels, resorts, or other lodging near any state park or forest can be obtained through the Wisconsin Department of Tourism's website at www.TravelWisconsin.com or by phone at (800) 432-8747.

A SHORT HISTORY OF WISCONSIN STATE PARKS AND FORESTS

Native American people were the first and longest-serving stewards of the land we know as Wisconsin or "Meskonsing" in the ancient Ojibwe dialect. These early people revered and held sacred many of the same beautiful landscapes, waterfalls, rivers, lakes, bluffs, and shorelines protected within Wisconsin state parks and forests today.

Their assistance in guiding early European explorers often led to a positive exchange of culture and friendly coexistence, especially during Wisconsin's fur trade period. Later exploitation of the state's natural resources by lead miners, lumber companies, and a wave of new immigrants in search of land to farm led to many broken treaty agreements and loss of land by indigenous people. Today, Wisconsin is still home to eleven active Native American tribes, each with its own distinct culture, heritage, and reservation lands.

Most of these treasured sites would have been lost to private development had it not been for a few far-sighted individuals. The effort to preserve these beautiful natural areas for public use and enjoyment was a true American idea. The story of how each property became a designated state park, forest, or recreation area is often a tale of ordinary people doing the extraordinary.

In the late 19th century, people throughout the nation began to petition their state and federal representatives to establish more parks and forests in rural areas, especially along scenic lakes and waterways. Most longed for a place where

3

average citizens could go to enjoy nature and find a peaceful haven from the stress of working in factories, offices, and the urban lifestyle brought on by the Industrial Age. Conservation leaders, outdoor writers, and influential American naturalists of this era, such as John Burroughs, encouraged and fought for the development of local, state and national parks. Wisconsin's own famed naturalist and founder of the Sierra Club, John Muir (1835-1914), wrote, "*Everybody needs beauty as well as bread; places to play in and pray in; where nature may heal and give strength to body and soul.*"

Wisconsin was the first state in the union to create its own state park. This occurred only a few years after Yellowstone National Park was established by an act of Congress and signed into law by President Ulysses S. Grant. Madison legislators passed a resolution in 1878 to develop a 760-square-mile state park in northern Wisconsin, primarily in Vilas County. This first-of-its-kind public recreation area was named simply, *The State Park*. Wisconsin's first attempt in setting aside public land to create a state park was doomed to failure however. Only about 10 percent or about 50,000 acres of the land within the park boundaries was actually owned by the state at that time. The effort to establish *The State Park* fell apart primarily due to the advent of the logging boom of the 1880s. Influential lumber companies purchased nearly all northern forest property during that era, including most of the state-owned property originally set aside for the development of *The State Park*. Undeterred, state legislators worked to

Interstate State Park

4

establish a new system of state parks beginning in 1899. Their efforts were realized when lawmakers authorized the purchase of scenic property along the Dells of St. Croix River in Polk County to establish *Interstate Park* in 1900—Wisconsin's first "official" state park.

In 1903, Wisconsin's progressive Governor Robert La Follette appointed the first State Parks Commission with the goal of establishing two additional state parks; one at Devil's Lake near Baraboo and another along the Wisconsin River near Kilbourn City, later renamed Wisconsin Dells. La Follette also appointed a State Forestry Commission for the purpose of developing a state forest system and appointed E.M. Griffith to serve as the first state forester in 1904.

In 1907, President Theodore Roosevelt called for a national conference of state governors to meet in Washington D.C. The purpose of the gathering was to urge each state to appoint a conservation commission to purchase and preserve state-owned public parks, forests, and recreations areas. After the conference, Governor James Davidson appointed the first Wisconsin State Park Board of Directors. One of the first acts of the Board was to commission John Nolen, a nationally renowned landscape architect, to help draft a plan for the "new" Wisconsin State Park System. Among the many suggestions he presented to the board, Nolen recommended four sites he felt should be considered for state park acquisition. Three of these sites—*Peninsula State Park* (1910), *Devils Lake State Park* (1911), and Nelson Dewey, later renamed *Wyalusing State Park* (1917), were acquired and opened to the public within a few years. The fourth site, the Dells of the Wisconsin River, never became a state park. However, a five-mile-long section of the river known as the *Dells of the Wisconsin River State Natural Area* (2,115 acres), was purchased by the state in 1994.

In 1923, C.L. Harrington was appointed as the first superintendent of the Wisconsin State Park System. Harrington employed park rangers to manage and oversee the development of each state park as it was acquired. He also established the criteria for the purchase or donation of land intended for state park use. Potential sites had to be large enough for outdoor recreational activities and must prove to have significant natural, historic or scenic potential before being considered for designation as a state park. The State Forestry Commission also began to establish the first state forests, including the *Northern Highland State Forest* in 1925 and *American Legion State Forest* in 1929. The cost to manage state forests was to be funded through timber harvest sales and (until recently) an annual statewide mill tax on all real estate.

Much of the early development of state parks and forests, including the construction of roads, parking areas, hiking trails, picnic areas, and service buildings, was accomplished with the help of federal Depression-era work programs established by President Franklin Roosevelt's New Deal Program.

Thousands of unemployed young men worked and lived in *Civilian Conservation Corps* (CCC) and *Works Progress Administration* (WPA) camps set up in dozens of state park and forest properties throughout Wisconsin from 1933 until 1941. Many of the original structures built by CCC and WPA craftsmen, including bluff side trails and staircases, stone and timber shelters, restrooms, and administrative facilities are still in use today in state parks and forests.

The period between 1960 and the late 1980s proved to be one of the most active periods for the expansion and funding of the Wisconsin State Park and Forest System. In 1962, Wisconsin legislators enacted a state park admission fee requirement for all vehicles entering most sites. Although controversial at first, this new pay-as-you-go policy allowed the state park system to become nearly self-funded and not dependent on tax revenues for operational budgets. Two other long-range finance programs for the acquisition and development of state-owned properties were established during this period as well; the federal *Outdoor Resources Action Program* (ORAP) in 1961 and Wisconsin's *Knowles-Nelson Stewardship Program* in 1989. As a result of these new funding sources, the State was better able to maintain existing properties and acquire additional lands to develop new parks, forests, and recreation. These funds were also used to develop biking and hiking trails along abandoned railroad grades throughout the state.

WISCONSIN STATE PARKS AND FORESTS TODAY

For more than 100 years, the Wisconsin State Park and Forest System has continued to grow and meet the expectations of its visitors. Management of this complex system relies on ongoing efforts to accommodate the ever-changing recreational needs of park users by updating existing park facilities and establishing new properties as they are acquired. Over time, this process has led to the development of 5,000 campsites, hundreds of picnic areas, playgrounds, swimming beaches, shelter facilities, and interpretive centers plus thousands of miles of hiking, biking, and nature trails. Today, Wisconsin has more than 60,000 acres of state park property and 470,000 acres of forest land in public ownership for everyone to enjoy.

Recent emphasis on modernizing state park and forest facilities has resulted in improvements to several properties, such as replacing outdated restrooms and shower facilities and providing electric service to many more campsites. In addition, dozens of hiking trails, picnic areas, restrooms, shelters, and other park facilities have been updated to accommodate mobility-impaired visitors. Although facility improvements and new construction projects are undertaken every year, the State's long-standing policy of leaving at least 90 percent of each property undeveloped remains intact. These woodlands, prairies, sand dunes, wetlands, and other natural areas at each park and forest allow native plants and wildlife to flourish relatively undisturbed.

Park and forest supervisors, rangers, wardens, clerical, interpretive, and maintenance staff strive to make each visit to a state park or forest enjoyable for everyone. They don't do it alone, however. Citizen volunteers donate more than 20,000 hours of service each year throughout the state park and forest system. Some volunteers work on maintenance and development projects, while others help serve park visitors at property entrance stations. Interpretive volunteers present evening programs, guided hikes, or serve as nature center hosts. Camp hosts take up residence at most state campgrounds for a month or more to help maintain campsites, restroom/shower facilities, and to provide assistance to fellow campers.

Many volunteers are also members of their local *Friends* group, a not-for-profit support organization for a state park or forest property. Friends groups raise funds through membership fees, donations and the sale of property merchandise and firewood to help fund interpretive programs and facility improvement projects. There are currently **89** *Friends* chapters at state parks and forests throughout the state. All are under the umbrella of the statewide *Friends of Wisconsin State Parks* organization, which actively promotes the preservation and enhancement of all the parks, forests and recreational areas. More information on these support organizations and volunteer programs can be found on the *Friends of Wisconsin State Parks'* website at www.fwsp.org .

Another non-profit organization, the *Natural Resource Foundation of Wisconsin*, provides funding for property and facility improvements for all state-owned lands, including state parks and forests. The Foundation's goal is to protect the state's lands, waters, and wildlife and to promote educational opportunities to connect people to nature.

More information regarding the Natural Resource Foundation of Wisconsin can be found at www.wisconservation.org.

EXPLORE, ENJOY, PLAY, LEARN, DISCOVER, AND RELAX

We hope this guide will inspire you to visit many, if not all, of Wisconsin's beautiful state parks, forests, recreation areas, and scenic water flowages. Whether traversing a hiking trail or enjoying time with family and friends at a beach, picnic area, or around a campfire, spending time at one of Wisconsin's crown jewels is sure to create treasured memories to last a lifetime. Take time to find out what so many have already discovered—a visit to any of these special natural areas can refresh the body, soul, and mind like no other place.

"I go to nature to be soothed and healed, and to have my senses put in order."
John Burroughs (1837-1921) American naturalist, writer,
early conservation activist

Rock Island 25

Peninsula 20
Newport 19

Copper Culture 2

Whitefish Dunes 26

Potwawatomi 23

Green Bay

Heritage Hill 42

Lost Dauphin 8 17

Hartman Creek 6

High Cliff 9

Point Beach 22

Oshkosh

Manitowoc

Fischer Creek 4

REGION 1

Kettle Moraine-North 10

Kohler-Andrae 12

Harrington Beach 5

Pike Lake 21

Lizard Mound 15

Cushing Memorial 3

Loew Lake 16

Havenwoods 7

Milwaukee

Lapham Peak 14

13 Lakeshore

Kettle Moraine-South 11 18

Mukwonago River

Richard Bong 24

Big Foot Beach 1

Region One — Eastern and Southeast Wisconsin: Lake Michigan, Winnebago, Lake Geneva, Green Bay and the Door County Peninsula, Oconto, Crystal and Fox rivers, Northern and Southern Kettle Moraine Forests and Greater Milwaukee Area.

1. BIGFOOT STATE PARK
Lake Geneva Park—A Natural Oasis

Ceylon Lagoon

PARK SNAPSHOT

Bigfoot Beach State Park is a small but attractive 271-acre property is located along the eastern shore of beautiful Geneva Lake in Walworth County. The park offers camping, hiking, nature study, picnicking, swimming, fishing, wildlife viewing and much more.

Bigfoot Beach is part of the community of Lake Geneva and is one of only two state parks in the system located entirely within a city limits. The other is Lakeshore State Park in Milwaukee. (See Chapter 13)

GEOLOGY AND HISTORY

The rolling landscape of the park and the Lake Geneva area was shaped by glacial activity during the last Ice Age. When the glaciers began to melt and recede northward about 12,000 years ago, they deposited sand, gravel, rocks and silt throughout the local area. During this period, a colossal block of ice became separated from the main glacier and was buried under hundreds of feet of glacial debris inside an ancient river valley just west of the park. When the underground ice finally melted, a deep water-filled basin (Geneva Lake) was formed.

The Geneva Lake region was used by several Native American tribes who hunted, fished and tended gardens here for generations. The Potawatomi called the lake *Kishwauketoe* or "Lake of Clear Waters." Later, early American settlers renamed it Big Foot Lake after a Potawatomi chief of the same name whose village was located along the south end of the lake. In 1833 the U.S. government negotiated the Treaty of Chicago, which required the Potawatomi to cede all their lands in southeast Wisconsin and leave the state. Many went to reservations in Kansas or moved to northern Wisconsin. Chief Bigfoot and his followers decided to relocate to the Platte River area of Missouri in 1836.

In 1835, John Brink, a government surveyor, arrived at Big Foot Lake. He took it upon himself to rename it Geneva Lake after the town of Geneva in New York where he grew up. Brink also named the nearby settlement Geneva but this was later changed to Lake Geneva by the U.S. Postal Service due to a conflict with another town of the same name in Illinois.

In the 1870s, European immigrant farmers began to clear the oak and hickory forests of the area to plant crops and raise cattle. Historical records indicate that cattle were still being pastured on farm land within the park as late as the 1940s. Beginning in the early 19th century, wealthy vacationers from nearby urban centers such as Chicago and Milwaukee built expensive summer homes along Geneva Lake. Many of these lakeside estates had well-known family business names such as Wriggly (chewing gum), Morton (table salt), Seipp (beer), and Borden (dairy products).

The rural beauty of the area and crystal-clear waters of Geneva Lake eventually transformed the local area into one of the most popular (and expensive) vacation areas in the country. In 1949, the Wisconsin Conservation Commission purchased land to establish Big Foot Beach State Park in part to secure public access to the Geneva Lake shoreline.

Swimming beach on Geneva Lake

BEACHES, PICNIC AREAS, AND SHELTERS

The sandy swimming beach at Big Foot State Park is the most popular area of the park during the summer season. The width of the beach fluctuates depending on the water level of Geneva Lake. The lake is known for its clear, clean water ideal for swimming, boating and fishing. Lakeshore Drive divides the swimming beach from the rest of the park. Visitors are advised to use caution when crossing this busy highway to access the beach. The park has a large picnic area with playground equipment, a volley ball court, horseshoe pits and an outdoor shelter building that can accommodate up to 125 people.

HIKING, BIKING AND NATURE TRAILS

There are six miles of hiking trails to explore in the park. All trail routes are identified by colored emblems on posts.

The Shady Oak Nature Trail (0.5 miles) is located on the west side of the park near the Ceylon Lagoon. This self-guided nature trail has interpretive signs which describe some of the more common wildlife species of the park area.

> **Ranger Tip**
>
> Walking the entire 26-mile path can take 8 to 10 hours. Many hikers enjoy taking the short walk from Big Foot Beach to downtown Lake Geneva and explore city's many unique shops, parks, and restaurants.

> **Ranger Trail Pick**
>
> The Green Trail (2.9 miles) loops through hardwood forests, wetlands, and restored prairie areas. A section of the trail runs parallel to the Ceylon Lagoon. A side trail crosses an arched bridge over the lagoon, offering a great view of nearby Geneva Lake.

The Geneva Lake Shore Path is a 26-mile-long footpath that circles the entire shoreline of Geneva Lake. The walkway leads past many expansive lakeshore estates and ornate mansions. Sections of this historic walking trail are believed to have been used by indigenous people as early as 2,500 B.C. In the early 19th century, lakeshore homeowners created a local ordinance requiring that a 20-foot-wide path around the lake must remain open for public use. The trail can be accessed from the park's swimming beach area.

WILDLIFE VIEWING

The park's wetlands, prairies, wooded swamps, and upland hardwood forests support a variety of wildlife. White-tailed deer, gray squirrels, raccoon, rabbits, and chipmunks are often seen by visitors but more secretive animals such as red fox, woodchucks and coyote live here as well.

Waterfowl and shorebirds are abundant along the shores of Geneva Lake and the Ceylon Lagoon. Songbirds including cardinals, blue jays and chickadees are commonly seen in woodland areas. More open grassy areas harbor bluebirds and tree swallows. Red-tailed hawks and occasionally bald eagles are seen soaring above the park by day, while both barred and great-horned owls can be heard at night.

INTERPRETIVE PROGRAMS AND FACILITIES

> **Ranger Trivia**
>
> The Ceylon Lagoon was built in the early 20th century by the son of F.L. Maytag, the founder of the popular washing machine company. The shape of the lagoon was designed to represent a miniature version of the nearby Geneva Lake. The Maytag family once owned much of the Big Foot State Park property and had a summer home here.

Interpretive panels are located along nature trails and at other key areas of the park. Signage at the Ceylon Fishing Lagoon explains how it was built and ongoing efforts to maintain the water quality of Geneva Lake.

BOATING AND FISHING

The park does not have a boat launch but a drop-off parking area for canoes and kayaks is available near the swimming beach along Lakeshore Drive. A public boat launch is located in nearby city of

Lake Geneva along Wriggly Drive. The park's Ceylon Lagoon is 6-acre pond that harbors bass, northern pike, and several species of panfish. The lagoon is also stocked with rainbow trout each spring as part of the Urban Waters Fishing Program.

Geneva Lake (5,231 acres) is an exceptionally clear water lake known for its excellent fishing and boating opportunities. The lake has good populations of walleye, lake trout, rainbow trout, northern pike, bass, and panfish.

ACCESSIBLE FACILITIES AND TRAILS
Most of the park's restrooms, shelters, and picnic areas are accessible to mobility-impaired visitors. Campsites 88 and 90 are designated accessible sites with electrical service and paved access to a shower/restroom facility. A wheelchair accessible fishing pier is available on the Ceylon Lagoon. A paved walkway from an accessible parking area provides access to the pier. Most of the park's hiking trails have a natural turf base and are generally not suitable for wheelchair use.

CAMPING
Big Foot has a 100-unit campground (34 electric) divided into 63 standard vehicle access campsites and 37 walk-in tent sites. Many of the campsites are well shaded by mature oak and hickory trees; others are located in more open sunny areas. The campground area has a central restroom/shower building and vending machines for snacks and refreshments. Firewood and ice sales are available from the camp host. An RV water fill and sanitary station is located near the park entrance area.

WINTER ACTIVITIES
Park staff groom six miles of inline cross-country ski trails in winter. The trail is relatively flat and suitable for beginners and families with small children. Ski rentals are available nearby in downtown Lake Geneva. Other off-season activities at Big Foot include hiking, birdwatching, snowshoeing, and winter camping.

> **Ranger Note**
>
> The **Black Point State Historic** site is accessed primarily through a popular tour boat operated by the Lake Geneva Cruise Line located at the Riviera Dock, 812 Wrigley Drive, Lake Geneva, WI 53147. (262) 248-6206 www.cruiselakegeneva.com

AREA ATTRACTIONS
The Lake Geneva area is one of the most popular tourist destinations in the Midwest with many attractions to explore. www.visitlakegeneva.com

Black Point Estate and Gardens State Historical Site is an attractive lakeside Victorian home built in 1888 by Chicago beer baron Conrad Seipp. The site is located west of Big Foot Beach State Park and operated by the Wisconsin Historical Society.
www.blackpointestate@wisconsinhistory.org (262) 248-1888.

DIRECTIONS TO THE PARK

Bigfoot is located about midway between Beloit and Kenosha near Wisconsin's border with Illinois. From the Highway 12 and 50 interchange take Highway 50 (W. Main St.) west into the city of Lake Geneva. Turn south onto S. Lakeshore Road and travel about one mile to the state park entrance.

Bigfoot State Park
1550 South Lakeshore Drive
Lake Geneva, WI 53147
(262) 248-2528

2. COPPER CULTURE STATE PARK
Wisconsin's Oldest Native American Burial Grounds

Oconto River at Copper Culture

PARK SNAPSHOT

Copper Culture is a 51-acre day-use state park located along the Oconto River just a few miles inland from the Green Bay shoreline. This historic park contains a hidden buried treasure like no other place in Wisconsin. Archeologists have determined the ancient Native American burial ground discovered here is the oldest of its kind in the state and one of the most important prehistoric sites in the United States.

The park offers hiking trails, picnic areas, wildlife watching, and river fishing opportunities. A small museum located near the park entrance showcases artifacts from the Copper Culture people that once lived here.

GEOLOGY AND HISTORY

Remnants of a prehistoric Native American village and burial ground site found at Copper Culture are located on an elevated ridge of sand, gravel and other glacial deposits. Geologists believe this ridge was part of a large sand bar at the mouth of the Oconto River along the shoreline of Green Bay. At the time, about 8,000 years ago, the water elevation of Green Bay was about 25 feet higher than it is today and was part of Lake Algonquin, a post-glacial lake that was much larger and deeper than Lake Michigan is today.

The prehistoric burial grounds were discovered in 1952 by a 13-year-old boy who found human bones while playing in an abandoned gravel quarry. Several archeological surveys and excavations were conducted by the Oconto County Historical Society and the Milwaukee Public Museum in the years that followed. Archeologists found 21 undisturbed burial pits that held at least 45 individuals. It's believed that as many as 200 Copper Culture people may have originally been interned here but most pits had been destroyed by quarrying operations at the site.

> **Ranger Note**
>
> It's often difficult to fully comprehend the age of historic sites. Put into context, the Copper Culture people built a large village community here at least 2,000 years "before" the construction of the pyramids of Giza in Egypt.

Items unearthed during site excavations included turtle shells, snail shell necklaces, stone arrow heads, spear points, knives, axes and many other artifacts. Some objects were trade items such as clam shells brought here from the Mississippi River and a rare whelk (conch) seashell from the Atlantic Ocean. The most significant discovery at the site were the many copper tools, weapons, ornaments and other artifacts found buried with the dead. Radiocarbon analysis of burial material indicated that the inhabitants of this site were from Late Archaic period (4000 to 2000 B.C). Today these early inhabitants are referred to as the Old Copper Culture People who lived in this region of Wisconsin more than 6,000 years ago.

Copper Culture people are considered to be the first known metalsmiths in America. They mined copper throughout the Lake Superior region, including the copper range of northern Wisconsin. Raw copper was transported to tooling villages such as the one discovered here. Copper was heated, hammered and cooled repeatedly by skilled craftsmen and then fashioned into knives, awls, spear points, fish hooks, bracelets and many other tools, ornaments, and weapons.

In the mid-19th century, an influx of European immigrants came to Oconto River area to work in logging camps and sawmills and to farm cut-over timber lands. One of the largest employers at that time was the Holt Lumber Company, which operated a sawmill in the nearby city of Oconto from 1847 to 1938. In its heyday, Holt Lumber was considered the largest hardwood sawmill in the Midwest.

The Oconto County Historical Society purchased the Copper Culture location soon after the discovery of the burial grounds to preserve this rare archeological site. In 1959, the site was transferred to the State of Wisconsin to develop Copper Culture State Park. Today, the park is jointly managed and operated by the State and the Oconto County Historical Society.

BEACHES, PICNIC AREAS AND SHELTERS

Copper Culture has a picnic area and a shelter pavilion adjacent to the park's welcome center and museum. There is no swimming beach in the park but a nice sandy beach is located nearby along the shores of Green Bay at Oconto City Park, 5182 County Highway N.

HIKING, BIKING AND NATURE TRAILS

There are three miles of hiking trails within the park. The **River Trail** follows the forested shoreline of the Oconto River. The **Bluebird Trail** leads through a 15-acre restored short-grass prairie with several species of prairie flowers and forbs that bloom throughout the summer season.

The **Copper Culture Biking and Hiking Trail** (4.0 miles) begins in the park and connects to the Oconto River State Trail (8 miles). This multi-use trail travels west along an abandoned railroad corridor through scenic farm fields and forests. The trail ends at the **Stiles Junction Railroad Station**, a Wisconsin Historic Registered site.

> ### Ranger Trail Pick
>
> The **Copper Culture Mound Trail** (0.26 miles) is a short gravel path that leads from the visitor center to the Native American burial grounds. At the end of the trail is a stone monument and historical plaque that describes the importance of this site.

WILDLIFE VIEWING

The park's forests, prairies, wetlands, and the Oconto River provide excellent habitat for white-tailed deer, squirrels, chipmunks, raccoons, rabbits, mink, muskrats and other small mammals. More than a hundred species of birds have been observed either nesting or migrating through this area, including woodland songbirds, waterfowl, shorebirds, and raptors such as red-tailed hawks and barred owls.

The nearby **Oconto and Rush Point Marsh** is a 928-acre state wildlife area that features a 220-acre open water impoundment and a 2.8-mile wildlife observation hiking trail.

The marsh is part of the **Great Wisconsin Birding and Nature Trail** network and is considered one of the best birding sites in the state. The property is located along County Highway Y (Main Street) just north of the nearby city of Oconto.

Copper Culture Museum

INTERPRETIVE PROGRAMS AND FACILITIES

The **Copper Culture Museum** is located inside a red brick, Belgium-style farm house adjacent to the park entrance. This historic structure was built in 1924 by Charles Werrebroeck, a skilled mason who immigrated to America from Belgium in 1911.

The museum houses interpretive displays of Copper Culture artifacts unearthed by archeological excavations in the park. On display are clam shell ornaments, copper tools, fish hooks, arrowheads, spear points and a very rare prehistoric whistle made from the wing bone of a swan. A short video explaining how archeologists conducted their work excavating and restoring the burial grounds can also be viewed. The museum is staffed by Oconto Historical Society volunteers and open weekends from Memorial Day to Labor Day.

BOATING AND FISHING

Canoes and kayaks can be launched along the Oconto River anywhere within the park. Larger watercraft can access the river and Green Bay at the **Breakwater Park** boat launch at 1301 Harbor Road in Oconto. The Oconto River is well known as having some of the best smallmouth bass fishing in the state. Walleye fishing can also be good at times, especially during the spring spawning run from the waters of Green Bay.

ACCESSIBLE FACILITIES AND TRAILS

The park's picnic area, restrooms, shelter, and the graveled path to the burial site are accessible for mobility impaired visitors. Most hiking trails have natural turf and are not wheelchair accessible.

CAMPING

Copper Culture is a day-use-only park so there are no campsites here. **Holtwood Park Campground** in the nearby city of Oconto has 130 campsites (rustic and full -service hookups) along the scenic Oconto River. The campground is located at 400 Holtwood Way, (920) 834-7732. www.cityofoconto.com.

North Bay Shore County Park has a 34-unit campground with modern amenities. The park is located north of Oconto along the Green Bay shoreline at 500 Bay Road, Oconto, WI 54153. (920) 834-6995 www.co.oconto.wi.us/park.

WINTER ACTIVITIES

Snowshoeing, hiking, cross-country skiing, and wildlife watching are favorite winter activities at Copper Culture. None of the park trails are groomed for skiing but the nearby Oconto City Park offers groomed and tracked cross-country ski trails.

AREA ATTRACTIONS

Oconto County is known for its outstanding historical sites, museums and its many outdoor recreational opportunities. (920) 834-6919 www.ocontocounty.org

The **Beyer Home and Carriage House Museum** is located in the nearby city of Oconto. The 1868 Beyer house was the first brick home built in Oconto County and is on the National Register of Historic Places. The **Carriage House Museum** has an extensive collection of antique cars, boats, sleighs, and artifacts from the Oconto Brewery and the Stanley Toy Factory. The museum is located at 917 Park Ave in Oconto. (920) 834-6206 www.ocontohissoc.org

DIRECTIONS TO THE PARK

Copper Culture is located in Oconto County along the north shore of the Oconto River. From Highway 41, take exit 191 to the roundabout to Hwy 22 / Charles Street. At the next roundabout, turn south on Copper Culture Way and then west on Mott Street to the park.

Copper Culture State Park
260 Copper Culture Way
Oconto, WI 54114   (715) 757-3979

3. CUSHING MEMORIAL (STATE) PARK

Hilltop Park Honors Civil War Heroes

Civil War Memorial Obelisk

PARK SNAPSHOT

Cushing Memorial Park is a day-use park dedicated to all who served in the Union Army during American Civil War, but especially to memory of three brothers (Alonzo, William and Henry Cushing) for whom the park is named. This small 8-acre park is nestled on a hill overlooking the scenic Bark River on the same plot of land that was once home to the Cushing brothers and their parents.

Cushing Park is located near the city of Delafield in northwestern Waukesha County. The park offers hiking, bird-watching, and fishing. It also has a picnic area and one of the best playgrounds in southeastern Wisconsin.

GEOLOGY AND HISTORY

Cushing Memorial Park is located on the crest of grass-covered knoll perched high above the Bark River. The park is part of the Kettle Moraine region, a vast landscape of rolling hills, kettle lakes, wetlands, and rivers left behind by receding glaciers more than 12,000 years ago.

3. Cushing Memorial State Park

The park's historic hillside was once known as **Peace Spring** and was used for untold generations of Native American people as a rendezvous gathering site to conduct trade and hold cultural ceremonies. The Bark River, which borders the south side of the park, once served as water highway for Native Americans traveling between Nagawicka Lake and Nemahbin Lake; two of the largest lakes in the region. The

Potawatomi established themselves as the dominant tribe of this area soon after they migrated to Wisconsin in the early 1600s after losing their homelands in southwestern Michigan due to conflicts with the Iroquois. The Potawatomi built a village that once housed over 1,000 people along the nearby Fox River. Their village site is now occupied by the city of Waukesha.

In 1839, Dr. Milton Cushing of New York purchased the Peace Spring hillside property. He built a homestead there for his family including his three sons, Alonzo, William, and Henry. Milton served as a physician in the Delafield area for many years before the family moved to Chicago. The family's hilltop property was once again used as a rendezvous for warriors, but time it was for veterans of the Civil War who gathered here for reunion celebrations. In 1915, the Waukesha County Historical Society erected a 75-foot-high memorial obelisk to honor Dr. Cushing's three sons, who distinguished themselves fighting for the Union Army.

At the same time, the property was transferred to the State and became *Cushing Memorial State Park*. Much of the early development and ongoing maintenance of the state park was accomplished by workers from the nearby *Delafield State Fish Hatchery*. In 1980, after 65 years as a state park, Cushing Memorial was deeded to the city of Delafield after the state fish hatchery closed.

BEACHES, PICNIC AREAS, AND SHELTERS

The park does not have a swimming beach but many visitors enjoy wading in the cool shallow waters of the Bark River in summer. The park's picnic area features a fortress-like playground facility known as **Fort Cushing** (see next page). The playground has towering wooden structures with many small rooms for children to explore.

Cushing Fort Playground

HIKING, BIKING, AND NATURE TRAILS

There are no marked hiking within the park but the property serves as an important trailhead for several other trails in the local area. The **Veteran's Memorial River Walkway** is a scenic paved trail and boardwalk that follows the shoreline of the Bark River from Cushing Park to the historic State Fish Hatchery property in Delafield. The trail features ten memorial plaques honoring the veterans who fought in each of our country's wars.

The park also serves as a parking hub for several bicycle/hiking trails including the **Cushing Park Road Trail** (3 miles), which connects to both the **Glacial Drumlin State Trail** (52 miles) and the **Lake Country Recreational Trail** (15 miles). The Cushing Park Road Trail also leads north to the **Lapham Unit of the Kettle Moraine State Forest.** (See Chapter 14)

WILDLIFE VIEWING

Bird-watching is a popular activity at Cushing Park. More than a hundred species of waterfowl, shorebirds, woodland songbirds, owls, hawks, and other raptors have been observed here. The Bark River shoreline, wetlands, and the upland forest of acorn-bearing oak trees attract white-tailed deer, wild turkey, chipmunks, raccoon, and squirrels.

3. Cushing Memorial State Park

INTERPRETIVE PROGRAMS AND FACILITIES

An interpretive panel at the park entrance area highlights the history of the property. Memorial plaques attached to the 75-foot obelisk monument in the upper picnic area provide a brief description of the Cushing brothers' military service during the Civil War. A more detailed look into their remarkable lives are as follows:

Ranger Trivia

In 2014, President Barack Obama presented a post-mortem Medal of Honor to Lt. Alonzo Cushing in honor for his bravery and sacrifice during the Civil War, more than 151 years after his death.

William Cushing was an officer in the US Navy. He is credited with planning and executing the sinking of the Confederacy's iron-clad ship, the CSS *Albermarle,* which he accomplished by ramming the enemy vessel with a small boat fitted with a torpedo. William detonated the torpedo himself, which not only sank the Confederate ship but nearly killed him in the explosion.

Howard Cushing was an artillery officer who was engaged in several battles throughout the south. At the end of the Civil War, he was promoted to a 1st Lieutenant with Troop F of the 3rd Calvary, which served in Texas and Arizona. He was involved in many skirmishes with Native American warriors, including Apache Chief Cochise. In 1871, Howard and his troop were ambushed by a band of Apache in Arizona, which resulted in his death during fierce hand-to-hand combat.

Alonzo Cushing attended West Point Military Academy, along with classmate George Armstrong Custer. During the Civil War, he served as a 1st Lieutenant and fought in several conflicts including the Battle of Chancellorsville. He is best remembered for his heroic efforts as the officer in charge of Battery A, 4th U.S. Company, which defended Cemetery Hill during the Battle of Gettysburg. Alonzo was wounded by shrapnel to his shoulder and later shot in the abdomen during a Confederate attack known as Pickett's Charge. Despite life-threatening injuries, he continued to rally his men on horseback but was eventually killed by a Confederate bullet that struck his head.

BOATING AND FISHING

Canoes and kayaks can be launched along the shoreline of the Bark River within the park. A public boat launch for motorized watercraft is located on **Nemahbin Lake** about a mile west of the park off Delafield Road.

The Bark River is a shallow connector stream that links **Nagawicka Lake** (981 acres) located east of the park to **Upper Nemahbin Lake** (277 acres) to the west. Anglers can expect to catch a variety of panfish, northern pike, bass, and walleye in these waters.

ACCESSIBLE FACILITIES AND TRAILS
Much of the park is accessible to mobility impaired visitors including sections of the **Fort Cushing Playground**. The nearby **Veteran's Memorial Community Walkway** features wheelchair-accessible paved trails and wooden walkways along the Bark River.

CAMPING
Cushing Memorial is a day-use-only park so there are no camping facilities here. **Ottawa Lake Campground** within the **Kettle Moraine State Forest—Southern Unit** has 100 campsites (65 electric) with restroom/showers and a swimming beach. (See Chapter 11) Ottawa Lake is located about 9 miles south of the park at S59 W36530 Hwy. ZZ, Dousman, WI 53118 (262) 594-6220.

WINTER ACTIVITIES
Cushing Park remains open in winter for hiking, wildlife watching, snowshoeing, and cross-country skiing. There are no groomed ski trails in the park.

AREA ATTRACTIONS
Waukesha County has many lakes and outdoor recreation attractions, making it one of the top tourist destinations in the state. www.visitwaukeshacounty.com

The **Retzer Nature Park** is a 450-acre Waukesha County nature preserve that offers several miles of hiking trails, a modern interpretive center, and planetarium. The park is located 9 miles southeast of Cushing Memorial Park at S14 W28167 Madison St. in Waukesha. (262) 896-8007.
www.waukeshacounty.gov

DIRECTIONS TO THE PARK
Cushing Memorial Park is west of the city of Delafield in Waukesha County. From Interstate I-43 take the Genesee Street exit and travel north to Delafield. At the intersection with Main Street, turn left (west) to North Cushing Park Road and then turn right (north) over the fieldstone Bark River bridge into the park.

Cushing Memorial State Park
775 N. Cushing Rd.
Delafield, WI 53018
(262) 646-6220

4. FISCHER CREEK STATE RECREATION AREA

Lakeshore Beauty on Lake Michigan

Lake Michigan beach

PARK SNAPSHOT

Fischer Creek Recreation Area is located in Manitowoc County along the Lake Michigan shoreline. This scenic park preserves nearly a mile of undeveloped Lake Michigan shoreline, forests, and wetlands plus a clearwater trout stream, which flows into lake.

Much of the recreation area is perched atop 40- to 80-foot-high clay-lined bluffs, which offer spectacular views of Lake Michigan. The property has hiking trails, picnic areas, and offers opportunities for bird-watching, fishing, and beach combing.

GEOLOGY AND HISTORY

Fischer Creek's shoreline landscape was shaped by immense glacial activity over time. Near the end of the last ice age about 12,000 years ago, massive glaciers retreating north to the Canadian Arctic left behind thousands of moraine hills, streams, and kettle lakes throughout eastern Wisconsin. One of the largest glaciers excavated a deep valley more than 100 miles wide and 300 miles long, which filled with water to become Lake Michigan.

Native American people found the shoreline and streams of Lake Michigan an ideal place to hunt, fish, and gather wild plants. They used the lake as a water highway that allowed them to travel great distances to conduct trade with other tribes. Several different Native American cultures lived along these shores for thousands of years, including the Ottawa, Sauk, Chippewa, Winnebago, and the Potawatomi.

> **Ranger Note**
>
> The bluff area overlooking Lake Michigan at Fischer Creek was once considered a sacred site by Native Americans. Several archeological sites, including a burial mound and remnants of an ancient Indian trail, have been discovered here.

French fur traders and missionaries were the first to explore the western shoreline of Lake Michigan in the 17th century. They were followed by East Coast fisherman who took advantage of excellent fisheries of the lake. In the mid-1800s immigrant farmers from Germany, Ireland and other west European countries arrived to carve out farmsteads in the fertile soils adjacent to the lake. A few sections of the original immigrant wagon route and military road that connected Milwaukee to Green Bay are still visible at Fischer Creek today.

> **Ranger Trivia**
>
> Fischer Creek was named after one of the earliest pioneer families of the area. August Wilhelm Fischer, along with his wife and children, immigrated to America from Prussia in 1863. They settled here in the township of Centerville area, so named because it was located between the communities of Manitowoc and Sheboygan.

In the 1990s, a Chicago developer purchased the property with the intension of building a private marina and several multi-story condominiums overlooking Lake Michigan. A local group of citizens opposed to the project lobbied local, state, and county officials to stop the destruction of this rare shoreline area and transform it into a public nature preserve instead. Their efforts paid off when private and government financial assistance allowed the State to purchase the 124-acre property in 1991.

> **Ranger Trivia**
>
> The story of the grassroots effort to save the Fischer Creek property from development was immortalized in a 1990's feature documentary film entitled, *Worth Fighting For: People Protecting the Great Lakes*. The film was narrated by musician James Taylor and featured original music composed and performed by Peter Buffet. The film won two Emmy awards.

4. Fischer Creek State Recreation Area

Today, Fischer Creek remains a quiet, peaceful green oasis of forests, streams, and sandy beaches. The property is part of the state park system and is maintained by the Manitowoc County Parks Department.

BEACHES, PICNIC AREAS AND SHELTERS

Picnic areas with grills, tables, and vault-type restrooms are located adjacent to parking areas located along County Highway LS about one mile north of the village of Cleveland.

A sand and pebble beach is located along the Lake Michigan shoreline adjacent to an historic iron bridge that spans Fischer Creek. The width of the beach varies with the water level of the lake.

HIKING, BIKING AND NATURE TRAILS

Fischer Creek has two hiking trails that are well maintained but are not marked with guide posts. Property maps showing trail routes are posted on bulletin boards at parking areas. The **Main Trail** connects the north and south use areas with outstanding views of Lake Michigan. The trail leads downhill from both parking areas and crosses Fischer Creek over an iron truss bridge built in the early 1900s. This part of the trail provides access to the beach along Lake Michigan.

Fischer Creek

Ranger Trail Pick

The Nature Trail is a near-level trail that begins near the north parking area and follows an abandoned historic roadway. The trail leads over a wooden foot bridge with a wildlife viewing shelter above a small creek. The trail ends above a deep ravine on the north boundary of the park. There are several rustic side trails off this trail as well. Some travel to bluff overlooks areas above Lake Michigan. Others lead to long-gone cottage sites and an old concrete silo from an abandoned farm.

WILDLIFE VIEWING

Fischer Creek's hardwood forests, wetlands, grassy fields, and clear-running streams provide excellent habitat for wildlife, including white-tailed deer, squirrels, chipmunks, rabbits, mink, and raccoon. The western shore of Lake Michigan is an important migratory route for thousands of waterfowl, gulls, terns, birds of prey, sandpipers, and dozens of woodland songbirds.

INTERPRETIVE PROGRAMS AND FACILITIES

There are no interpretive signage or self-guided nature trails at Fischer Creek but many visitors come here for birdwatching or just to enjoy the sights and sounds of nature.

BOATING AND FISHING

Fischer Creek is a popular fishing spot for many anglers. The creek is considered a class II steelhead (rainbow trout) stream. Steelhead and white suckers migrate upstream from Lake Michigan each spring to spawn. In fall, brown trout and coho (king) salmon can often be found here as well. Fischer Creek is also known for its rock bass, panfish, northern pike, and smallmouth bass fishing.

Ranger Trivia

The stream and property are named after the family name, Fischer, a German surname meaning "one who fishes" by profession.

There is no boat launch at Fischer Creek. A small boat launch is located at Hika Bay Park along Lake Michigan in the village of Cleveland about two miles south of the park.

CAMPING

Fischer Creek is a day-use-only property so there are no campsites here. **Kohler-Andrae State Park**, located south of Sheboygan, has **141** sites (52 electric) with restroom/showers. (See Chapter 12)

4. Fischer Creek State Recreation Area

WINTER ACTIVITIES
Fischer Creek is open all year for hiking and wildlife viewing. Snowshoeing and cross-country skiing are allowed but trails are not groomed.

AREA ATTRACTIONS
Manitowoc County has many popular tourist attractions, parks, and museums. www.manitowoc.info

The **Farm Wisconsin Discovery Center** is an interesting farm education center. The center is filled with fun, interactive displays that highlight the many different modes of agriculture in producing food, fiber, and fuel on Wisconsin farms. The center features a one-of-kind dairy cow birthing barn where visitors can often watch the live birth of calf. A visit to the center includes a bus tour to a nearby modern dairy farm operation. Farm Wisconsin is located about nine miles north of Fischer Creek at 7001 Gass Lake Road, Manitowoc, WI 54220. (920) 726-6000   www.farmwisconsin.org

DIRECTIONS TO THE PARK
Fischer Creek Recreation Area is located about midway between Manitowoc and Sheboygan. From Highway I-43 take the County Hwy XX (North Ave.) exit and the travel east into the Village of Cleveland. Turn left (north) on County Hwy LS (Lakeshore Drive) and travel about 1.8 miles to one of the parking areas.

Manitowoc County Park System
3500 Hwy 310
Manitowoc, WI 54220
(920) 683-4185
www.manitowoccountywi.gov/parks

5. HARRINGTON BEACH STATE PARK

Shimmering Jewel of Lake Michigan

Lake Michigan beach

PARK SNAPSHOT

Harrington Beach State Park is located along the shores of Lake Michigan in northern Ozaukee County. For more than 40 years, this scenic park was a lesser-known day-use property, but with the addition of a new campground in 2009, Harrington Beach has become one of the most popular state parks in Wisconsin.

The park conserves a mile-long stretch of the Lake Michigan shoreline along with its wide, sandy beaches. Harrington has several picnic areas, hiking trails, and offers both drive-in and walk-in campsites. **Quarry Lake**, a scenic, crystal-clear, 26-acre man-made lake popular with anglers, hikers, and photographers, is also located here.

GEOLOGY AND HISTORY

Harrington Beach's landscape was shaped by both natural and man-made forces. About 600 million years ago, a warm, shallow inland ocean known as the Silurian Sea covered much of Wisconsin. Over the course of millions of years, deep deposits of sea life, sand, and silt beneath this sea were transformed into hard dolomite limestone; the bedrock of the park.

During Wisconsin's last Ice Age about 30 million years ago, a mile-high glacier excavated a massive river valley to the east of Harrington Beach. When the glaciers began to melt and retreat north again about 12,000 years ago, they left behind thousands of moraine hills, streams, and lakes, including Lake Michigan. Several Native American cultures are known to have hunted, fished, and lived along the lakeshore area for generations, including the Menominee, Winnebago, and Potawatomi people.

> **Ranger Trivia**
>
> The park was named in honor of C.L. Harrington who served as the state's first Superintendent of State Parks and Forests from 1923 to 1950.

In mid-1800s, European immigrants primarily from Belgium, Luxemburg and Germany moved to the lakeshore region of Ozaukee County to establish farmsteads and towns. In 1890, a businessman from Milwaukee began to acquire lakeshore property from local farmers to develop a limestone quarrying operation here called the **Northern Stone Company**. Crushed limestone was in demand for use in road building, concrete, mortar, and many other uses. Quarrying rock in the early 19th century was dangerous work, as it required the use of dynamite, stone crushers, and mule-pulled carts. Limestone was loaded onto steamships docked along a 700-foot-long pier built into Lake Michigan and shipped to distant ports.

In 1901, the quarry was purchased by the **Lake Shore Stone Company**, which built several houses along with a company store near the quarry to house immigrant workers from Italy, Luxemburg, and Austria. This small company town, known as **Stonehaven**, remained active until the quarry closed in 1925. In 1966, the Wisconsin Conservation Department purchased the quarry, along with 715 acres of private land, to develop Harrington Beach State Park.

> **Ranger Note**
>
> Harrington's beach, Quarry Lake, and most picnic areas are closed to motor vehicle traffic to maintain the quiet atmosphere of the park. Visitors can walk to any of these areas or catch a ride on the park shuttle, which operates on summer weekends and holidays. Many guests bring along a coaster wagon, hand truck, or bicycle to transport coolers, folding chairs, or other equipment to day-use areas.

BEACHES, PICNIC AREAS, AND SHELTERS

The park's mile-long wide, sandy beach is the main draw for many park visitors. Lake Michigan water temperatures are usually very cold but occasionally, offshore winds push warmer surface water towards the beach. The park has several picnic areas, including a few adjacent to the Lake Michigan shoreline.

The **Puckett's Pond Picnic Area**, located near the center of the park, has an open-air shelter with picnic tables. Also located here is a fishing pier on the pond and the **Plunkett Observatory**. Evening stargazing programs are offered at the observatory from time to time.

The **Ansay Welcome Center** is an indoor shelter located adjacent to the main beach parking area. The center has a larger gathering room with tables and chairs and kitchenette. The welcome center facility and all open-air shelter facilities can be reserved for gatherings in advance.

HIKING, BIKING, AND NATURE TRAILS

Harrington Beach has several miles of hiking trails to explore. The **White Cedar Nature Trail** (0.8-miles) is a self-guided interpretive trail that leads through a lowland white cedar and yellow birch forest.

The **Bobolink Nature Trail** (0.5 miles) loops through a restored prairie area with interpretive panels that highlight grassland flora and fauna of the area. Birds such as bluebirds, tree swallows, meadowlarks, and bobolinks are often spotted along this trail.

The **Stonehaven Historical Interpretive Trail** begins at the Ansay Welcome Center and follows the shuttle road to Quarry Lake. Interpretive panels

Quarry Lake

> **Ranger Trail Pick**
>
> The Quarry Lake Trail (0.75 miles) circles the perimeter of Quarry Lake with scenic views of the lake, limestone ledges, and a seasonal waterfall. Interpretive panels along the route describe the quarrying operations that once operated here and how the lake was created. Exposed rock along this trail reveals ancient fossils imbedded in the limestone and scrape marks cut into the rock by glaciers during the last Ice Age.

along this route recount the history of the quarry company community of Stonehaven. A kiosk at the **Quarry Lake Picnic Area** feature historic photos of limestone quarry operations.

Bicycle riding is allowed on all park roads including the two-mile-long shuttle road. The **Ozaukee Interurban Bicycle Trail** (30 miles) can be accessed in the nearby village of Belgium. The trail follows a former railroad grade south to the cities of Port Washington, Grafton, and Cedarburg. www.interurbantrail.us.

WILDLIFE VIEWING

The park's diverse mix of grasslands, forests, wetlands, and Quarry Lake attract a wide array of wildlife, especially white-tailed deer, squirrels, rabbits, and raccoon. Harrington Beach is favorite stop for bird watchers in spring and fall when many species of shorebirds, waterfowl, woodland songbirds, hawks, falcons, and other raptors use the Lake Michigan shoreline as migration route.

INTERPRETIVE PROGRAMS AND FACILITIES

Nature hikes, programs, and special events are offered from time to time throughout the summer season in the park. Members of the **Northern Cross Science Foundation** present several stargazing programs throughout the year at the **Plunkett Observatory** located at the Puckett's Pond Picnic Area. Participants can view the night sky using the group's stargazing equipment, which includes a powerful 20-inch telescope.

An interpretive panel at the **Point Picnic Area** recounts the loss of 60 passengers and crew members on the steamship *Niagara*, a passenger vessel that caught fire and sank just offshore from the park in 1856. The original 3,000-pound anchor from the *Niagara* is on display here.

BOATING AND FISHING

There is no boat launch onto Lake Michigan in the park but a public ramp is available in the nearby city of Port Washington. Watercraft are not allowed on Quarry Lake. Shore fishing in the quarry is popular for bluegill, perch, bass, and rainbow trout. Fishing piers are available on both Quarry Lake and Puckett's Pond.

ACCESSIBLE FACILITIES AND TRAILS

Most buildings, restrooms, picnic areas, and shelters are accessible for mobility-impaired visitors. A log cabin in the campground has a bedroom, kitchen, living room, and a wheelchair-accessible shower and restroom. The cabin is reserved for campers with physical disabilities. A paved walkway from Ansay Welcome Center provides wheelchair access to the beach area. The Quarry Lake Trail is surfaced with packed limestone to accommodate wheelchair use.

CAMPING

Harrington Beach has 73 campsites (33 electric) including five walk-in tent sites. The campground has a central restroom building with private shower rooms. The park also has an outdoor group camp with a capacity of up to 30 people. The group camp sites are limited to tents only but one wheeled camping unit is allowed in the parking area.

A canoe/kayak campsite is available along the Lake Michigan shoreline to provide overnight camping for paddlers along Wisconsin's **Great Lakes Kayak Trail.**

WINTER ACTIVITIES

Park staff groom three miles of inline cross-country ski trails and a separate trail for snowshoeing and walking. Ice fishing is popular at Quarry Lake. All campsites are closed for the season beginning the last weekend of October

AREA ATTRACTIONS

Ozaukee County and the Belgium area have many visitor attractions to explore. www.ozaukeetourism.com, www.belgiumchamber.org

Riveredge Nature Center is a 379-acre woodland preserve located along a mile-long undeveloped stretch of the Milwaukee River. The center offers several miles of hiking trails and a modern nature center with interpretive displays. Riveredge is located about 14 miles southwest of Harrington Beach at **4458** County Road Y, Saukville. www.riveredgenaturecenter.org

DIRECTIONS TO THE PARK

Harrington Beach is located east of the village of Belgium in Ozaukee County. From Interstate I-43, exit onto County Highway D and travel east towards Lake Michigan to the park's entrance.

Harrington Beach State Park
531 Highway D
Belgium, WI 53004   (262) 285-3015

6. Hartman Creek State Park

Chain O' Lakes Park Has It All

Allen Lake

PARK SNAPSHOT

Hartman Creek is often described as state park that has something for everyone. The 1,500-acre park offers camping, swimming, picnicking, fishing, paddling, nature study, and miles of hiking, biking, and equestrian trails. Hartman Creek has several spring-fed lakes that are part of the Chain O' Lakes system of 22 interconnected lakes, one of Wisconsin's most popular tourist destination areas. The park is conveniently located near the center of the state within Waupaca and Portage counties.

GEOLOGY AND HISTORY

The park's rolling moraine hills, sandy plains, and clear springs are the work of glacial activity that occurred near the end of the last Ice Age about 12,000 years ago. Many Native American cultures are known to have lived in central Wisconsin for

> **Ranger Trivia**
>
> Waupaca is a Menominee term meaning "pale or clear waters." The nearby city of Waupaca and Waupaca County were both named after the Waupaca River. It was also the name of the local Potawatomi leader, Chief Waupaca, who lived along the river when the first white settlers arrived.

thousands of years, including the Effigy Mound builders, known for their elaborate burial mounds built in this region as early as 650 A.D. More recent inhabitants of the local Waupaca area included the Potawatomi, Ojibwe (Chippewa), HoChunk, and Menominee people.

Immigrants, mostly from Western Europe, arrived in the area in the 1830s to cut timber and clear farmland. Although many early attempts to farm the sandy soils of the local area failed, the introduction of growing **hops** as a cash crop was a commercial success. Hops are perennial vines that are grown on pole and string structures. The plant's cone-like flowers are used as a bittering or flavoring agent in brewing beer. One of the earliest pioneers in the area, the Allen family, grew hops in the local area beginning in 1874 and eventually became the largest hop growers in Waupaca County.

In 1925, George Allen started a brook trout hatchery within the park area by installing several dikes and a dam across Hartman Creek to create three manmade lakes. The hatchery was later purchased by the Wisconsin Conservation Commission to raise smallmouth bass and other native fish used to restock degraded river systems around the state in 1939. When the state fish hatchery closed in 1960, the state purchased additional private land in the area to develop Hartman Creek State Park, which opened to the public in 1966.

Hartman Lake swimming beach

BEACHES, PICNIC AREAS, AND SHELTERS

The Hartman Lake Beach and Picnic Area has a 300-foot swimming beach and a well-stocked concession where snacks, refreshments, and camping supplies are sold. Bicycles, canoes, kayaks, paddle boards, and floatation mats can also be rented here.

The **Middle Picnic Area**, located between Hartman Lake and Mid Lake, has an attractive six-sided indoor shelter that can accommodate up to 50 people. The park's outdoor amphitheater is also located here. The **Allen Lake Picnic Area** north of the park's campground is a popular fishing spot. A floating pier is located here along with canoe/kayak racks for campers.

The **Whispering Pines Marl Lake Day-Use Area** is a quiet, picturesque area of the park located about two miles east of the main park area along Whispering Pines Road. The site has some of the oldest white pines of the area and overlooks Marl Lake, a brilliant, blue-colored, spring-fed lake. A stone staircase descends to a floating pier on the lake.

HIKING, BIKING, AND NATURE TRAILS

Hartman Creek has more than 28 miles of hiking trails to explore. The **Oak Ridge Trail** (4.87 miles) leads through tall grass prairies, sandy hills, and mature oak and pine forests. The **Windfelt Trail** (2.0 miles) travels through similar topography and leads to **High Point**, a scenic knoll with a beautiful panoramic view of the surrounding countryside.

> **Ranger Note**
>
> Whispering Pines was once a private amusement park owned by Christ and Emma Hyldgaard in the 1950s and '60s. The site had a gift store, picnic area, playground, several flower rock gardens, a museum, and a concession stand that sold food, refreshments and bait for trout fishing on Marl Lake. In 1977, descendants of the Hyldgaad family donated the Whispering Pines property to the state as part of Hartman Creek State Park.

> **Ranger Note**
>
> Many visitors enjoy exploring the beautiful Waupaca area Chain O' Lakes by bicycle. An excellent publication entitled, **Hartman Creek Suggested Area Bike Routes Guide** is available at the park office.

> **Ranger Trail Pick**
>
> The **Deer Path Trail** (1.0 miles) loops around Allen Lake and is one of the most scenic hikes in the park. The trail offers great views of the lake and the headwater springs that feed the clear waters the park's man-made lakes. The trail can be accessed from the **Allen Lake Picnic Area**.

The **Dike Hiking Trail** (1.0 miles) encircles Hartman Lake and crosses the dam over Hartman Creek. **Coach Road Trail** (1.0 miles) is located south of the Hartman Lake dam. This trail follows a 19th-century stagecoach line route that once provided transportation between the cities of Stevens Point and Oshkosh. A three-mile segment of the **Ice Age National Scenic Trail** passes through Hartman Creek and the **Emmons Creek Fishery and Wildlife Area** in the southeast section of the park.

An asphalt-paved hiking/biking trail connects the campground to the Hartman Lake swimming beach area. A challenging 13.5-mile single-track, off-road bike trail is also located in the park.

WILDLIFE VIEWING

The lakes, streams, wetlands, prairies, and oak/pine forests found at Hartman Creek provide excellent wildlife habitat. Park staff actively thin oak stands and restore prairies to provide habitat for the state- threatened Karner Blue butterfly.

White-tailed deer, gray squirrels, rabbits, raccoons, chipmunks, and many other mammals are commonly seen in the park. Nearly 200 species of woodland and grassland songbirds, waterfowl, shorebirds, hawks, eagles, and owls have been recorded in the park.

The **Laedtke Memorial Butterfly Garden**, located adjacent to the park office, has several native plants on display and identified. A publication entitled *Wildflowers of Hartman Creek State Park* can be obtained at the office. The brochure lists most of the wildflowers found in the park and what time of the year they bloom.

INTERPRETIVE PROGRAMS AND FACILITIES

Nature hikes are offered from June through Labor Day weekend. Interpretive programs are held at the park's outdoor amphitheater located at the **Middle Picnic Area.**

The **Hellestad House**, located at the **Allen Lake Picnic Area**, serves as the park's interpretive center. This Norwegian-style log cabin was built in 1864 by Ole Olson Hellestad on his farm near the Little Wolf River. The historic structure was moved to Hartman Creek in 1997.

BOATING AND FISHING

There are no boat launches in the park but carry-in non-motorized watercrafts are allowed on all the lakes. Fishing piers are located at both Allen and Marl lakes. Each lake has good populations of bass, northern pike, and panfish. Trout fishing opportunities are available at **Emmon's Creek State Fishery Area** located adjacent to the park. Fishing equipment can be borrowed from the park office and bait can be purchased at the concession stand.

Public boat launch facilities are available on most lakes within the Chain-O-Lakes system. A small canoe/ kayak landing is located on Marl Lake (28 acres) along **Whispering Pines Road.**

ACCESSIBLE FACILITIES AND TRAILS

Most of the park's buildings, shelters, and restroom facilities have paved walkways to accommodate mobility-impaired visitors. Campsite #47 is a wheelchair-accessible site with an asphalt walkway to a restroom/shower facility. An accessible fishing pier is located on Allen Lake adjacent to the picnic area.

> **Ranger Tip**
>
> Over the years, Hartman Creek State Park has become known as the unofficial state park employee "getaway" camping spot. The reputation of this park as quiet vacation destination by many off-duty park staff is very telling as to the quality of the outdoor experience to be found here.

CAMPING

Hartman Creek has 110 campsites (27 electric) all within well-shaded forest areas. The campground has two restroom/shower facilities. An RV water fill and sanitary station is located across the road from the campground entrance.

The park has five tent-only group campsites located along the perimeter of a beautifully restored prairie area in the southwest section of the park. Each site can accommodate up to 50 campers. One wheeled camping unit is allowed in each campsite parking area.

WINTER ACTIVITIES

Park staff groom nine miles of cross-country ski trails for both inline and skate skiers. A separate one-mile snowshoe and hiking trail is also available. Ice fishing is popular on the park's lakes and throughout the Chain O' Lakes system. Hartman Creek's campgrounds are closed from December 1st through early April.

AREA ATTRACTIONS

Waupaca County has long been one of Wisconsin's top tourist destinations, especially the beautiful Chain O'Lakes and Crystal River area. www.waupacaareachamber.com

The **Ding's Dock Crystal River Canoe and Tubing Trips** are popular, family-friendly canoe and tubing trips along the length of the beautiful Crystal

River. Ding's Dock is located about three miles north east of Hartman Creek at E1171 County Road Q, Waupaca, WI 54981. (715) 258-2612. www.dingsdock.com

DIRECTIONS TO THE PARK

Hartman Creek is located about six miles west of the city of Waupaca. From Highway 54, exit on Hartman Creek Road and drive two miles south to the park entrance.

Hartman Creek State Park
N2480 Hartman Creek Road
Waupaca, WI 54981-9727
(715) 258-2372

7. HAVENWOODS STATE FOREST

Wisconsin's Only Urban State Forest

Environmental Awareness Center

FOREST SNAPSHOT

Havenwoods State Forest is 237-acre nature preserve located within the city of Milwaukee; Wisconsin's largest city. The property is a popular outdoor recreational destination for many, especially urban residents who often lack access to traditional state parks, forests and recreations areas.

Havenwoods offers hiking trails, picnicking, and wildlife viewing, especially birdwatching. It is also has one of the finest environmental learning centers in the state and provides year-round nature-based learning experiences for visitors and school groups.

GEOLOGY AND HISTORY

The property's rolling landscape of prairies, wetlands, creeks, and ponds has been altered many times by both natural and man-made conditions. Retreating glaciers near the end of the last Ice Age 12,000 years ago left behind moraine hills, wetlands, rivers, and excavated Lake Michigan a few miles to the east.

Several Native American cultures are known to have inhabited the Milwaukee area for thousands of years, including the Meskwaki (Fox), Sauk, Menominee, Potawatomie, and the Ho Chunk (Winnebago) people. One of the first Europeans to explore the local area was French missionary Fr. Jacques Marquette, who arrived by canoe at the mouth of the Milwaukee River along Lake Michigan in 1674.

Ranger Trivia

The name Milwaukee is attributed to the Algonquin word *Miloke,* meaning "pleasant land". The Potawatomi referred to the area at the confluence of Lake Michigan and the Milwaukee River as *Minwakin,* a "gathering place by the water."

Large numbers of European immigrants arrived in the Milwaukee area in the early 1880s. Many built farmsteads to raise crops and cattle, while merchants and tradesmen added to the growing city of Milwaukee. As the local population grew, the need for public services such as law enforcement grew as well. In 1917, Milwaukee County built a House of Corrections prison on what is now the Havenwoods property. During World War II, the county prison was taken over by the U. S. Army to incarcerate mostly German prisoners-of-war and court-marshaled American soldiers.

In 1956, during the Cold War Era, the U.S. Army built an underground launch silo for a Nike Ajax missile here. They also installed eight other Nike missile sites encircling the Milwaukee area to shield the city from long-range Soviet bombing attacks. These sites soon became obsolete due to the development of intercontinental missiles.

After the World War II, the property was used for Army Reserve training exercises. When the federal government abandoned the site in 1974, a group of real estate developers proposed building houses, factories, and apartment housing on the site. A local group of citizens and politicians apposed more residential development. They envisioned turning the property into an urban green space where citizens could enjoy nature and learn about the environment. This grassroots effort led the way for the state to develop Havenwoods State Forest as an outdoor recreation area and environmental learning center in 1980.

Ranger Trivia

The name Havenwoods was chosen in a local naming contest for the new state forest. The winning entry was submitted by a student from Custer High School in Milwaukee.

PICNIC AREAS

A picnic area is located adjacent to the environmental center and next to the visitor parking area. Water fountains and restrooms are located inside the nature center.

HIKING, BIKING AND NATURE TRAILS

Havenwoods has six miles of hiking trails that are marked with color-coded emblems. Trail maps are available inside the Environmental Center and online. A property map is also posted on the kiosk outside the center. Leashed pets are allowed on most trails but are prohibited in wildlife refuge areas off marked trails. A designated two-mile **Pet Trail** has been set aside specifically for pet owners. A two-mile biking/hiking route follows the outer perimeter of the property.

The **Trek Through Time Trail** is a fun route for those interested in geocaching. A guide book for this trail and GPS unit can be borrowed at the Environmental Center. The **People and the Land Trail** (1.0 miles) is a self-guided nature trail that highlights the history of the property and the ongoing efforts to restore its flora and fauna.

Ranger Trail Pick

To experience the diverse landscape of Havenwoods, a combination of various trails is recommended. Begin west of the nature center and follow the trail about a quarter mile to the 120-foot-long **Lincoln Creek Bridge**. Beyond the bridge are several looped trails to explore through restored prairie areas. Cross back over the bridge and take the first trail to the right (south), which has interpretive panels and views of several man-made ponds. A short side trail leads to a boardwalk and pier at one of these ponds. Farther down this trail are several looped trails that cross over **Intermittent Creek** on the south end of the forest. From here a wide gravel roadway leads north back to the **Environmental Center**. Hikers who wish to explore further can hike a series of side trails through a forested area on the east side of the property that also lead back to the center.

WILDLIFE VIEWING

The once barren, degraded landscape at Havenwoods has been transformed into a mix of attractive wetlands, ponds, creeks, restored prairies, and small woodlots where native plants and animals once again thrive. More than 160 different species birds have been observed here, including hawks, owls, waterfowl, great blue herons, tree swallows, and bluebirds. White-tailed deer, squirrels, chipmunks, rabbits, and raccoon are commonly seen along hiking the trails. More secretive animals such as red fox and coyotes are also spotted here as well.

INTERPRETIVE PROGRAMS AND FACILITIES

Havenwood's Natural Resource Educators offer environmental education and outdoor learning programs for visitors throughout the year including interpretive hikes, educator workshops and children's nature activities. Forest staff also

Pond and prairie area

provide indoor and outdoor classes for thousands of elementary students each year.

As part of the property's "green" educational efforts, several showcase gardens are maintained at Havenwoods, including a **Kid's Garden**, **Naturalist's Backyard Garden**, and **Heritage Vegetable Gardens**. The property also has a butterfly garden, rain garden, and an urban arboretum, which features native wildflowers and trees.

BOATING AND FISHING

Havenwoods State Forest does not have any boating or fishing opportunities.

ACCESSIBLE FACILITIES AND TRAILS

The Environmental Awareness Center, restrooms, and auditorium are accessible for mobility-impaired visitors. The **People and the Land Trail** (1.3 miles) is a designated wheelchair-accessible route and self-guided interpretive trail. An interpretive brochure for this trail is available at the Environmental Center.

CAMPING

Havenwoods is day-use-only property so there are no camping facilities here. Two state park campgrounds located about 30 miles from Milwaukee include Harrington Beach State Park in Ozaukee County (see Chapter 5) and Pike Lake State Park in Washington County (see Chapter 21).

WINTER ACTIVITIES

Cross-country skiing, snowshoeing, hiking, and wildlife watching are popular winter activities. The property's trails are not groomed but the parking area is plowed. Snowshoes can be borrowed from the Environmental Awareness Center.

AREA ATTRACTIONS

The city of Milwaukee ranks as the top tourist destination in the state due to its multitude of museums, theaters, parks, and other world class attractions. www.visitmilwaukee.org

The **Schlitz Audubon Nature Center** is a 185-acre nature preserve located along the Lake Michigan shoreline. The property has miles of hiking trails and is home to the nationally recognized **Vallier Environmental Learning Center,** which features live raptor displays and offers interpretive programing throughout the year. Schlitz Audubon is located about seven miles east of Havenwoods at 1111 E. Brown Deer Rd. in the community of Bayside. www.schlitzaudubon.org

DIRECTIONS TO HAVENWOODS STATE FOREST

Havenwoods is located on the north side of Milwaukee. From Interstate I-41 or Hwy 145, exit east onto W. Silver Spring Drive and turn north onto N. Sherman Blvd. Exit onto W. Douglas Ave. to the state forest entrance.

Havenwoods State Forest
6141 N. Hopkins Street
Milwaukee, WI 53209
(414) 527-0232

Ranger Note

State park entrance fees are not required at Havenwoods State Forest. The property is open every day from 6 a.m. to 8 p.m. The Environmental Awareness Center is open most days except Sundays and holidays.

---

8. HERITAGE HILL STATE PARK

Living History Park—A Step Back in Time

Belgian-American farmhouse

PARK SNAPSHOT

Heritage Hill State Park is perched on a hillside overlooking the Fox River near the city of Green Bay. The park has a collection of 24 historic buildings that house thousands of rare artifacts, some dating as far back as 1672. The property operates as a living history park. Interpreters dressed in period costumes offer presentations about everyday life in early Wisconsin and provide guided tours at several buildings.

Park buildings and displays are organized by historical eras ranging from the Canadian-French fur trading posts of the mid-17th century all the way to the 19th century, when thousands of European immigrants arrived to set up farmsteads and establish towns throughout Wisconsin.

GEOLOGY AND HISTORY

Heritage Hill is located on a glacial moraine hill that slopes west towards the Fox River. As the last Ice Age came to an end about 12,000 years ago, massive geologic forces caused a large region of northeastern Wisconsin to tilt towards the east. As a result, the Fox River, which once flowed to the west, changed course and began to flow east, and carved its river channel ever deeper over time. Today, the Fox River continues its fast-paced 49-mile descent from Lake Winnebago to the waters of Green Bay and Lake Michigan.

In 1634, French-Canadian explorer Jean Nicolet arrived by canoe at Red Banks, a few miles northeast of the Fox River in Green Bay. Although he was the first European ever to step foot in Wisconsin, only a small band of Ho-Chunk people was there to greet him. At that time, at least 6,500 other Native American people lived in the nearby area, including the Menominee, Meskwaki (Fox), and the Ojibwe (Chippewa). Nicolet's most important "discovery" was the well-established Native American canoe trade route west to the Mississippi River from the Fox River at Green Bay.

Heritage Hill is located on part of the original site of **Camp Smith**, an upriver wilderness military post of **Fort Howard** in Green Bay. Both forts were built around **1816** to defend U.S. interests along the important Fox River trade route and to provide military protection for settlements in the western frontier.

The effort to develop Heritage Hill State Park began in the early 1960s by members of the Brown County Historical Society. The historical society was in the process of saving several historic buildings from demolition in the Green Bay area and wanted to relocate them to a site along Fox River. At the time the Heritage Hill property was owned by the Green Bay State Reformatory and was used to grow vegetable crops and apples for use at the reformatory. In 1975, a six-lane freeway (Highway 172) was built over the Fox River, cutting off the reformatory from its farm fields. In 1977, the State, in partnership with the Brown County Historical Society, acquired the property and established Heritage Hill State Park.

> **Ranger Trivia**
>
> Jean Nicolet was the first to name the bay area *La Baie Verte*, "the Green Bay" due to the greenish color of the water. He also claimed the entire Fox River region for the King of France in his misguided belief that it was a river trade route to China. Nicolet never got a chance to explore his fabled route to the Orient. He drowned at age 44 when his boat capsized during a storm while traveling the St. Lawrence River in 1642.

BIKING, HIKING, AND NATURE TRAILS

A section of the 25-mile **Fox River State Recreational Trail** leads along the western boundary of Heritage Hill State Park. This popular asphalt-paved, biking/hiking trail follows an abandoned railroad grade adjacent to the Fox River. Several interesting interpretive panels along the route highlight the history of the area.

The historic buildings and outdoor displays at Heritage Hill are connected throughout the site by developed foot trails and roadways. Interpretive signs are posted along walking trails and at most buildings.

1851 Moravian Church

INTERPRETIVE PROGRAMS AND FACILITIES

Interpreters dressed in period costumes representing of various historical eras conduct tours at several of buildings. The **LaBaye Area** portrays the fur trade era from the mid-1600s to the early 19th century. On displays here are several pioneer log cabins including an original fur trading post built in the early 1800s. Inside the trading post are various goods that would have been used in trade with Native American trappers. A 1830s log structure representing Wisconsin's first courthouse, which was located at this site, is also on display.

The **Fort Howard Area** portrays a time period after the War of 1812 when the U.S. government built military forts such as **Camp Smith** along the Fox River to defend the western frontier. The site preserves some of the original buildings of Camp Smith. It also has an Army officer's quarters, a company kitchen, a civilian schoolhouse, and a military post hospital. The **Cotton House** is beautiful Greek revival-style home built in 1840 for the John Cotton Family. The house is on the National Register of Historic Places and is open for guided tours.

The **Growing Community Area** features structures built from the 1850s to the end of the 19th century, including a beautifully restored Moravian church built in 1851. Other buildings include a blacksmith shop, a printer's shop, and the Franklin Hose Firefighter Company shed. One structure, the **1835 Baird Law Office**, has been designated as a National Register of Historic Places.

The **Ethnic Agricultural Area** interprets a period starting around 1853 when French-speaking Belgium immigrants and other ethnic groups from Western Europe arrived in the local area. This area features a cheese factory, a roadside chapel, and several other structures.

> **Ranger Pick**
>
> The Belgian Farmstead showcases an impressive red and white-brick farmhouse, which has been meticulously restored and furnished with artifacts from the 1800s. The farmstead includes a barn with several outbuildings and live farm animals.

BOATING AND FISHING

There are no boating and fishing opportunities at Heritage Hill. The nearby Fox River has several public boat launches, including the **Bomier Boat Landing** and the **Fox Point Boat Launch** located on the east side the river in the nearby city of DePere.

The Fox River is known for its excellent walleye fishing, especially during the early spring spawning period. The river also has good populations of panfish, northern pike, white bass, and smallmouth bass.

ACCESSIBLE FACILITIES AND TRAILS

Most of the facilities at Heritage Hill are accessible for mobility-impaired visitors. Wheelchair ramps and walkways (some paved) are located throughout the park. Golf cart trams are available most days to transport guests throughout the property.

CAMPING

Heritage Hill is a day-use-only park so there are no camping facilities here. The **Brown County Fairgrounds** offer 57 urban-type campsites with electric service and restroom/shower facilities. The fairground is located a few miles southeast of Heritage Hill at 1500 Fort Howard Avenue in De Pere. www.browncountyparks.org

WINTER ACTIVITIES

Heritage Hill State Park remains open in winter from Tuesday through Friday of each week. None of the historic buildings are open and interpreters are not available during this period.

AREA ATTRACTIONS

The Green Bay area is one of the most popular tourist destinations in the state due to its many attractions, including the Green Bay Packers Hall Fame, the Neville Public Museum, and the Bay Beach Amusement Park. www.greenbay.com

Bay Beach Wildlife Sanctuary is a 600-acre wildlife refuge and rehabilitation center. The site has live birds of prey displays and several open water lagoons that attract hundreds of geese and ducks. A modern nature center and the Wisconsin Wildlife Zoo are also located here. Bay Beach is located about two miles north of Heritage Hill along Green Bay's south shoreline at 1660 E. Shore Drive in Green Bay. www.baybeachwildlife.com

DIRECTIONS TO THE PARK

Heritage Hill is located on the southeast side of Green Bay. From U.S. Interstate I-43 or I-41 exit onto Highway 172. Turn north onto Webster Avenue off Highway 172 to the park entrance.

Heritage Hill State Historical Park
2640 S. Webster Avenue
Green Bay, WI 54301
(920) 448-5150 or (800) 721-5150
www.heritagehillgb.org

9. HIGH CLIFF STATE PARK
Majestic Park Overlooks Lake Winnebago

View of Lake Winnebago

PARK SNAPSHOT

High Cliff State Park is located along the northeast shoreline of Lake Winnebago, Wisconsin's largest freshwater lake. The park was named after the scenic limestone bluffs that tower more than 200 feet above the lake.

This popular 1,147-acre state park offers camping, picnicking, swimming, boating, nature study, wildlife viewing, and miles of hiking, biking, and equestrian trails. It also has one of the finest marina and boat launch facilities in the state.

GEOLOGY AND HISTORY

The impressive bluffs at High Cliff are composed of dolomite limestone. This hard rock was formed from the sediments of sand, mud, and ancient sealife laid down more than 400 million years ago beneath ancient seas that once covered most of Wisconsin. The park's limestone bluffs are part of the Niagara Escarpment, a massive 1,000-mile-long dolomite rock ridge that stretches from northern Illinois north along the eastern shore of Lake Winnebago, through Door County and upper Michigan, all the way to its namesake at Niagara Falls in New York State.

51

The tall limestone bluffs at High Cliff resisted most of the crushing effect of Ice Age glaciers which pushed through this region of Wisconsin 30,000 years ago. The base of the bluffs were altered somewhat by ice and were also eroded by wave action of glacial Lake Oshkosh; now called Lake Winnebago.

Many Native American people lived along the shoreline of Lake Winnebago for hundreds of years. More recent tribes included the Fox (Meskwaki), Sauk, and the Winnebago (Ho-Chunk) people who lived, hunted and raised crops in this area for generations. The park preserves several ancient burial mounds, many in the shape of animals, which are believed to have built by Effigy Mound people as far back as 1,500 years ago.

> **Ranger Note**
>
> Lake Oshkosh was an Ice Age lake that formed behind a receding glacier about 14,000 years ago. This massive lake was 65 feet higher than the current water level of Lake Winnebago and sub-merged a large area of the state from Portage to Green Bay.

> **Ranger Note**
>
> A 12-foot-tall bronze statue of Chief Red Bird, a prominent leader of the Winnebago Nation, is located at an overlook area in the Upper Picnic Area. Red Bird was a trusted ally of the U.S. government in the early 1820s, but changed his allegiance and became a leader in what would be called the "Winnebago war against the United States." He surrendered in 1827 in hopes of avoiding an all-out war against his people, but died in prison while awaiting his trial.

In 1855, a mining company began excavating deposits of clay from the base of the park's bluffs for use in making bricks. This led to the establishment of a new mining town called "Clifton." This small settlement was later renamed **High Cliff** and was composed of several homes, a company office, post office, a general store, and a tavern. In 1870, quarry operations shifted to mining limestone in the bluff areas. Rock was blasted apart, broken into smaller chunks and deposited into the top of wood-fired kilns. Powdered lime was extracted from the bottom of the kilns and sold for use in the manufacture of plaster, cement, and to enrich agricultural soils. The limestone quarry closed in 1957 and was sold to the State to develop High Cliff State Park.

BEACHES, PICNIC AREAS, AND SHELTERS

High Cliff's shady forests, bluffs and cool refreshing breezes along Lake Winnebago have attracted visitors to this area for generations. The **Lower Picnic Area** has a popular sandy swimming beach and bathhouse along the Lake Winnebago shoreline.

The **Upper Picnic Area**, located on top of the bluff area, features a 40-foot wooden observation tower that offers panoramic views of Lake Winnebago. An attractive pavilion with an indoor central fireplace and a large open-air shelter are also located here. Both facilities can be reserved in advance.

HIKING, BIKING AND NATURE TRAILS

High Cliff has nine miles of hiking trails to explore. Most routes are easy to hike but some trail sections are close to bluff edges with steep drop-offs. Hikers are strongly advised to stay on the marked trails at all times when hiking along bluff areas.

The **Lime Kiln Trail** (1.7 miles) starts near the historic lime kiln interpretive area in the lower picnic area. The trail leads south parallel to Lake Winnebago shoreline and ascends to the upper Niagara Escarpment bluff area. This trail is very scenic but is rocky in areas and has a few wet spots near hillside springs.

> **Ranger Note**
>
> The forested bluff area along the **Lime Kiln Trail** is a designated State Natural Area and contains several rare plant, ferns, and outstanding rock formations. Watch for tall, ancient-looking trees with distinctive "bumps" on their trunks along this trail. These are **hackberry** trees, a species not often found in this part of Wisconsin.

The **Indian Mound Nature Trail** (0.6 miles), located in the upper park area, leads past several ancient effigy burial mounds. Archeologists believe these

Rock outcrops along Lime Kiln Trail

53

> **Ranger Trail Pick**
>
> The **Red Bird Trail** (3.8 miles) begins in the upper picnic area and leads south past a statue of Chief Red Bird to several overlook areas with panoramic views of Lake Winnebago. A staircase near the campground provides access the lower bluff area for a close up view of limestone rock formations.

animal-shaped effigy mounds built in the shape of panther water spirits, deer, and buffalo most likely had important spiritual significance for these prehistoric people.

The **Horse and Off-Road Bike Trails** (7.75-miles) are located east of the campground area. These are the longest trails in the park and loop through open grasslands and upland hardwood forests.

WILDLIFE VIEWING

High Cliff's varied landscape of rocky bluffs, wetlands, grasslands and forests provide ideal habitat for wildlife such as white-tailed deer, raccoon, opossum, squirrels, woodland songbirds, bluebirds, tree swallows, and fox sparrows. Turkey vultures, hawks, and bald eagles are occasionally seen soaring above bluff areas. The Lake Winnebago shoreline and inland ponds provides excellent viewing for purple martins, ducks, geese and shorebirds.

INTERPRETIVE PROGRAMS AND FACILITIES

Nature hikes and evening programs are presented throughout the season by park naturalists and guest speakers. The **General Store Museum** located in the lower picnic area, serves as the park's interpretive center. This historic red brick building was once a post office, telegraph office, and supply store for the mining company town in the 1890s. The general store is open on weekend afternoons during the summer season.

BOATING AND FISHING

Lake Winnebago (138,000 acres) is the largest inland lake in Wisconsin and attracts sail and motorboat enthusiasts from all over the Midwest. Anglers can expect to catch walleye, northern pike, white bass, perch, crappie and several types of panfish in the lake. Fishing from the lakeshore is also popular and at **Butterfly Pond**, a small man-made pond located near the park office.

High Cliff Marina has several public boat launch lanes. Boaters can spend the night at the marina or moor their watercraft for the entire season in one of the 100 boats slips for rent. The marina store sells a variety of items, including snacks, clothing, boating supplies, and camping items. A lakeside launch area adjacent to the marina allows paddlers to launch smaller watercraft such as kayaks, canoes, and wind-surf boards.

ACCESSIBLE FACILITIES AND TRAILS

The park's visitor station, shelter facilities, restrooms and bathhouse are accessible for mobility-impaired visitors. Campsites 26 and 86 are wheelchair accessible and have electrical service. A modern cabin with an accessible kitchen, living room, bedroom, restroom, and a wheel-in shower can be reserved in advance for campers with physical disabilities.

The **Butterfly Pond Trail** (1.5 miles) is an asphalt-paved accessible walkway located in the lower area of the park just west of the park office. The trail has interpretive signs that describe flora and fauna of area.

CAMPING

High Cliff has 121 campsites (32 electric) located within a mature, well-shaded forest. Campsites are well spaced and most have adequate room for either tents or wheeled camping units. A flush toilet/shower facility is located near the entrance to the campground.

The park also has an outdoor group camp with eight mostly shaded campsites. The group camp has vault-type restrooms and can accommodate up to 255 people. An RV water fill/ sanitary station is located along the main road east of the group campground.

WINTER ACTIVITIES

Park staff grooms six miles of cross-country ski trails for inline skiing. Snowshoes can be checked out at the park entrance station and used on any trail not groomed for skiing. Fat tire biking is allowed along a section of the Red Bird Trail.

Lake Winnebago is a premier ice fishing destination for thousands of anglers in search of walleye, perch, and white bass. The lake is also known for the dozens of ice shanty towns that pop up all across the frozen lake each February during the annual lake sturgeon spearing season.

AREA ATTRACTIONS

Calumet County and the Fox River Area of Outagamie County have many popular tourist attractions. www.travelcalumet.com   www.foxcities.org

The **1,000 Islands Environmental Center** is a 350-acre nature preserve located along the scenic Fox River. The site has a modern nature center which has several interpretive and live animal displays. The preserve has 10 miles of hiking trails, including several wooden walkways adjacent to the fast-flowing Fox River. Bald eagles are often seen along the Fox River here, especially in winter. The center is located about eight miles north of High Cliff off of Highway 55 at 100 Beauliea Court, Kaukauna, WI 54130   (920) 766-4733. www.1000islandsenvironmentalcenter.org

DIRECTIONS TO THE PARK
From U.S. Hwy. 10 southeast of the Appleton/Menasha area, exit onto Hwy 114 toward the village of Sherwood. Exit onto Pigeon Road south and follow the directional signs to the park entrance visitor station.

High Cliff State Park
N7630 State Park Road
Sherwood, WI 54169
(920) 989-1106

10. KETTLE MORAINE STATE FOREST–NORTHERN UNIT

Heart of Wisconsin's Ice Age Legacy

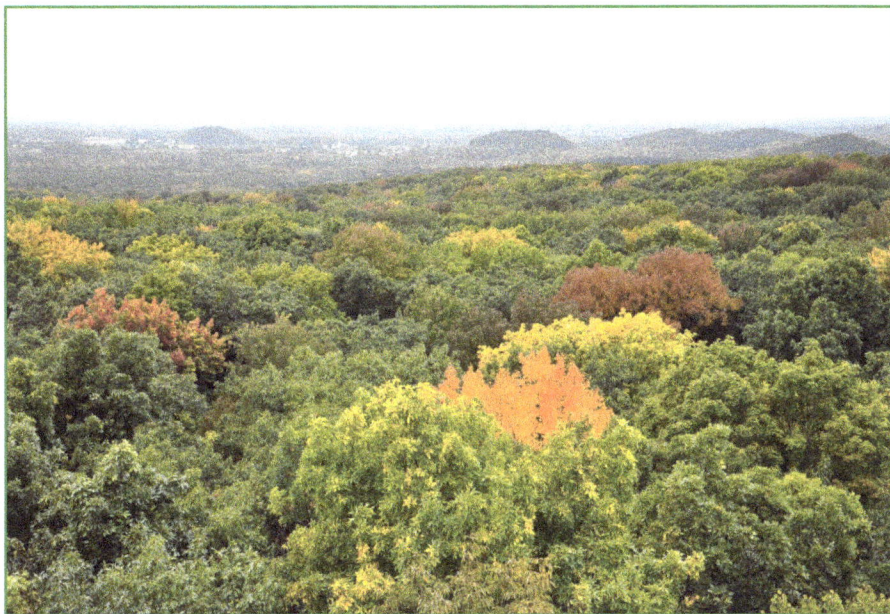

Parnell Observation Tower view

FOREST SNAPSHOT

The Kettle Moraine State Forest-Northern Unit is the largest state-owned property in southern Wisconsin. The 30,932-acre forest stretches for 25 miles from Sheboygan and Fond du Lac counties south into Washington County.

The forest offers a multitude of outdoor recreational activities, including camping, hiking, swimming, picnicking, biking, horseback riding, boating, fishing, hunting, cross-country skiing, and nature study. The Kettle Moraine is a unit of the Ice Age Reserve, a National Park Service program established to preserve Wisconsin's outstanding glacial features.

GEOLOGY AND HISTORY

The Kettle Moraine Forest is a labyrinth of rolling hills, lakes, wetlands and streams. The rugged landscape of the forest was shaped during the last Ice Age about 25,000 years ago when two massive glaciers collided here. As the glaciers began to recede northward about 12,000 years ago they deposited towering hills of sand, gravel, rock, silt and other glacial till at the base and outer edges of retreating ice fields. Geologists refer to these elevated hills as **moraines**. Retreating glaciers also left behind thousands depressions in the landscape called **kettles**. Kettles were formed when large chunks of glacial ice broke off the main ice sheet and became buried beneath glacial till. When these pockets of underground ice melted, deep holes or depressions in the landscape appeared. Some of these kettles remain dry, while others closer to the water table hold seasonal ponds.

Larger and deeper depressions filled with water and became **kettle lakes**.

> ### Ranger Trivia
> Many kettle formations are circular in shape. Early settlers to the region thought these round depressions resembled the iron caldrons or "kettles" they used for cooking food or making soap. This descriptive term continues to be in used by scientists and the general public alike.

Native American people including the Ojibwa (Chippewa), Ho-Chunk (Winnebago) and the Potawatomi lived within the Kettle Moraine region for generations prior to European settlement. In 1833, the Potawatomi were forced to cede all of their land in eastern Wisconsin to the U.S. government and relocate to a reservation in the Iowa Territory. Many years later, displaced Potawatomi people were able to return to Wisconsin when the tribe obtained ownership of their own 15,000-acre reservation in Forest County.

In the 18th and 19th centuries, thousands of European immigrant farmers settled in the Kettle Moraine area. Thousands of acres of forested land were leveled to create open farm fields to raise cattle and grow crops. Within a few decades, more than half of the forest trees of the Kettle Moraine area were cleared for agriculture. In the 1920s, a grassroots effort of concerned citizens lobbied lawmakers in Madison to purchase more land for public recreation in eastern Wisconsin. At the same time, legislators were also being urged to protect the Milwaukee River watershed and enhance timber production. All these efforts eventually led to establishment of the Kettle Moraine State Forest in 1937.

BEACHES, PICNIC AREAS, AND SHELTERS

The forest has two main recreation areas at Mauthe Lake and Long Lake. Both have picnic areas, shelters, swimming beaches, and bathhouses. Mauthe Lake also has a concession stand that offers prepared food, beverages, and canoe/kayak rentals.

HIKING, BIKING, AND NATURE TRAILS

The Northern Unit State Forest has more than 80 miles of hiking trails to explore, including a 31-mile segment of the **Ice Age National Scenic Trail**. The **Greenbush Recreational Trails** (13.5 miles) and **New Fane Trails** (5.5 miles) have multiple loops of hiking routes that lead through mature hardwood forests and pine plantations.

Butler Lake Trail (1.5 miles) overlooks a small kettle lake atop a steep-sided glacial formation called an **esker**. An esker is a long, narrow, snake-like ridge composed of sand, gravel, and silt deposited by a stream that flowed beneath glacial ice. Butler Lake is located east of the Long Lake Recreation Area. The **Zillmer Trail System** (11.0 miles) is great for hiking any time of the year but the trails here are especially popular with cross-country skiers in winter.

The **Tamarack Trail** (2.0 miles) circles the entire perimeter of Mauthe Lake. The trail leads through oak, cedar, tamarack, and pine forests and crosses a bridge spanning the Milwaukee River. **Spruce Lake Bog Nature Trail** (0.25 miles) located west of Long Lake, offers one of the most unique hikes in the forest. This short trail follows a raised boardwalk through a spruce bog that harbors rare bog plants such as sundews, lady slipper orchids, and pitcher plants.

Summit Trail (1.0 mile) is a self-guided nature trail located adjacent to the Long Lake campground. This trail leads uphill 250 feet to the crest of Dundee Mountain, a cone-shaped moraine hill called a **kame**. Interpretive panels along the trail describe how glacial activity created the kame and surrounding landforms.

Ranger Trail Pick

The **Parnell Tower Trail** (3.5 miles) begins with a 362-step staircase climb to the top of a moraine hill. Here a 60-foot-high observation tower provides beautiful panoramic view of the Kettle Moraine's glacial landscape. The trail then leads through a rolling landscape of oak and maple forests before looping back to the tower. Parnell Tower is located adjacent to County Highway U in the northern section of the forest.

The **Lake to Lake Bike Trail** (6.5 miles) is a crushed limestone biking/hiking trail that connects the Long Lake and Mauthe Lake recreation areas. Off-road bikers have 14 miles of rugged trails to ride. Most routes are located within the Greenbush Trail area.

The forest maintains 42 miles of equestrian trails. Horseback riders can camp overnight at the 22-unit equestrian campground located along County Road SS near New Prospect.

WILDLIFE VIEWING

The large expanse of forests, lakes, streams, wetlands, and grasslands of the Kettle Moraine provide ideal habitat for many wildlife species. Common mammals here include white-tailed deer, gray and fox squirrels, raccoon, woodchucks, rabbits, and chipmunks. More secretive animals such as the flying squirrel, skunk, mink, otter, fox, and coyote can be found here as well.

The Wisconsin Society for Ornithology has listed the Kettle Moraine Forest as one of top 50 birding sites in the state. More than 260 species of birds have been identified within the forest. During the spring and fall migration period many species of hawks, waterfowl, warblers, and woodland songbirds can be seen here.

The Northern Unit of the Kettle Moraine has twelve State Natural Areas. One of them, the **Haskell Noves Memorial Woods**, has never been logged out and preserves many old-age oak and maple trees. This site is also a designated *National Watchable Wildlife Area*. Haskell Noves Woods is located along County Highway GGG north of the Mauthe Lake Recreation Area.

INTERPRETIVE PROGRAMS AND FACILITIES

Interpretive programs and nature hikes are presented by forest naturalists and guest speakers throughout the year. The **Henry S. Reuss Ice Age Visitor Center,** located off Highway 67 near Dundee, is one of the most visited sites in the forest. The center has exhibits highlighting the glacial geology and human history of the area. A 20-minute movie shown in the center's auditorium describes how glacial landscapes were created during the last Ice Age.

View from the Greenbush Kettle wayside

The **Moraine Nature Trail** (0.75 miles) located near the Ice Age Visitor Center, has interpretive signs identifying native trees and explains how each was used by Native American people and early pioneers.

The **Greenbush Kettle Wayside** offers a scenic view of a large **kettle pond.** The site has a viewing platform with interpretive panels and short trail that descends to edge of the pond. The site is located along Kettle Moraine Drive about one mile north of Highway 67.

BOATING AND FISHING

The forest has 12 kettle lakes, several clear-running creeks, and a section of the East Branch of the Milwaukee River. **Mauthe Lake** (78 acres) has a boat launch and canoe/kayak rentals. Boating is restricted to electric motors only. **Long Lake** (423 acres) is the largest and most popular lake in the forest for fishing, swimming, boating, and waterskiing. A boat launch is available at the Long Lake Recreation area. Anglers can expect to catch blue gill, sunfish, crappie, northern pike, and largemouth bass in most lakes in the forest.

ACCESSIBLE FACILTIES AND TRAILS

Most of the forest's buildings, including the Ice Age Visitor Center, are accessible for mobility-impaired visitors. Both Mauthe Lake and Long Lake recreation areas have accessible campsites with electric service and wheelchair-accessible asphalt walkways to picnic areas and fishing piers.

CAMPING

The forest has 371 campsites. **Mauthe Lake Campground**, located in the southern section of the forest off County Highway GGG, has 135 campsites (51 electric). This campground is very popular due to its close proximity to the swimming beach and picnic areas. **Long Lake Campground** is located in the northern section of the forest off Division Road. The campground has 200 campsites (no electric). Both Mauthe and Long Lake campgrounds have multiple restroom/showers facilities.

The **Greenbush Group Campground** has nine well-spaced campsites (tent only) with access to vault-type restrooms. Each campsite can accommodate up to 40 people.

There are five **Backpack Shelter Campsites** located along the Ice Age National Scenic Trail throughout the forest. Each backpack shelter area has a fire ring and a vault-type restroom. Campers must pack in drinking water.

WINTER ACTIVITIES

Forest staff and volunteers of the Kettle Moraine Nordic Ski Club groom 25 miles of trails for both inline and skate skiing. The Greenbush and Zilmer ski areas have heated indoor shelters located at their trailheads.

Other off-season activities include snowshoeing, hiking, wildlife watching, and ice fishing. Winter camping is available at the Mauthe Lake Campground. The forest has 60 miles of snowmobile trails linked to county trails.

AREA ATTRACTIONS

There are several small towns, museums, and other visitor attractions within the Fond du Lac and Kewaskum area. www.fdl.com   www.kewaskum.org

The **Wade House Historical Site and Wesley Jung Carriage Museum** is a 240-acre outdoor museum that features an historic 1850s stagecoach inn. The site also has a world-class antique carriage museum and a water-powered sawmill. The Wade House is managed by the State Historical Society and is located on the north end of the forest at W7965 Highway 23, Greenbush, WI 53026. (920) 526-3271. www.wadehouse.org

DIRECTIONS TO THE FOREST

The Northern Unit of the Kettle Moraine is located in parts of Sheboygan, Fond du Lac and Washington Counties. The forest headquarters is located along County Highway G south of U.S. Highway 67.

Kettle Moraine State Forest–Northern Unit
Forest Headquarters
N1765 County Highway G
Campbellsport, WI 53010-3303
(262) 626-2116

11. Kettle Moraine State Forest– Southern Unit

Wisconsin's Own "Smoky Mountains"

Fall colors at Kettle Moraine State Forest

FOREST SNAPSHOT

A French map from the 1820s identified a region of towering hills located in the southeastern Wisconsin as the *Smoky Mountains*. Today, this 200-year-old depiction of what is now part of the Kettle Moraine Forest is still an apt description of the property, especially on cool summer mornings when a haze of fog cloaks the hills and valleys of the forest.

The Kettle Moraine Sate Forest–Southern Unit encompasses **22,300** acres that stretch for **40** miles through Walworth, Jefferson, and Waukesha counties. More than **1.2** million visitors from throughout the Midwest flock to this popular state forest each year to camp, swim, picnic, hike, bike, hunt, fish, boat, ride horses, ski, or just relax and enjoy nature at its best.

GEOLOGY AND HISTORY

The rolling landscape of the forest was shaped by continental glaciers that pushed through this region of the state about 30,000 years ago. The Kettle Moraine area and much of southeastern Wisconsin are considered some of the best sites in the world to view glacial formations. The most noticeable of these are the many **moraine** hills found throughout the forest. Moraine hills were built up layer by layer from deposits of sand, gravel, rocks, and other glacial debris at the base and sides of retreating glaciers near the end of last Ice Age about 12,000 years ago. The glacial activity also left behind thousands of **kettle** depressions. Kettles were formed when massive chunks of ice broke off larger sheets of glacial ice and became buried beneath glacial till. When the underground ice melted, large, often round *kettle-like* depressions in the earth were exposed. Most kettles remained high and dry but some depressions held seasonal or permanent ponds. Many large and deep kettles filled with water and are now permanent lakes.

Several Native American cultures have hunted, fished, and lived in the Kettle Moraine region over time including the Ojibwe, Sauk, Meskwaki, Ho-Chunk, and the Potawatomi people. The first permanent European immigrants arrived here in the early 19th century to set up farmsteads. The new settlers were welcomed at first by local people but over time, disagreements arose over hunting and land rights. When the *Black Hawk War* broke out in 1832, the Potawatomi sided with the U.S. Army against the Sauk people. Despite their allegiance to the U.S. government during the war, Potawatomi leaders were themselves forced to cede all their lands in southeast Wisconsin through the Treaty of Chicago in 1833.

> **Ranger Note**
>
> The Black Hawk War of 1832 started when Black Hawk, a respected Sauk leader, along with 1,600 of his followers , attempted to reclaim their former village sites in Illinois. The Sauk had been forcefully removed from Wisconsin and Illinois to the Iowa Territory in 1830. During the conflict, Black Hawk eluded U.S. government troops twice by hiding in the wooded hills of what is now the Kettle Moraine State Forest. The war ended tragically when most of Black Hawk's followers were killed during the Battle of Bad Axe along the Mississippi River.

By the early 20th century, nearly seventy percent of forest land in the Southern Kettle Moraine had been leveled for agriculture. Conservation groups such as the Izaak Walton League began to lobby lawmakers to preserve the remaining forests of the area for public recreation, timber protection, and to safeguard water resources. In 1937, the State established the Kettle Moraine State Forest to fulfil these needs.

BEACHES, PICNIC AREAS, AND SHELTERS

The Ottawa Lake Recreation area has the most popular swimming beach and picnic area in the forest. A smaller beach is located along Whitewater Lake. Both sites have picnic areas and open-air shelter facilities. The forest has two indoor facilities, the **D. J. Mackie** and **Nordic** shelters. Both have electric service, fireplaces, and can be reserved in advance.

HIKING, BIKING, AND NATURE TRAILS

There are 127 miles of hiking trails in the forest, including a 30-mile section of the Ice Age National Scenic Trail. In addition, the forest has 87 miles of equestrian trails, 54 miles of off-road and tour bike trails, and a marked canoe trail on Ottawa Lake.

The **John Muir Trail** (12 miles) and the **Emma Carlin Trail** (8 miles) have trail routes of varying distances that lead through oak, hickory, and white pine forests. Off-road bicycle riding is allowed these trails as well.

The **Scuppernong Springs Nature Trail** (1.5 miles) is a popular trail near the Ottawa Lake Recreation Area. Several interpretive stops feature historic sites along the trail, including a former trout hatchery, cheese factory, a tourist hotel, and a former mining site for *marl*, a calcium-rich soil found in wetlands once used to improve farmland.

Paradise Springs

> ### Ranger Trail Pick
>
> The **Paradise Springs Nature Trail** (0.5 miles) has a paved walkway that leads to a beautiful spring-fed trout pond. In the 1920s, a horse race track and golf course were built here by Louis Petit, a Milwaukee millionaire. In the late 1940s, a fashionable resort was built adjacent to the pond, as well as a commercial spring water bottling plant. At the end of the trail are the remains of a springhouse. This 100-year-old fieldstone structure was built to shelter the spring outlet, which releases 30,000 gallons of cold water per hour into the adjacent trout pond.

Bald Bluff Trail (0.5 miles) is a short but strenuous 1,050-foot hike to the top of one of the tallest hills in the forest. Early pioneers named it *Bald Bluff* because of the lack of trees on its crest. Archeologists believe this open hilltop was used by Native Americans for ceremonial gatherings and to send smoke signals during daylight hours. Large beacon fires lit at night could be seen for miles around from this hill.

There are five self-guided nature trails throughout the forest. The **Rice Lake Nature Trail** (0.5miles) features wetland flora and fauna. The **Stute Springs Trail** (1.0 mile) leads to an historic farmstead with a spring house and other farm structures built by the Stute family in the 1850s. A side trail leads to the top of *The Big Hill*, the highest point in the Kettle Moraine with an outstanding view of the forests and farms below.

WILDLIFE VIEWING

There are nine State Natural Areas located throughout the property. The hardwood forests, wetlands, and lakes of the state forests provide ideal habitat for a variety of wildlife including white-tailed deer, fox, raccoon, coyotes, squirrels, rabbits, hawks, owls, wild turkey, and dozens of species of woodland songbirds.

The 3,500-acre **Scuppernong Marsh** is one of the best birding sites in the forest. This marsh is the largest wet-mesic prairie areas of its kind in Wisconsin. Many species of waterfowl, cranes, herons, shorebirds, amphibians, turtles, and other wetland species can be found here.

> ### Ranger Trivia
>
> *Scuppernong* is a Ho-Chunk term that means "sweet-scented land."

INTERPRETIVE PROGRAMS AND FACILITIES

Interpretive programs, hikes, and special events are offered throughout the year. A favorite stop is the **Kettle Moraine Natural History Museum**, located at the forest headquarters just off Highway 59. Exhibits here describe the geologic, animal, and human history of the forest. Wildlife displays feature mounts of mammals and birds of the forest, including animals no longer present such as elk, bison, and the passenger pigeon. An informative 20-minute movie featuring the Kettle Moraine Forest can be viewed in the auditorium upon request.

The **Stoney Ridge Nature Trail** (1.0 miles), located just outside the Natural History Museum, leads through an oak, cherry, and hickory hardwood forest. Along the trail are interpretive signs that explain the geologic formations of the Kettle Moraine region.

BOATING AND FISHING

Ottawa Lake (17 acres) and Whitewater Lake (625 acres) are the most popular lakes in the forest for fishing, boating, and paddling. Both have boat launch facilities. Ottawa Lake is limited to electric motors only. Anglers can expect to catch northern pike, panfish, largemouth bass, and occasionally walleye on these lakes. Catch-and-release fishing for rainbow trout is available at Paradise Springs Pond and Scuppernong Springs Creek.

ACCESSIBLE FACILITIES AND TRAILS

All forest offices, the history museum, and restroom/shower facilities are accessible for mobility-impaired visitors. The Paradise Springs Nature Trail (0.5 miles) has an asphalt-paved walkway along its route. Wheelchair-accessible fishing piers are located at Ottawa Lake, Whitewater Lake, and the Paradise Fishing Pond.

All forest campgrounds have one or more accessible campsites that offer electrical service and paved access to restrooms. An accessible cabin within the Ottawa Lake Campground has kitchen, living room, bedroom, and a wheel-in restroom/shower. The cabin overlooks Ottawa Lake and can be reserved in advance for campers with physical disabilities.

CAMPING

The forest has three drive-in campgrounds and two group campgrounds with a combined total of 336 campsites. The **Pinewoods Family Campground** is located in the northern section of the forest and has 101 campsites (no electric). This campground is designated "quiet" camping area, which means pets, radios, or any other electronic devices that may produce noise are prohibited. The nearby **Pinewoods Group Camp** can accommodate groups of up to 40 people. The campground is served by a central shower facility and vault-type restrooms.

Ottawa Lake Campground is located about two miles south of Pinewoods along Highway 67 and has 100 campsites (49 electric), including five walk-in sites. This popular campground offers easy access to the Ottawa Lake swimming beach and picnic area. Both the Ottawa Lake and Pinewoods campgrounds have restrooms/shower facilities and RV water fill and sanitary stations.

Whitewater Lake Campground, located at the far south end of the forest, has 63 sites (no electric) and vault-type toilets. The nearby Rice Lake/Whitewater Lake Picnic Area has a boat launch and a small, sandy swimming beach. The **Hickory Woods Group Camp,** located adjacent to the Whitewater Campground, can accommodate up to 20 campers.

There are three backpack shelter campsites along the Ice Age Trail, which leads through the forest. Each site has a shelter, fire ring, and a pit toilet. Backpackers must reserve the shelter campsites in advance and pack in their own drinking water.

WINTER ACTIVITIES

The cross-country ski trails at the southern unit are one of the most popular in the state. Forest staff grooms 51 miles of trails for both classic and skate skiing. Snowshoeing is allowed on any trail not maintained for skiing. Winter camping is available at the Ottawa Lake Campground. The forest has 54 miles of snowmobile trails that connect to local county trail networks.

AREA ATTRACTIONS

There are many unique shops, museums, and other visitor attractions throughout Walworth, Waukesha, and Jefferson counties.

www.walworthcounty.com

www.waukeshacounty.com

www.enjoyjeffersoncounty.com

Old World Wisconsin is one of the largest outdoor museums of its kind. The site has more than 60 historic homes, barns, shops, schools, churches, and other structures built by ethnic groups who immigrated to Wisconsin between 1840 and 1910. Costumed interpreters are available at many of sites. Old World Wisconsin is operated by the Wisconsin Historical Society and is located off of Highway 67 near the community of Eagle, about two miles east of the forest headquarters. www.oldworldwisconsin.org

DIRECTIONS TO THE FOREST

Kettle Moraine State Forest–Southern Unit is located about 40 miles southwest of Milwaukee and 60 miles east of Madison within Walworth, Waukesha, and Jefferson counties. A detailed map of the forest can be downloaded from the state park website or can be obtained at the forest headquarters, located about three miles east of the Village of Eagle along Highway 59.

Kettle Moraine State Forest–Southern Unit

S91W39091 State Road 59

Eagle, WI 53119 (262) 549-6200

12. Kohler-Andrae State Park
Beach Park offers Sand, Surf, and Fun

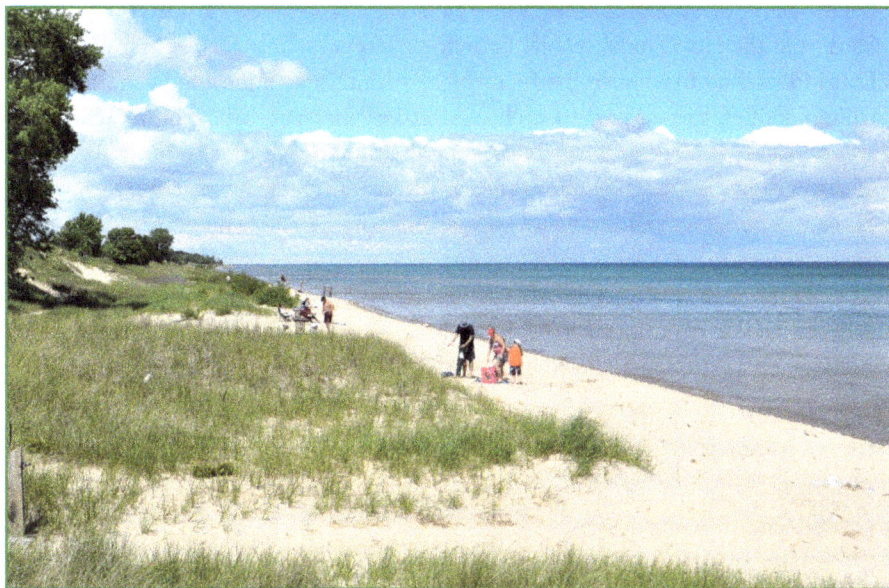

Lake Michigan beach

PARK SNAPSHOT
Kohler-Andrae State Park is located along the beautiful shoreline of Lake Michigan in Sheboygan County. This popular 1,028-acre park offers a wide array of outdoor activities including camping, hiking, biking, swimming, cross-country skiing, picnicking, wildlife viewing, nature study, and much more. Native white pine forests, wind-swept sand dunes, river marshes, abundant wildlife, and more than two miles of wide, sandy beaches await those who explore this natural gem.

GEOLOGY AND HISTORY
The park's sand dunes and wide beachfront are continually altered and reshaped through the action of wind, waves, and the fluctuating water levels of Lake Michigan. The 300-mile-long Lake Michigan basin was excavated by massive glaciers that advanced and retreated between Wisconsin and the state of Michigan near the end of the last Ice Age, about 12,000 years ago. Water levels of this ancient lake would rise and fall several times over millennia. About 7,500 years ago, Lake Nippissing, the last of the post-Ice Age lakes, submerged the park area

under 25 feet of water. A remnant of its shoreline can be seen along the **Ancient Shores Trail** near the west boundary of the park.

Several Native American cultures have hunted, fished, and lived along the Lake Michigan shoreline over time including the Hopewell, Ottawa, Ojibwe, Sauk, and the Ho-Chunk people. French explorers Father Jacques Marquette and Louis Joliet may have been the first to make contact with local indigenous people while paddling the western shoreline of Lake Michigan by canoe in 1673. The area's first fur trading post was built along the Sheboygan River a few miles north of the park by Jacques Vieau for the Northwest Fur Company in 1795. Soon after, Yankee fishermen from the East Coast set up commercial fishing operations along the lakeshore to net lake trout and white fish.

By the mid-1800s, thousands of immigrants, mostly from Germany, Ireland, and Holland, arrived to clear land and build farmsteads along the lakeshore and inland areas. Port cities like Sheboygan grew quickly during this period to serve the ever-growing commercial and passenger ships sailing the waters of the Great Lakes.

> **Ranger Note**
>
> Elsbeth Andrae was a civic-minded environmental activist who was years ahead of her time. Reflecting on her donation of land for the state park she said, "Public parks have inestimable psychologic value in a way we do not ordinarily estimate. They are social correctives. They serve to stabilize."

Terry Andrae State Park was established in 1928 thanks to generosity of Mrs. Frank (Elsbeth) Andrae of Milwaukee. Elsbeth and her husband, businessman Frank Theodore (Terry) Andrae, purchased 122 acres of lakeshore property here to build a summer home overlooking Lake Michigan. When Mr. Andrae passed away, Elsbeth donated their sand dune property to the state under the condition that it be named Terry Andrae State Park as a memorial to her husband.

J. M. Kohler State Park was established through a donation of 280 acres of lakeshore property just north of Terry Andrae State Park by the Kohler Company in 1966. The park was named in memory of the plumbing company's founder, John Michael Kohler. Today, both properties are managed together and are referred to as Kohler-Andrae State Park.

BEACHES, PICNIC AREAS AND SHELTERS

The natural cooling effect of the Lake Michigan has attracted beachgoers to this shoreline area for generations. On warm summer days, hundreds of people flock to the park's expansive beaches to enjoy a refreshing swim, bask in the sun, or just stroll barefoot through the cool surf.

The **South Beach and Picnic Area** has a bathhouse facility with a foot-washing station and a vending machine concession. Also located here is a modern playground, a baseball diamond and volleyball court, a gazebo-style open-air shelter, and an historic indoor shelter facility with a fieldstone fireplace. Both shelters can be reserved in advance.

HIKING, BIKING, AND NATURE TRAILS

Kohler-Andrae has 12 miles of hiking trails to explore. The **Black River Trail (2.5 miles)** is a multi-use trail system that leads through open meadows, a hardwood forest, and a pine plantation adjacent to the Black River. Off-road bike riding and horseback riders are allowed on this trail. The **Interurban Bike Trail (30.0 miles)**, located west of the park, is a multi-county bike route that connects several communities from the city of Sheboygan to the Milwaukee metro area. www.interurbantrail.com

The **Ancient Shores Interpretive Trail** (1.2 miles) leads to an extinct shoreline of Lake Nippissing, a post-glacial lake that submerged much of eastern Wisconsin 7,500 years ago. The trail begins near the **Friends Fishing Pond** along the western edge of the park. The **Black River Boardwalk** (0.25 miles) follows an elevated boardwalk through a cattail marsh along the Black River just west of the campground area. Interpretive panels along the boardwalk highlight the flora and fauna the river and wetlands area.

The **Creeping Juniper Nature Trail** (0.5 miles) begins at the **Sanderling Nature Center** with interpretive stops that describe the flora, fauna, and geology

Dunes Cordwalk Trail

> **Ranger Trail Pick**
>
> The **Dunes Cordwalk Trail** (2.0 miles) leads through the 286-acre Kohler Dunes State Natural Area with scenic views of park's largest dunes and Lake Michigan. The entire trail follows a cordwalk, a wood-and-cable walkway designed to protect the fragile dune ecosystem. Many visitors enjoy hiking the entire length of the cordwalk from either the north or south parking areas followed by a cool, refreshing walk along the Lake Michigan shoreline on the return route.

of Lake Michigan and the park's sand dune formations. The **Woodland Dunes Nature Trail** (1.0 miles) begins in the south picnic area and has interpretive signs that explain how the micro-climate created by Lake Michigan has influenced the diversity of trees that grow here.

WILDLIFE VIEWING

The river wetlands, forests, dunes, and the Lake Michigan shoreline provide ideal habitat for wildlife. The park has over 30 species of mammals, including white-tailed deer, coyote, fox, raccoon, 13-lined ground squirrels, chipmunks, and squirrels. A large multi-unit bat house is located near the park's RV sanitary station, along with an interpretive panel describing each bat species found in Wisconsin.

> **Ranger Note**
>
> A **Wildlife Garden**, featuring a small pond, waterfall, birdbath, and bird feeders is located at the park entrance station. An indoor viewing room for the garden is located inside the office.

More than 200 species of birds have been identified in the park. The Lake Michigan shoreline is an important migratory route for many species of hawks, songbirds, waterfowl, and shorebirds. A purple martin nesting house is located near the **Sanderling Nature Center.**

> **Ranger Trivia**
>
> The Sanderling Nature Center was named after a small shoreline bird called a "sanderling." These tiny sandpipers are often seen along the beach in small flocks probing the wet sand with their tweezer-like beak in search of food.

INTERPRETIVE PROGRAMS AND FACILITIES

Guided nature hikes are offered throughout the use season. Interpretive and other evening programs are presented in the nature center auditorium and at the park's outdoor amphitheater.

The **Sanderling Nature Center**, located along the Lake Michigan shoreline, is one of the finest interpretive centers of its kind in the state. The center has a 99-seat auditorium and many interpretive displays that feature the park's human history, geology, and wildlife. A staircase from inside the center leads to an observation deck with panoramic views of the surrounding dunes, beach and Lake Michigan.

BOATING AND FISHING

Canoes, kayaks, and other small watercraft can be launched from any beach in the park. Larger watercraft can access public launch facilities at the Sheboygan Harbor, located about three miles north of the park. Lake Michigan is known for its world-class trout and salmon fishing opportunities. Charter boat fishing outings can be booked at the Sheboygan River harbor area. Shoreline fishing is available from breakwater piers at the harbor.

The **Friends Fishing Pond**, located west of the campground along Old Park Road, is a fun place to fish for both kids and adults. The pond is stocked annually with rainbow trout and bluegills.

ACCESSIBLE FACILITIES AND TRAILS

The park's office, nature center, bathhouse, and most other buildings are accessible for mobility-impaired visitors. The **Woodland Dunes Accessible Trail** (0.25 miles) starts on an asphalt-paved walkway from the south picnic area and leads to a level, compacted limestone interpretive trail. The **Pond Trail** (0.30 miles) is an asphalt-paved walkway that circles the Friends Fishing Pond along Old Park Road. A wheelchair-accessible fishing pier is also located here.

A log cabin with a wheelchair-accessible shower/restroom, kitchen, bedroom, and living room is located in the south campground. The cabin can be reserved by any physically disabled campers. The **Marsh Trail** (0.5 miles) adjacent to the cabin leads to the **Black River Marsh Boardwalk** (0.25 miles). Both trails are wheelchair accessible. Campsites 14 and 60 are wheelchair accessible. Both have electric service and are located adjacent to a showers/restroom facilitiy.

> **Ranger Tip**
>
> An accessible beach wheelchair can be checked out free of charge at the Sanderling Nature Center. The chair's unique balloon-type tires and lightweight PVC frame allow it to be pushed over sandy areas along the beach with ease.

> **Ranger Note**
>
> All of Kohler-Andrae's camping facilities are very popular and usually fill to capacity daily in summer and on most weekends from late spring to the end of October. Advance campsite reservations are recommended.

CAMPING

The park has 141 campsites (52 electric). The **North Campground** has well-shaded campsites with electric service. The campground has a central shower/restroom facility, an indoor shelter, and a coin-operated laundry.

The **South Campground** is located within a mature forest area. Most sites in this campground do not have electrical service but it does have showers and a flush restroom facility.

The group campground is located adjacent to the Kohler Dunes State Natural Area. There are two group sites that can accommodate up to 25 campers each.

WINTER ACTIVITIES

Park staff groom six miles of inline cross-country ski trails and maintain a one-mile snowshoe trail. Winter camping is available in the North Campground. A frost-free drinking water tap is located at the park's RV/water fill station.

AREA ATTRACTIONS

Sheboygan County is a popular tourist destination that offers several world-class attractions including the **Road America Racetrack**, the **Elkhart Lake Resort Area** and the **Whistling Straits Golf Course.** (920) 457-9491
www.sheboygan.org

The **Riverfront Boardwalk, Blue Harbor Resort,** and **Harbor Center Marina** are located along the Lake Michigan shoreline in the nearby city of Sheboygan. New England-style fish shanties are located along a pedestrian boardwalk along the Sheboygan River. These structures house restaurants, ice cream parlors and several unique shops. Lake Michigan charter fishing boats are docked here as well.

DIRECTIONS TO THE PARK

Kohler-Andrae State Park is located four miles south of the city of Sheboygan. From Interstate Highway I-43, take Exit 120 and then turn east on County Highway V. Follow the directional signs to the park entrance.

Kohler-Andrae State Park
1020 Beach Park Lane
Sheboygan, WI 53081
(920) 451-4080

13. LAKESHORE STATE PARK

Wisconsin's Only Urban State Park

Lakeshore State Park

PARK SNAPSHOT

Lakeshore State Park is one of the smallest properties in the state park system. Despite its diminutive size (22 acres), the park's location along the shoreline of Lake Michigan near the heart of downtown Milwaukee, often attracts more visitors than other much larger state parks. Visitors to this peaceful green space find Lakeshore Park a welcome oasis from the nearby urban landscape of steel, glass, concrete, and traffic.

The park offers scenic views of Lake Michigan and has restored prairie areas, hiking trails, interpretive signs, and paved walkways for biking, jogging, and inline skating. Lakeshore also has a sheltered lagoon perfect for fishing, canoeing, or kayaking, plus a boat dock with a marina for overnight boat camping.

GEOLOGY AND HISTORY

Lake Michigan encompasses an incredible 22,000 square miles of open water, making it the second-largest Great Lake. Its deep basin was excavated by massive glaciers near the end of the last Ice Age about 12,000 years ago. Glacial activity also spawned three nearby streams that flow into the lake; the Milwaukee, the

Kinnickinnic, and the Menomonee rivers. This fertile land of this three-river region was home to several Native American cultures, including the Ojibwa, Menominee, Oneida, Ho-Chunk, and Potawatomi who hunted, fished, and built villages here.

French explorer Robert LaSalle was one of the first Europeans to visit the Native American villages in the Milwaukee area in 1679. Much later in 1795, another French-Canadian, Jacques Vieau, set up a fur trading post and built the first cabin in Milwaukee. Vieu summoned Solomon Juneau, a clerk with the American Fur Company, to manage the trading post in 1818. Juneau helped found the city of Milwaukee and served as its first mayor.

> **Ranger Trivia**
>
> Milwaukee was named after a Potawatomi term, *Minwaking*, which means "gathering place by the water." Lake Michigan was derived from an ancient Ojibwa word *Michi-gami*, meaning "great water."

By 1850, Milwaukee was still a small community of 20,000 people but after the turn of the century, the city had grown to almost 300,000 citizens, primarily due to the influx of European immigrants. To make room for the city's ever-growing population, the lakeshore's wetland areas were filled with sand, soil, rock, and even garbage, expanding the shoreline of Milwaukee more than 1,000 feet into Lake Michigan.

In 1977 the Milwaukee Metropolitan Sewerage District (MMSD) began construction of a 17-mile underground excavation project called the Deep Tunnel. The purpose of the tunnel was to trap water overflow during heavy rainstorms and store it underground until it could be treated and safely released into Lake Michigan. The excavation resulted in the removal of millions of cubic yards of waste rock. Much of this rock was deposited into the waters of Lake Michigan to create a man-made peninsula called **Harbor Island City Park**. In 2001, the city of Milwaukee transferred ownership of Harbor Island to the state to develop Lakeshore State Park, Wisconsin's first urban state park.

BEACHES, PICNIC AREAS, AND SHELTERS

Lakeshore State Park does not have a swimming beach or picnic area but two nearby Milwaukee county parks, McKinley and Bradford, offer both of these amenities. The park also does not have restrooms or water fountains but visitors are allowed to use these facilities at the **Discovery World Science Museum** located adjacent to the park on the north end of the property.

HIKING, BIKING, AND NATURE TRAILS

Visitors to Lakeshore are often surprised that there are no shade trees in the state park. The landscaping plan for the property intentionally left out tree planting due to an agreement with the city of Milwaukee, which wanted to maintain an unobstructed view of Lake Michigan from the city's shoreline. Park staff have planted several shortgrass prairie areas that bloom from early spring to fall with a variety of colorful, native grassland plants, including purple coneflower, orange butterfly weed, and prairie smoke. The prairie areas showcase native grasses, such as big blue stem and Indian grass.

The park has 1.5 miles of paved walkways and gravel trails to explore. Several interpretive panels along walkways highlight the park's flora and fauna and explain the human history and development of the shoreline of Milwaukee.

> **Ranger Note**
>
> The **Hank Aaron State Trail** (14 miles) is a paved hiking/biking route that heads west from Lakeshore State Park to the Waukesha County line. This trail leads past several attractions, including the Urban Ecology Center, the Menomonee River Valley, Mitchell Park Domes and the American Family Baseball Stadium. The trail has connector routes to Milwaukee County's **Oak Leaf Trail** (108 miles).

The **Northern Entrance Walkway** is located near the Discovery World Science Museum. This section of the trail spans the park's lagoon and marina over a beautiful arched metal-sculptured bridge. The **South Entrance Walkway** begins on the south side of the Henry Maier Festival (Summerfest) Grounds. Bicycles, scooters, and inline skates are allowed on all paved walkways. Bike rentals are available nearby at Bradford Beach and Veterans Park, located north of Lakeshore Park.

WILDLIFE VIEWING

Lakeshore Park is home to a surprising number of wildlife species, considering it location adjacent to Wisconsin's largest city. Gray and red fox have been known to raise their kits (young) in underground dens in the park. Other wildlife includes white-tailed deer, mink, woodchucks, rabbits, and ground squirrels.

The Lake Michigan shoreline is an important flyway for migratory birds. Over 70 species of birds have been recorded in the park, including shorebirds, waterfowl, songbirds, and birds of prey. Red-breasted mergansers, goldeneye, scaup, bufflehead, and mallard ducks are often seen on Lake Michigan and the park's lagoon. In summer, dozens of tree swallows make use of nesting boxes installed throughout the park grounds.

View of the Milwaukee skyline and the Discovery Center

INTERPRETIVE PROGRAMS AND FACILITIES

Park naturalists, volunteers, and guest speakers offer interpretive programs and hikes from time to time for park visitors. School groups and community organizations can schedule field trips to the park. Special events are held throughout the season, including kid's fishing clinics, family-night-out programs, and art-in-the-park displays. Each fall, a Sturgeon Fest Celebration is held at Lakeshore Park, where up to 1,000 lake sturgeon fingerlings are released into Lake Michigan by local grade-school children.

BOATING AND FISHING

Lakeshore Park has a 20-slip marina for mooring boats up to 60 feet long. Slip rentals are available for both daily and overnight stays. The park does not have a boat launch but a public boat landing is available just north of the property at the Milwaukee County McKinley Marina. Canoes and kayaks can be launched from the watercraft beach area located along the north lagoon.

Fishing is allowed anywhere along the shoreline of the park. Casting for trout and salmon is popular throughout the year in the park. A fishing pier is located along the lagoon in the north end of the park. Anglers can expect to catch a variety fish here including northern pike, bass, walleye, bullhead, carp, perch, bluegill, and crappie. Non-native invasive fish such as round goby and Eurasian ruffe are also caught here.

ACCESSIBLE FACILITIES AND TRAILS

All paved walkways, the arched entrance bridge, and the fishing pier along the north lagoon are accessible for mobility-impaired visitors. Wheelchair-accessible restrooms and drinking fountains are available inside the nearby Discovery World Museum.

CAMPING

Lakeshore is a day-use-only state park so there are no camping facilities here except for registered boaters moored overnight in the marina.

> **Ranger Note**
>
> Lakeshore does not have a parking area but several public and private parking lots are located nearby. There is no admission fee for Lakeshore State Park. The park is open all year from 6 a.m. to 10 p.m.

The **Wisconsin State Fair RV Park** has paved campsites with full service (water, sewer, electric) and shower/ restroom facilities. The campground is located six miles west of the park at 601 South 76th Street in Milwaukee. www.wistatefair.com/rv-park

Cliffside County Park, located about 18 miles south of the park, has 92 campsites with electric service, restrooms and showers. The park is along the Lake Michigan shoreline at 7320 Michna Road near Caledonia. www.racine-county.com/cliffside-park

WINTER ACTIVITIES

Snowshoeing, hiking, and wildlife watching are popular activities at Lakeshore State Park in winter. Birdwatching can be exceptional at times during the off-season, especially for migrating waterfowl, birds of prey, or Arctic visitors such as snow buntings and snowy owls. Anglers enjoy ice fishing on the southern lagoon when ice conditions allow.

AREA ATTRACTIONS

The Milwaukee area has many world-class museums, parks, restaurants, theatres and other tourist attractions to enjoy. www.visitmilwaukee.org

The **Discovery World Science Museum** is a fun science and technology center that offers many interactive exhibits. The center features a popular Great Lakes fish aquarium and a large-scale working model of the Great Lakes watershed. The Discovery Center is also the home docking station for the *Denis Sullivan*, a 19th-century reproduction of a Great Lakes schooner. Sailing trips aboard the schooner into Lake Michigan are offered from time to time. Discovery World is located adjacent to Lakeshore State Park at 500 N. Harbor Dr., Milwaukee. www.discoveryworld.org

DIRECTIONS TO THE PARK

Lakeshore State Park is located along Lake Michigan just west of downtown Milwaukee. From Interstates I-94 or I-43, exit onto I-794 east towards the lakefront. Take the N. Lincoln Memorial Drive exit and travel north to E. Clybourn to the north entrance of Lakeshore State Park.

Lakeshore State Park
500 North Harbor Drive
Milwaukee, WI 53202
(414) 274-4280

14. LAPHAM PEAK–
KETTLE MORAINE STATE FOREST SOUTH

Top-of- the-Hill Property—A Scenic Gem

View from observation tower

FOREST SNAPSHOT

At an elevation of 1,233 feet above sea level, Lapham Peak seems like it's on top of the world. The property is located on the highest hill of Waukesha County within the Kettle Moraine Forest. On a clear day, the view from atop the 45-foot observation tower here can be spectacular. Below, as far as the eye can see, are hardwood forests and scenic kettle lakes. On the distant horizon, the northern boundary of Illinois comes into view.

Lapham Peak's 1,022 aces of oak forests, ponds, wetlands, hills, and prairies are a hiker's paradise. The property has more than 30 miles of hiking and off-road biking trails. In addition, visitors will find several picnic areas, shelter facilities, and an attractive nature center.

GEOLOGY AND HISTORY

The rolling landscape of Lapham Peak are the work of retreating glaciers, which left behind thousands of moraine hills and kettle depressions throughout this region of the state near the end of the last Ice Age 10,000 years ago. Lapham's "peak" is actually a very tall cone-shaped moraine hill that geologists call a *moulin kame*. A kame is formed when melt water falls through a vertical shaft or open hole (moulin) on the surface of a glacier. The downward spiraling water drains away beneath the glacier, but the sand, gravel and rock it carried builds up layer by layer into a cone-shaped pile similar sand in the bottom of an hour glass.

Several different Native American people lived within the Kettle Moraine area for generations, including the Sauk, Meskwati, Ho-Chunk, and the Potawatomi people. Indigenous people lived alongside white settlers in relative peace at first, but as more and more immigrants arrived to cut timber, build dams, and farm the land, conflicts over land use was inevitable. In 1833, the Potawatomi were induced to cede all their lands in southeastern Wisconsin to the U.S. government through the Treaty of Chicago, opening the entire region to the European immigrant settlement.

The towering hill and forest area of Lapham Peak was a popular recreation area long before it became part of the state park system. In 1851, an enterprising land owner, Charles Hanson, built a 20-foot observation tower on what was then called Stoney Hill. Hanson charged a fee for visitors to climb the tower or have a picnic at his private park.

Stoney Hill was later renamed Government Hill by U.S. government surveyors, who used the hill as a reference point to run land surveys in the surrounding area. The U.S. Army Signal Corps also used the hill to relay storm warnings of approaching severe weather from as far away as Pikes Peak, Colorado to the Great Lakes shipping ports to the east.

Increase Lapham, Wisconsin's foremost scientist and naturalist, conducted weather observations from Government Hill for several years. Under his guidance, the first National Weather Service was established in 1870.

In 1939, the State Conservation Department acquired the Lapham Peak property to create Lapham Peak State Park. Crews from the Depression-era WPA (Works Projects Administration) built many of the property's roads and trails and erected a new observation tower. Today, Lapham Peak is managed as a unit of the Southern Kettle Moraine State Forest.

> **Ranger Trivia**
>
> In 1916, the Waukesha County Historical Society renamed the hill "Lapham Peak" in honor of Increase Lapham (1811-1875). Lapham wrote more than 80 scientific books and articles on Wisconsin's geology, minerals, native wild plants, and fossils. He also carried out the first comprehensive survey of Native American effigy burial mounds sites in the state.

BEACHES, PICNIC AREAS AND SHELTERS

The **Homestead Hollow Picnic Area** is the property's largest day-use area. It has an open-air shelter and an indoor shelter facility, which can reserved in advance. The **Evergreen Picnic Area** has an open air shelter and an indoor lodge facility. The **Summer Stage Outdoor Amphitheater** is also located here. Local talent performs theatrical and musical performances here on weekends throughout the summer season.

The **Observation Tower Picnic Area** provides access to Lapham Peak's 45-foot high observation tower. From atop the tower, visitors can enjoy panoramic views of the kettle moraine landscape below. On a clear day, the iconic **Holy Hill Basilica and National Shrine** can be seen twenty miles to the north perched on the tallest hill in southeast Wisconsin.

There are no lakes or rivers at Lapham Peak. A swimming beach is available at **Naga-Waukee County Park,** located about four miles northwest of Lapham Peak along Highway 83 in the nearby city of Hartland.

HIKING, BIKING, AND NATURE TRAILS

Lapham Peak has 31 miles of hiking trails to explore including a 5.6-mile section of the **Ice Age National Scenic Trail.**

The **Kettle View Trail** (5.8 miles) and the **Moraine Ridge Trail** (6.6 miles) both lead through mature red oak forests and the unique glacial formations of the Kettle Moraine region. The **Kame Terrace Trail** (2.0 miles) is an easy walking trail that loops through woodlands and open grassy areas in the eastern section of the property. A side trail leads to the **Friends of Lapham Peak Butterfly Garden.**

The **Prairie Path Trail** (4.5 miles), located across County Highway C in the northwest section of Lapham Peak, is an equestrian and off-road bike route. The **Cushing Park Bike Trail** (3.0 miles), located adjacent to Highway C, leads north to the city of Delafield with connections to the **Waukesha County Lake Country Recreational Trail** (15.0 miles). The trail ends at **Cushing Memorial Park** (see Chapter 3). A two-mile section of the Cushing Trail also travels south from Lapham Peak to the **Glacial Drumlin State Bike Trail** (52.0 miles).

Ranger Trail Pick

The **Plantation Path Trail** (1.8 miles) is an asphalt-paved walkway that leads through prairies, oak savannas, hardwood forests, and conifer plantations. The trail has interpretive panels that highlight the flora, fauna, human history and geology of the area. Side trails lead to the **Hausmann Nature Center** and the **Observation Tower.**

WILDLIFE VIEWING

Lapham Peak's varying landscape of red oak forests, wetlands, ponds, and restored prairies provide ideal habitat for hundreds of species of birds, mammals, reptiles, and amphibians. Gray and fox squirrels, rabbits, raccoon, chipmunks, white-tailed deer, hawks, owls, and dozens of woodland bird species make their homes in forested areas.

Grassland birds such as bluebirds, tree swallows, meadowlarks, and bobolinks can be spotted in the restored prairie areas in summer. During spring and fall migration periods, dozens of species of songbirds, shorebirds, and waterfowl, plus red-tailed hawks and other birds of prey, can be seen throughout the property.

INTERPRETIVE PROGRAMS AND FACILITIES

Several interpretive panels are located along the **Plantation Path Trail** and highlight Lapham Peak's history, geology, and wildlife.

The **Hausmann Nature Center** has interpretive displays, a gift shop, and offers an outstanding view of a long-grass, restored prairie. The nature center is generally open on weekend afternoons in summer. The building can also be reserved for private gatherings.

> **Ranger Note**
>
> The nature center was originally the home of Dr. Paul and Bernice Hausmann of Delafield. In 1984 the couple generously donated 35 acres of land and their private home to the state of Wisconsin to serve as Lapham Peak's nature center.

Hausmann Nature Center

BOATING AND FISHING

There are no lakes or streams at Lapham Peak. Several nearby lakes offer boating and fishing opportunities including **Nagawicka Lake** (981 acres) and **Pewaukee Lake** (2,437 acres). Both lakes are located about four miles north of Lapham Peak. Anglers can expect to catch panfish, northern pike, walleye, bass, and muskie in these lakes.

ACCESSIBLE FACILITIES AND TRAILS

The visitor entrance station, restrooms, and most picnic areas are accessible for mobility-impaired visitors. Paved walkways provide access to the indoor shelter lodge facilities and the Hausmann Nature Center. The 1.8-mile Plantation Path Trail is one of the finest wheelchair-accessible trails in the park system.

CAMPING

Lapham Peak is a day-use property so there are no campground facilitates here. There are two large campgrounds within the Kettle Moraine State Forest–Southern Unit, located about eight miles southwest of the property. (See Chapter 11)

> **Ranger Note**
>
> A walk-in backpack campsite is located along the Ice Age Trail at Lapham Peak. The site has a fire ring, primitive toilet, and a small camping shelter.

> **Ranger Note**
>
> Lapham Peak is the only state park or forest to have snow-making equipment to supplement winter snowfall on the lighted section of the ski trails.

WINTER ACTIVITIES

Lapham Peak is known as having some of the best cross-country ski trails in the state. Hundreds of skiers flock to Lapham Peak daily to take advantage of the 17 miles of ski trails offered there. The **Evergreen Picnic Area** lodge serves as a warming house for skiers. Property staff groom trails regularly for both inline and skate skiing. A **2.5** mile section of the trail is lighted for night skiers

AREA ATTRACTIONS

Northwestern Waukesha County is known as the **Lake Country Area** due to the 20 named lakes of the area. This popular tourist destination has many resorts, museums, and other attractions to visit www.visitwaukeshacounty.com

Ten Chimneys National Historic Landmark is a historic Scandinavian-style mansion that once served as the summer home of theater legends Lynn Fontanne and Alfred Lunt. This grandiose house was built in 1915 and features ten working chimneys. The home and grounds is now a museum with many interactive displays that highlight the golden age of the theater before the dawn of movies and television. Ten Chimneys is located about seven miles south of Lapham Peak at S43 W31575 Depot Road, Genesee Depot, WI 53127 (262) 968-4161. www.tenchimneys.org

DIRECTIONS TO LAPHAM PEAK

Lapham Peak is located just south of the city of Delafield and about 25 miles west of Milwaukee. From Interstate I-94, exit onto County Hwy C (Genesse Street) and head south about a half mile to entrance of the forest on the east side of the road.

Lapham Peak-Kettle Moraine State Forest–Southern Unit
W329N846 County Highway C
Delafield, WI 53018
(262) 646-3025

15. LIZARD MOUND STATE PARK

Sacred Site and Archeological Treasure

Interpretive Center

PARK SNAPSHOT

Lizard Mound State Park is one of the most significant archeological locations in Wisconsin and is listed on the National Register of Historical Places. The site preserves 28 effigy burial mounds, many four feet high with some nearly 250 feet in length.

The mounds were built by the Late Woodland people from 800 to 1,500 years ago. Many are shaped in the outline of a mammal, bird, or other animal, both real and mythical.

This small 22-acre historic park is located northeast of the city of West Bend in Washington County. Lizard Mound offers hiking, nature study, and birdwatching. It also has a picnic area and an open-air shelter with interpretive displays.

GEOLOGY AND HISTORY

Lizard Mound is perched atop a moraine plateau a few miles east of the Milwaukee River. The area's rolling hills, wetlands, and numerous springs were formed by glacial activity near the end of the last Ice Age about 12,000 years ago.

Effigy Mound Builders lived in this area during the Late Woodland Period between 700 A.D and 1200 A.D. At that time, this region of Wisconsin would have been mostly long-grass prairies, spring-fed marshes, and small oak savanna woodlands.

Archeologists believe there were at least 60 mounds here when it was still an active burial site. The first field investigation of the Lizard Mound group was conducted by Professor Julius Torney of Milwaukee in 1883. His site map showed 47 mounds still intact at that time. Unfortunately, by the time the area was protected as a state park in 1950, only 28 of the mounds remained. Southern Wisconsin is considered the epicenter of the Effigy Mound Builders' culture. Archeologists estimate that at least 20,000 mounds were built in this region. Of these, fewer than 4,000 remain. Most were destroyed by farming practices, road building, quarrying operations, and urban development.

Excavations conducted by state archeologists in 1960 indicated that the Effigy Mound Builders were nomadic travelers who hunted, fished, and gathered plants in more open grassland areas in spring and summer. During the winter, they retreated to the cover and shelter found in lowland forest areas. The mound builders made use of clay pots, stone axes, and spears. They may also have been the first people to use bow and arrows.

Most effigy mounds contain one or more burials. Those buried here were likely revered men and women who were either tribal leaders or important clan shamans.

> **Ranger Note**
>
> Lizard Mound Park is a sacred burial site to all Native People and is protected by state law. Several tribal nations, including Ho-Chunk representatives, are working in tandem with the state to ensure that the mounds are properly maintained, interpreted, and preserved for future generations.

The significance of effigy mound shapes and what they represent may never be totally understood. Like all indigenous people, effigy mound builders were closely tied to the physical world they were familiar with. Both their daily and spiritual lives revolved around the three natural entities: earth, sky, and water. Earth-related effigy mounds were often built in the shape of bears, deer, bison, or snakes. Conical mounds are thought to represent a beaver, muskrat, or a turtle. Sky-related mounds were portrayed as falcons or eagles with outspread wings. Mounds built in reverence to water deities included geese, swans, herons, and cranes.

One of the most common effigy mounds built throughout southern and eastern Wisconsin was a long-tailed mythical, underground creature known as the water panther. One of the mounds discovered at this site was thought to resemble a four-legged, long-tailed lizard, which led to the naming of the park. New

Water panther effigy mound

evidence, however, suggests the mound may actually have been intended to represent a water panther built in a spread-eagle pose instead.

Lizard Mound opened to the public in 1950 and was managed as a state park until 1980, when ownership of the property was turned over to Washington County. The site was re-deeded back to the Wisconsin State Park System again in 2021. Ongoing efforts are underway to improve preservation of the mounds and restore the site back to its original oak savanna landscape.

BEACHES, PICNIC AREAS, AND SHELTERS
Picnic tables are available in the mowed area adjacent to the property's open-air shelter facility. There are no lakes or rivers in the park. A public swimming beach and additional picnic areas are can be found at nearby **Sandy Knoll County Park**. (See Attractions)

HIKING, BIKING, AND NATURE TRAILS
The park has a one-mile limestone-surfaced trail that loops around many of the 28 mounds preserved here. Off-trail hiking is allowed in the park except visitors are asked to respect the burial sites and not walk on the mounds.

The **Eisenbahn State Bike Trail** (24 miles) is located about four miles west of the park. The trail leads past an historic railroad depot in the city of West Bend and travels north adjacent to the Milwaukee River to the village of Eden in Fond du Lac County.

WILDLIFE VIEWING
The park's open fields, hardwood forests, and oak savannas attract a variety of wildlife including white-tailed deer, squirrels, chipmunks, rabbits, raccoon, fox, woodpeckers, blue jays, cardinals, and many songbirds, hawks, and owls.

INTERPRETIVE PROGRAMS AND FACILITIES
An open-air shelter near the entrance parking area serves as the park's interpretive center. Inside the shelter is a unique solar-powered interpretive display with a diorama of the park's mound group. The display has an interactive push-button question-and-answer system. Adjacent to the open-air shelter are several Native American-stylized metal kiosks. Each has an interpretive panel explaining the history and archeological findings of effigy mound building in Wisconsin.

BOATING AND FISHING
There are no boating or fishing opportunities at Lizard Mound.

ACCESSIBLE FACILITIES
Walkways to the interpretive center, restroom, and hiking trails are surfaced with crushed limestone. Most trails are level and accessible for mobility-impaired visitors with assistance. An accessible picnic table is located adjacent to the main parking area.

CAMPING
Lizard Mound is a day-use-only park so there are no camping facilities here. **Pike Lake State Park,** located about 17 miles southwest of the park, has 33 campsites (12 electric) with a shower/restroom facility. (See Chapter 21)

WINTER ACTIVITIES
The entrance road to Lizard Mound is closed from November 1st through March 31st. Visitors can still walk or ski into the park during this period by parking at the gate along County Road A.

AREA ATTRACTIONS

Washington County has a wide array of parks, museums and other attractions to explore. www.visitwashingtoncounty.com (262) 677-5069

Sandy Knoll County Park is 275-acre county facility located about four miles south of Lizard Mound. The park has picnic areas, a fishing pond, and a swimming beach. The park also has several miles of hiking trails that lead through restored prairie areas and mature hardwood forests. Sandy Knoll is located along Wallace Lake Road near West Bend.

DIRECTIONS TO THE RECREATION AREA

Lizard Mound is located northeast of the city of West Bend. From Highway 144 exit onto County Road A to the park entrance about one mile to the east.

Lizard Mound State Park
2121 County Hwy A,
West Bend, WI 53095
(262) 670-3400
(Pike Lake State Park)

16. LOEW LAKE UNIT
KETTLE MORAINE STATE FOREST SOUTH
Wilderness Lake—A Rare Oasis

Loew Lake

FOREST SNAPSHOT

The Loew Lake Unit of the Kettle Moraine State Forest is a serene, near-wilderness property nestled within a forested valley only eleven miles west of the Milwaukee Metro area. This rare, undeveloped 2,273-acre natural area offers hiking, hunting, fishing, horseback riding, canoeing, kayaking, birdwatching, and many other outdoor activities.

GEOLOGY AND HISTORY

Geologists consider Loew Lake's landscape to be a "true valley" due to its location between the towering hills that surround both the lake and the eastern branch of the Oconomowoc River, which flows through the property. Loew Lake is a mix of moraine hills, open grasslands, wetlands, and wooded swamps. Upland forests are composed of mature sugar maple, hickory, aspen, cherry, and red and white oak trees.

The landscape of Loew Lake was formed by glacial activity near the end of the last Ice Age about 12,000 years ago. As glaciers receded northward out of Wisconsin, several large chunks of ice broke off and became buried beneath deposits of sand, gravel, and other glacial debris. When these underground pockets of ice melted, water-filled depressions immerged throughout the landscape. Shallow depressions developed into wetlands or small ponds while larger deeper ones became permanent lakes such as Loew Lake. Retreating glaciers also sent vast amounts of melt water down the Oconomowoc River, cutting the steep-sided valley seen at Loew Lake today.

Several Native American tribes hunted, fished, and gathered wild plants in Loew Lake area for generations. The Potawatomi moved to Door County and then south along the Lake Michigan shoreline from their homeland in Lower Michigan in the 17th century due to an ongoing war with the Iroquois over fur trade and hunting rights. By 1800, the Potawatomi claimed ownership of all of northern Illinois and southeastern Wisconsin, but in 1833, they were forced to cede all their lands, including the Oconomowoc River region, to the U.S. government.

In the 1850s, a large influx of Irish Catholic immigrants settled in the Loew Lake area to carve out farmsteads in the fertile soils of the region. The Irish culture and traditions continue to permeate this area even today. Local road names such as Shamrock Lane and Emerald Drive and the local township of Erin, a Welch term for Ireland, attest to the strong Irish connection to the area.

In 1987, the State of Wisconsin began purchasing property to establish Loew Lake. The goal of acquiring this unique, undeveloped area was to provide more outdoor recreation opportunities for people living in the Milwaukee area and to protect the Loew Lake and Oconomowoc River watershed.

> **Ranger Trivia**
>
> The Oconomowoc River is named after the Potawatomi term Oconomowoc, which means "rivers of lakes." The term is an accurate description of the Oconomowoc River, which links 17 lakes along its 49-mile-long route to the Rock River in Jefferson County.

> **Ranger Trivia**
>
> Loew Lake was named after the Loew family, who operated a summer resort and cabin rental business along the lake in the early 1900s. The Loews also operated a popular log cabin tavern here. This historic cabin can still be seen along the east side of the lake and currently used as lodging for state employees.

BEACHES, PICNIC AREAS, AND SHELTERS

Loew Lake is managed as a wilderness area so there are no developed picnic areas or swimming areas and no restrooms or drinking water onsite. These amenities can be found at Pike Lake Unit of the Kettle Moraine about 19 miles north of Loew Lake. (See Chapter 21)

HIKING, BIKING, AND NATURE TRAILS

There are ten miles of hiking trails at Loew Lake including a 5.5-mile equestrian trail. Some trails are marked by directional emblems but most are not. A property map indicating established trails is posted on a kiosk located at the entrance parking lot.

A four-mile section of the **Ice Age National Scenic Trail** is located near the western boundary of the property. This trail leads north through the **Basilica of Holy Hill National Shrine** property and the **Pike Lake Unit** of the Kettle Moraine State Forest. The trail also heads south to the **Lapham Peak Unit** of the Kettle Moraine.

Ranger Trail Pick

From the main parking area off Emerald Drive, take the horse trail south to its junction with the Ice Age Trail. Continue south on the horse trail and cross the iron bridge over the scenic Oconomowoc River. From here, the trail loops through a restored prairie area and crosses a gravel drive. A short walk north on the gravel drive leads to an historic log cabin and small boat landing located on the east side of Loew Lake. From the cabin, return to the horse trail and continue east up a heavily forested hill to an open field where the route ends at parking area along St. Augustine Road.

WILDLIFE VIEWING

The property's hardwood forests, pine plantations, aspen groves, and prairie areas provide ideal habitat for white-tailed deer, raccoon, gray and red squirrels, rabbits, fox, coyote, and wild turkey. The lake, river, ponds, and wetland areas of the property are home to muskrat, mink, otter, and many species of shorebirds, herons, and waterfowl. Reptiles such as water snakes, garter snakes, painted turtles, snapping turtles, and several species of frogs, toads and salamanders are commonly seen near waterways as well.

Loew Lake is well known as one of the best birding areas in southeast Wisconsin. During the spring and fall migration periods, dozens of species of waterfowl, hawks, cranes, and both woodland and grassland birds can be observed here.

Oconomowoc River—Eastern Branch

INTERPRETIVE PROGRAMS AND FACILTIES

There are no interpretive programs or marked interpretive trails at Loew Lake.

ACCESSIBLE FACILITIES AND TRAILS

Boaters with physical impairments can obtain a permit from the state to use electric motors on Loew Lake. There are no wheelchair-accessible trails or facilities at this property.

BOATING AND FISHING

Loew Lake (24 acres) has an average depth of 11 feet and an undeveloped shoreline. Anglers can expect to catch primarily bluegill and crappie but the lake is also known for its northern pike and bass fishing opportunities. Only non-motorized watercraft are allowed on Loew Lake.

A public boat launch with access to the eastern branch of the Oconomowoc River is located on the far southwest section of the property along County Highway Q. The launch is designed primarily for small craft such as canoes and kayaks. From the landing, paddlers can head north on the scenic Oconomowoc River, which leads to Loew Lake. Paddling this tranquil lake is a great way to explore its rich wetlands and beautiful shoreline.

CAMPING
Loew Lake is a day-use property so there are no camping facilities here. The nearby Pike Lake Unit of the Kettle Moraine State Forest has 32 campsites (11 electric) with restrooms and shower facilities. (See Chapter 21)

WINTER ACTIVITIES
Hiking, hunting, snowshoeing, wildlife watching, and cross-country skiing are popular off-season activities at Loew Lake. None of the trails are groomed for skiing. Anglers enjoy ice fishing on the lake in winter as well.

AREA ATTRACTIONS
Washington County has many small, quaint towns, parks, museums, and other visitor attractions to explore. www.visitwashingtoncounty.com

Glacier Hills Park is 140-acre Washington County park located about two miles northeast of Loew Lake. The park offers fishing, kayak rental, picnic areas, and miles of scenic trails that wind through oak forests, restored prairies, and the rolling hills of the Kettle Moraine area. The park entrance is off Friess Lake Road near the community of Hubertus. www.webplan@co.washington.wi.us

DIRECTIONS TO LOEW LAKE
Lowe Lake is located 11 miles west of the cities of Menomonee Falls and Germantown in southern Washington County. From Highway 167, exit onto County Highway K and turn left (east) onto Shamrock Lane. At the junction of Shamrock Lane and Emerald Drive, turn right (south) to the Loew Lake parking area.

Loew Lake Unit–Kettle Moraine State Forest
Emerald Drive
Town of Erin, WI 53027
(262) 670-3400

17. Lost Dauphin State Park

The Lost Prince of France

Entrance sign

PARK SNAPSHOT

Lost Dauphin State Park is home to one the most peculiar legends ever told in Wisconsin— the tale of the lost heir to the throne of France. All that remains today of this incredible saga is an historical marker in a small park along the western bank of the Fox River in Brown County. Lost Dauphin is a former 19-acre state park that offers hiking, picnicking, fishing, and birdwatching.

GEOLOGY AND HISTORY

Lost Dauphin is perched on a moraine hill overlooking the Fox River valley. The landscape of the area was formed near the end of the last Ice Age about 12,000 years ago. Retreating glaciers during that period sent a flood of meltwater thundering down the Fox River, cutting the valley ever deeper before the river empties into Green Bay.

Native American people had hunted, fished, and gathered wild rice in the Fox River area long before French explorer Jean Nicolet "discovered" Wisconsin near Green Bay in 1634. By that time, thousands of Menominee, Ho-Chunk (Winnebago), and Meskwaki (Fox) people were already living in permanent settlements here. In 1673, Father Jacques Marquette and Louis Jolliet paddled up the Fox River past what is now Lost Dauphin and then took the Wisconsin River west to the Mississippi River. Although they didn't find the Northwest Passage to the Indies, they did discover a well-established Native American river highway connecting the Great Lakes to the Mississippi River. This water route became an important asset to the fur trade industry and was fought over between the French, English, and the American military for decades.

> **Ranger Note**
>
> The term *Dauphin* was a royal title given to the eldest son of the king that entitled him to be heir to the throne of France.

The legend associated with Lost Dauphin Park begins when King Louis XVI and his wife Marie Antoinette were executed by guillotine during the French Revolution in 1793. By French monarchy tradition, their eight-year-old son, Prince Louis Charles XVII, was then considered the Dauphin of France.

Revolutionary guards held the boy in prison, where he reportedly died from tuberculosis at age ten. Other accounts claim the boy was secretly sent in exile to Canada, where he was given the name Eleazer Williams and adopted by a Mohawk family. Williams later attended a missionary school (now Dartmouth College) in New Hampshire and served as a Protestant missionary among several displaced Native American tribes in New York State.

> **Ranger Trivia**
>
> William's claim of being of French royal ancestry seems to have been debunked through modern ancestor DNA testing, but many still believe that he really was the legitimate lost Dauphin of France. A book, *The Lost Prince*, written in 1853 by John H. Hanson, is often cited as proof regarding William's claim.

In 1822, Williams led a large group of Oneida and Stockbridge people west through the Great Lakes region in hopes of finding a new homeland for them. He managed to secure 4,800 acres of land from the U.S. government along the western bank of the Fox River, where he built a cabin with his Menominee-French wife. Rev. Williams intended to establish a Christian Indian Empire on this property and declare himself ruler but his plan was rejected by tribal leaders. The Oneida and Stockbridge people did remain in eastern Wisconsin, however, eventually settling on their own lands near Green Bay.

In 1841, Prince de Joinville, son of the newly restored King of France, traveled to America and met with Rev. Williams in Green Bay. According to Rev. Williams, it was at this meeting that the prince informed him that he was the long-lost Dauphin and rightful heir to the French throne. News of this meeting with French royalty spread far and wide, which led to Williams becoming a celebrity of sorts. Newspapers and magazines published fantastic stories of his life but

Williams never sought to claim his title. He eventually left Wisconsin and moved back to New York, where he died alone in 1858.

Lost Dauphin State Park was established in 1947 to preserve Rev. William's original log home and the legend of his claim to fame. Unfortunately, the cabin was destroyed by fire in the 1970s. Soon after, the state turned the property over to the Town of Lawrence, which maintains Lost Dauphin Park today.

BEACHES, PICNIC AREAS, AND SHELTERS
Lost Dauphin has a picnic area with tables and grills, a playground, and an open-air shelter facility. There is no swimming area along the Fox River.

HIKING, BIKING, AND NATURE TRAILS
The **Lost Dauphin Nature Trail** (0.25 miles) is gravel-surfaced and leads though the scenic hardwood forest and rolling hills of the property. Nearby is the **Fox River State Trail** (25.0 miles), an asphalt-paved biking/hiking trail that leads through the nearby city of De Pere to downtown Green Bay. Along the route are interpretive panels that describe the early history of this region. The trail is located along the east side of the Fox River and can accessed from Voyager's Park in De Pere, about five miles north of the park.

WILDLIFE VIEWING
Birdwatching is popular activity at Lost Dauphin, especially during the spring and fall migration periods. The area's woodlands attract songbirds such as cardinals, blue jays, chickadees, nuthatches, and woodpeckers. Bluebirds and tree swallows make use of nesting boxes in the grassland areas of the park.

Other wildlife seen in the park includes white-tailed deer, raccoon, rabbits, chipmunks, squirrels, red fox, and coyote. The Fox River attracts many species of waterfowl, sandpipers, herons and white pelicans, as well as mink, muskrats, turtles and an occasional otter

INTERPRETIVE PROGRAMS AND FACILITIES
A Wisconsin Historical Marker highlighting the peculiar life of Rev. Eleazer Williams is located in the picnic area. The narrative describes William's failed effort to establish a Christian Indian Empire here and his claim to fame as the Lost Dauphin of France.

BOATING AND FISHING
Shore fishing is available along the Fox River adjacent to County Highway D across from the park entrance. Anglers can expect to catch a variety of fish including bluegill, perch, northern pike, muskie, white bass, walleye, and bass.

Wisconsin historical marker overlooking the Fox River

The park does not have a boat launch but several are located on the east side of the river, including the **Bomier Boat Landing** at 700 Fox River Drive and **Fox Point Boat Launch** at 1075 North Broadway.

ACCESSIBLE FACILITIES AND TRAILS
The park's picnic area and shelter are accessible for most mobility-impaired visitors. The **Lost Dauphin Nature Trail** is surfaced with crushed limestone but some sections of the trail may be too steep for many wheelchair users without assistance.

CAMPING
Lost Dauphin is a day-use property so there are no campsites here. The nearby **Apple Creek Campground** offers 135 campsites with full hookups. The campground is located about six miles south of the park at N **3831** County Road U. www.applecreekcamping.com

WINTER ACTIVITIES
Snowshoeing, hiking, birdwatching, and cross-country skiing are the popular winter activities. The trails are not groomed.

AREA ATTRACTIONS

The Brown County and the Green Bay area is one of the most popular tourist destination areas in Wisconsin. The area has many parks and historical attractions to explore including the **Bay Beach Wildlife Sanctuary** and **Heritage Hill State Park** (See Chapter 8). www.greenbay.com

The **De Pere Lock and Dam Historic District** was one of the first lock and dam ever built in Wisconsin. The lock was opened in 1836 to provide passage around the De Pere rapids while boating the Fox River between Green Bay and Lake Winnebago. The lock and dam is still in service today and is open for public viewing. A perched walkway here provides a close up view the De Pere dam and a chance to see pelicans and cormorants that often congregate here to fish. The lock and dam are located adjacent to **Voyageur Park** at 100 William St. in De Pere. www.de-pere.org .

DIRECTIONS TO THE PARK

Lost Dauphin Park is located in Brown County about four miles south of the city of De Pere. From U.S. Highway 41, take the County Highway S exit and follow this road east to Hickory Road. Continue on Hickory Road for about a mile and then turn north on County Highway D (Lost Dauphin Rd.) one mile to the park entrance.

Lost Dauphin Park
2821 Lost Dauphin Road (County Hwy. D)
De Pere, WI 54115
(920) 336-9131
www.townoflawrence.org

18. MUKWONAGO RIVER UNIT—
KETTLE MORAINE STATE FOREST

A Diamond in the Rough

Rainbow Springs Lake

FOREST SNAPSHOT

The Mukwonago River is a semi-wilderness unit of the Southern Kettle Moraine State Forest located in Waukesha and Walworth counties. The property was named after the Mukwonago River, which flows through the property. The recreation area offers hiking, fishing, boating, hunting, birdwatching, and nature study.

GEOLOGY AND HISTORY

The property's landscape is made up of moraine hills, upland hardwood forests, pine plantations, creeks, ponds, restored prairies, and beautiful Rainbow Springs Lake. Adjacent to Mukwonago is **Lulu Lake State Natural Area**. Both Rainbow and Lulu lakes are considered kettle lakes by geologists.

Several Native American cultures lived in the Kettle Moraine area over time, including the Potawatomi people who built a large village called *Mequanego* or "Place of the Bear" along the Mukwonago River. This important Native American town was home to at least 600 people and the tribal seat of the Potawatomi Bear

Clan for decades. In 1833, the Potawatomi were forced to cede all their lands in southeast Wisconsin to the U. S. government and moved onto reservation land in Kansas. Some did leave the state but many other tribal members decided to move north into Forest County.

European immigrant farmers, merchants, and lumbermen transformed much of the Mukwonago River area into farm fields, roads, and villages during the 19th century. Unlike settlers in other regions of Wisconsin at the time, the new caretakers of this land preserved many of the forests and clean waters of the area. They even adopted the Potawatomi town's name, *Mequanego*, for their own village.

In the early 1960s, a Milwaukee businessman purchased the Mukwonago River property to develop Rainbow Springs Golf Course Lodge and Resort. Plans for this upscale resort called for the construction of two 18-hole golf courses, several multi-level hotel units, a central gathering lodge, and a convention center that would house 10 dining rooms and 13 bars. The golf course and the central lodge were completed and operated by several different owners over a 40-year period but the venture never became profitable. In 2008, the 913-acre Rainbow Springs property was sold to the State of Wisconsin to develop the Mukwonago River Recreation Area.

BEACHES, PICNIC AREAS, AND SHELTERS

Mukwonago River is currently managed as a semi-wilderness property so there are no developed picnic areas, restrooms, or swimming areas here. All these amenities can be found at **Mukwonago County Park** located a few miles northeast of the property (see Camping section)

HIKING, BIKING, AND NATURE TRAILS

Mukwonago has several miles of trails and abandoned roadways to explore but none are marked or signed. Some walking paths follow asphalt-paved roads or

> **Ranger Hiking Route Pick**
>
> Begin at the north parking lot along County Highway LO and head south along the asphalt road across the Mukwonago River. A side road to the left (east) leads to an abandoned boat landing along the shores of Rainbow Spring Lake. Another spur road dead-ends near the remains of the Rainbow Springs Resort hotel and the site of the abandoned lodge. Return to the main road to access the south section of the property, which has scenic ponds, streams, wetlands, open fields, and woodland areas to explore.

former golf cart paths. Others lead down old farm lanes, gravel driveways, or mowed hunter access trails.

WILDLIFE VIEWING

The Mukwonago River property is gradually being transformed from a golf course back into the prairies, wetlands, and hardwood forests that once thrived here. These restoration efforts have brought back many wildlife species to the site. Ironically, in areas where golfers once scored a one-under par birdie, visitors can now observe real bird life, including bluebirds, tree swallows, ring-necked pheasants, ruffed grouse, owls and hawks and many woodland and grassland songbirds.

Other wildlife commonly seen here are white-tailed deer, raccoon, cottontail rabbits, gray squirrels, woodchucks, chipmunks, coyote, and red fox. Bald eagles are occasionally seen soaring in the skies or snatching a fish from Rainbow Springs Lake.

BOATING AND FISHING

There are no boat launch facilities on Rainbow Springs Lake or the Mukwonago River. Carry-in canoe and kayak access is available from the parking area along County Highway LO. Rainbow Springs Lake (35 acres) has a maximum depth of 16 feet and is linked to other lakes in the area via the Mukwonago River and other connector streams. Anglers can expect to catch largemouth bass, northern pike, and several species of panfish.

The Mukwonago River is one of the cleanest rivers in southeast Wisconsin and is a Class II brook and brown trout stream. Restoration projects, such as rerouting the river bed back to its original course, have improved the area's water quality and

> **Ranger Note**
>
> Due to its high water quality and diversity of rare fish species, the state chapter of **The Nature Conservancy** has designated the Mukwonago River as one of the "Last Great Places" in Wisconsin.

recreation. As a result, fish can now freely swim upstream and paddlers are able travel on the river without having to portage around culverts. The river supports at least 53 species of fish, including several rare, state-threatened fish such as the long-ear sunfish and the starhead topminnow.

CAMPING

The Mukwonago River is a day-use property so there are no camping facilities here. The nearby **Ottawa Lake and Pinewood Campgrounds** in the Southern Unit of Kettle Moraine State Forest have 200 campsites (49 electric) with restrooms and shower facilities (See Chapter 11)

Mukwonago Park offers 35 campsites (no electric) with showers, restrooms, and a swimming beach. The park is operated by Waukesha County and is located four miles northeast of the Mukwonago River at S100 W31900 County Highway LO. www.waukeshacounty.gov/mukwonago

WINTER ACTIVITIES

Ice fishing on Rainbow Springs Lake is popular in winter along with hiking, birdwatching, snowshoeing, and cross-country skiing. Trails are not groomed.

AREA ATTRACTIONS

Waukesha and Walworth counties are popular vacation destinations for many. Both have many fine parks, museums, lakes, and other attractions to explore. www.waukeshacounty.com   www.walworthcounty.com

Lulu Lake State Natural Area

Lulu Lake State Natural Area (1,660 acres) is known for having some of the most diverse wetlands and upland glacial topography in the state. The site is located along Lulu Lake (95 acres) and shares a property boundary with Mukwonago River. The property protects rare sedge meadows, fen wetlands, southern bogs, and oak savannas.

The natural area has 2.6 miles of hiking trails and interpretive panels that describe the unique geography, history, flora, and fauna of the area. Lulu Lake is located north of County Highway J. www.dnr.wi.us/org/land/er/sna

DIRECTIONS TO THE FOREST

The Mukwonago River Unit is located about five miles northwest of East Troy. From Highway 67 near the Kettle Moraine State Forest or Highway 83 in the village of Mukwonago, exit onto County Highway LO to the main parking area located on the northwest section of the property.

Mukwonago River Unit
Kettle Moraine State Forest–Southern Unit
County Hwy LO
Mukwonago, WI 53149
(262) 594-6204 (Kettle Moraine Headquarters Office)

19. NEWPORT STATE PARK

Wisconsin's Only Designated Wilderness Park

Newport Bay beach

PARK SNAPSHOT

Newport State Park is located near the tip of Door County along the pristine shoreline of Lake Michigan. This scenic property offers miles of forested hiking trails, backpack camping, nature study, and wildlife viewing, plus picnic areas and a sandy swimming beach.

Newport was designated as Wisconsin's first **Wilderness State Park**. It also became the state's first **International Dark Sky Park** due to the exceptional star gazing opportunities found here.

GEOLOGY AND HISTORY

The varied landscape of Newport includes conifer forest lowlands, upland hardwoods, and both sandy and rocky Lake Michigan shoreline. All were shaped by wind, water, and glacial activity near the end of the last Ice Age about 12,000 years ago. Retreating glaciers on both sides of the Door County Peninsula during that period were responsible for excavating the Green Bay and Lake Michigan basins.

Newport has one of the best-recorded histories of human occupation in Door County. Several Native American cultures are known to have hunted, fished, and built semi-permanent villages in this area, including the Ho-Chunk, Menominee, Ottawa, Sauk, Ojibwe, and Potawatomi. French explorer Robert LaSalle was most likely the first European to set foot on this shoreline in the 17th century. LaSalle was fluent in several indigenous languages and visited a number of Native American encampments along the western shoreline of Lake Michigan by canoe in 1679.

Ranger Trivia

The park was originally named Europe Bay State Park when it opened to the public in 1966 but was later renamed Newport State Park in 1970.

European settlement in the local area was much slower than other regions of northeastern Wisconsin. The thin soils and shallow limestone outcrops found in much of Door County made farming difficult here. The dense stands virgin forests here did not go unnoticed, however. In 1881, Scandinavian immigrant Hans Johnson, along with his partner Peter Knudsen, built a shipping pier, general store, and a lumber mill along what is now the state park shoreline. The new settlement, called Newport, became a thriving logging town and shipping harbor for the next 40 years. Sailing vessels and steamboats docked at the Newport Bay harbor to load lumber, posts, railroad ties, and firewood for transport to Milwaukee, Racine, Chicago, and other major ports on the Great Lakes.

The logging industry attracted workers from Scandinavia, Germany, and Bohemia, but by the early 20th century, most of Door County's forests were gone, bringing the local lumber industry to end. Over time, the once-thriving community of Newport was abandoned and eventually became a ghost town. In 1919, a Chicago investor purchased much of the land here with the intension of selling house lots. His plan never materialized due to onset of the Great Depression and poor employment opportunities at the time.

In 1946, the State of Wisconsin offered to purchase the property to establish a new state park along Lake Michigan. Local citizens strongly opposed this plan, fearing their quiet neighborhood would be overrun by tourists and campers. Twenty years later, the state finally managed to get local approval to acquire the 2,373-acre property, but with the restriction that the park be a low-impact wilderness park with minimal development.

BEACHES, PICINIC AREAS, AND SHELTERS

Newport has a picnic area along the Lake Michigan shoreline. An open-air shelter facility with a fireplace is located here and can be reserved in advance. Nearby is a wide sandy swimming beach.

HIKING, BIKING, AND NATURE TRAILS

The park has **33** miles of hiking trails to explore, including **15** miles of off-road bike routes. The **Newport Trail** (5.0 miles) begins at the picnic area and loops south through a rare boreal forest of hemlock, spruce, balsam fir, and white cedar trees.

The Newport Trail has several side trails, some of which lead to one of the **11** shoreline backpack campsites located here. The **Sand Cove Trail** (1.0 miles) follows the lakeshore to a secluded sand beach and past several attractive rock ledge formations.

> **Ranger Note**
>
> Boreal forests are usually only found in far northern Wisconsin along Lake Superior and in upper Michigan. The cooling effect of Lake Michigan has allowed a remnant boreal forest to thrive at Newport State Park.

> **Ranger Tip**
>
> The **Hotz Trail** can also be accessed by vehicle north of the park along Europe Bay Road. This road also provides access to a local township park along Lake Michigan that has swimming beach and picnic area.

The **Lynd Point/Fern Trail** (2.2 miles) is a side trail off the Europe Bay Trail that follows the rocky shoreline of a peninsula jutting into Lake Michigan. The trail leads through a lowland conifer forest and past interesting limestone formations. The **Hotz History Trail** (1.3 miles) leads through an upland forest of oak, pine, and beech trees to the northern tip of the park with sweeping views of beautiful **Europe Lake**, a 297-acre inland lake.

> **Ranger Trail Pick**
>
> The **Europe Bay Trail** (7.0-mile loop) begins at the picnic area and leads to the far northern boundary of the park. This historic trail was once a wilderness access road to the village of Newport in the late 1800s. The trail leads through a heavily forested area adjacent to Lake Michigan's Europe Bay.

WILDLIFE VIEWING

The park's varied landscape of upland hardwoods, conifer swamps, grasslands, and the Lake Michigan shoreline provides ideal habitat for many wildlife species. White-tailed deer, raccoon, rabbits, porcupine, red and gray squirrels are often encountered in the park. Other more secretive mammals, such as coyote, bobcat, and black bear are occasionally spotted in the park.

More than **170** species of birds are known to nest or migrate through Newport State Park. Chickadees, nuthatches, blue jays, woodpeckers, wild turkeys, and evening grosbeaks can be observed at the park office bird feeding

Rocky Lake Michigan shoreline

stations. In spring and fall, waterfowl, shorebirds, hawks, and many other types of birds can be observed along the Lake Michigan shoreline, an important migration route for thousands of birds.

INTERPRETIVE PROGRAMS AND FACILITIES

Park naturalists and volunteers offer nature hikes and evening programs throughout the use season. A nature center room with interpretive displays and creative learning stations is located inside the park office. Newport's remote location in the northeast corner of Door County gives the park the distinction of being one of the darkest nighttime spots in Wisconsin. As a result, the park has been designated as a **Dark Sky Park** by the International Dark-Sky Association; one of only **18** such sites in the United States. Visitors can obtain a brochure that lists the best locations in the park to experience a nighttime view of the sky at the park office.

Fern Trail (.75 miles) leads through a mature spruce forest and wetland area featuring several types of ferns, including ostrich, bracken, sensitive, and maidenhair ferns. A mobile audio-guide can be checked out at the park office for this trail. The audio-guide has sensors that read interpretive panels and play recordings describing the flora, fauna, and history of the site.

BOATING AND FISHING

Canoes, kayaks, or other carry-in watercraft can access Lake Michigan along the beach area. Motorboats can be launched at the **Sister Bay Marina,** located on the Green Bay side of the peninsula southwest of the park. Anglers interested in Lake Michigan fishing can book a charter boat in the nearby villages of Gills Rock and Ellison Bay.

A public boat launch is located on **Europe Lake** (297 acres), an inland lake that borders the northwestern boundary of the park. This scenic lake has good populations of panfish, smallmouth bass, northern pike, and walleye.

ACCESSIBLE FACILITIES AND TRAILS

The park's visitor entrance station, nature center room, restrooms, picnic areas, and shelters are all accessible to mobility-impaired visitors. Due to the rustic wilderness characteristic of Newport, most hiking trails and backpack campsites are not wheelchair accessible.

CAMPING

Newport has 17 backpack (walk-in) campsites and a group site. Most sites are situated along or near the Lake Michigan shoreline except for the two campsites located at Europe Lake in the north section of the park. Each site has a fire ring, bench, and a steel food storage box. Vault-type restrooms are available near each camping area. Drinking water must be carried in. All campsites at Newport are in high demand throughout the warm season. Advance reservations are recommended.

WINTER ACTIVITIES

Hiking, birdwatching, winter camping, and exploring the ice-encrusted shoreline of the park are popular off-season activities. Park staff groom 21 miles of inline cross-country trails and two miles for skate skiing. The park also has five miles of trails for snowshoeing, hiking, and fat-tire bicycle use.

AREA ATTRACTIONS

Door County is one of the state's top tourist destination areas. The county has many parks, museums, orchards, wineries, and picturesque villages to explore. www.doorcounty.com

Death's Door Maritime Museum has many interesting displays and artifacts related to the commercial fishing and maritime history of the area. The museum is located along the shoreline of what early French explorers called *Porte des Mortes* or "Door of Death," a treacherous sailing route linking Lake Michigan to

Green Bay. The museum is within the village of Gills Rock at 12724 W. Wisconsin Bay Road, Ellison Bay. www.dcmm.org

DIRECTIONS TO THE PARK
Newport is located about three miles east of the village of Ellison Bay in northern Door County. From Highway 42 north of Ellison Bay, take County Highway NP south and east and then south on Newport Lane to the park entrance.

Newport State Park
475 County Road NP
Ellison Bay, WI 54210
(920) 854-2500

20. PENINSULA STATE PARK

Crown Jewel of Door County

Nicolet Bay

PARK SNAPSHOT

Peninsula State Park is often described as the most scenic park in Wisconsin. More than a million people flock to the park each year to enjoy the stunning views from roadside bluff overlooks and the parks observation tower.

This popular 3,776-acre park is located in northern Door County along a forested peninsula encircled by the waters of Green Bay. The park offers camping, swimming, hiking, picnicking, boating, biking, nature study, and an 18-hole golf course.

GEOLOGY AND HISTORY

The rocky bluffs of Peninsula State Park are composed of dolomite limestone. This hard rock was formed by crystallized sand, calcium carbonite, and decomposing sea life that settled beneath an ancient ocean that once covered this area. The entire park is part of the **Niagara Escarpment**, a massive stretch of dolomite rock that begins at its namesake, the Niagara Falls in New York. From here, the escarpment heads to upper Michigan and then south down the backbone Door County and along the east shore of Lake Winnebago before terminating in northern Illinois.

The towering rock bluffs at Peninsula resisted the advance of mile-high glaciers that excavated the Green Bay and Lake Michigan basins near the end of the last Ice Age 12,000 years ago. The bluffs were also eroded by wave action from Lake Nippissing, a massive post-glacial lake much deeper than Lake Michigan and Green Bay are today.

Native American people have hunted, fished, gathered wild plants, and lived within the park region for thousands of years.

In 1634, French explorer Jean Nicolet became the first known European to set foot in Wisconsin and make contact with indigenous people. During his exploration of Lake Michigan and Green Bay, Nicolet reportedly landed on **Horseshoe Island**. This island is located just offshore from Nicolet Bay and is now part of the state park.

During the early 1800s, the Ottawa, Ojibwe, and Potawatomi people were engaged in trading furs for ironware, rifles, blankets, and other items with French, English, and American fur companies. Conflicts over land use arose when permanent settlers arrived to set up farmsteads and began cutting trees for the timber industry. This tense situation led to President Andrew Jackson signing the Indian Removal Act of 1830, which forced all Native American tribes to cede their lands to the U.S. government and relocate west of the Mississippi River.

> **Ranger Note**
>
> Archaeological excavations near Nicolet Bay in the park revealed stone tools, arrowheads, and other artifacts used by early Woodland and Oneonta people who lived in a village here from about 400 BC until 1300 AD.

People were drawn to the forested bluffs and clear waters of Peninsula long before it became a state park. In 1909, Tom Reynolds, a local politician, along with other influential citizens, urged Wisconsin legislators to purchase land in the Fish Creek area to develop a state park. Their efforts paid off in 1919 when Peninsula State Park was established as Wisconsin's second state park after **Interstate Park** in 1900. (See Chapter 70)

BEACHES, PICNIC AREAS, AND SHELTERS

The **Nicolet Bay Beach and Picnic Area** is the most popular day-use area in the park. On warm summer days, hundreds of sunbathers and swimmers flock to the sandy beach along Nicolet Bay. The concession stand here offers prepared food, beverages, clothing, and camping supplies and also rents bikes, kayaks, canoes, and paddleboats.

The park has six picnic areas overlooking the scenic waters of Green Bay. **Weicker's Point** and **Weborg Point Picnic Areas** have indoor shelter facilities that can be reserved in advance. The **Eagle Terrace Picnic and Day Use Area** has a 60-foot-tall observation tower that offers panoramic views over the blue waters of Green Bay.

The park's 18-hole **Peninsula Golf Course** is located near the east entrance. The golf course has the distinction of being the only one of its kind in the state park system.

HIKING, BIKING, AND NATURE TRAILS

There are 20 miles of hiking and biking trails in the park. The **Sunset Biking/ Hiking Trail** (10.0 miles) is surfaced with crushed limestone and leads to many of the park's most popular sites, including the nature center, **Nicolet Beach**, the **Eagle Bluff Lighthouse**, and most campgrounds. Off-road bicycle riders have 12 miles of challenging trails located in the southeastern section of the park.

The **Sentinel Nature Trail** (2.0 miles) loops through a mixed forest of maple, red pine, and beech trees. Interpretive signs along the route explain the forest ecology of the park. This trail can be accessed from the **Eagle Bluff Tower** area. **Hemlock Trail** (1.8 miles) leads through a white cedar and hemlock grove to an upland hardwood forest. This trail connects to **Skyline Trail** (3.0 miles) with scenic views from **Sven's Bluff** and then loops back down to the trailhead along the western lakeshore of the park.

Eagle Cave

> **Ranger Trail Pick**
>
> **Eagle Trail** (2.0 miles) offers some of the best hiking and scenic views in the park. The trail begins in a hardwood forest on a bluff area and then descends a series of limestone rock outcrops to the shoreline of Eagle Bay. From there, the trail follows the shoreline through old-growth cedar and over tiny streamlets beneath a 150-foot limestone bluff. Several small caves carved by wave action from an extinct post-glacial lake can be seen along this route. One of the most impressive is **Eagle Cave** perched in a towering bluff 30 feet above water's edge. The trail ends with a hike back up to the top of the bluff.

> **Ranger Trivia**
>
> Research biologists have discovered that some of the stunted white cedar trees that grow along the rocky bluffs at Peninsula are more than 500 years old.

The **White Cedar Nature Trail** (0.5 miles) begins near the nature center and follows an extinct 5,000-year-old post-glacial lake shoreline. Interpretive signs along the trail describe how the flora and fauna of the park provide food and shelter for white-tailed deer.

> **Ranger Note**
>
> A bat nursery and roosting house, located at Welcker's Point Picnic Area, was erected to help offset the nationwide decline in bat populations. Visitors can watch bats as they exit the roosting house at sunset to feed on insects.

WILDLIFE VIEWING

Peninsula State Park is a designated **National Migratory Bird Stopover Site**, which makes it a premiere destination for birdwatchers, especially in spring and fall. The park is home to at least 125 known species of birds that nest here in spring and summer.

The hardwood forests, wetlands, rocky bluffs and shoreline of the park provide habitat for many animal and plant species. White-tailed deer, chipmunks, squirrels, raccoon, and rabbits are often seen in the park. More secretive animals such as coyote, fox, black bear, and bobcat are occasionally observed here.

Peninsula has a variety of plant life, including many species of wildflowers seen along woodland trails, roadsides, bluffs, and wetlands. The park's diverse forest areas contain nearly every tree species native to Wisconsin.

INTERPRETIVE PROGRAMS AND FACILITIES

The **White Cedar Nature Center** has interpretive displays highlighting the flora, fauna and history of the park and several kid-friendly learning stations. Park naturalists and volunteers present nature hikes, evening programs, and outdoor skills workshops throughout the season.

The **Eagle Bluff Lighthouse** is located in the northeast section of the park. This historic brick structure was built in **1864** to guide sailing vessels through the shallow and rocky areas of Green Bay. Guided tours are presented by members of the Door County Historical Society. www.eagleblufflighthouse.org

The nearby **Ridges Sanctuary** is a designated **National Natural Landmark** that features rare sand ridge and swale wetlands and an historic **1869** range lighthouse. The 1,600-acre natural area is located eight miles southeast of Peninsula Park along Hwy 57 in Baileys Harbor. www.ridgessanctuary.org

BOATING AND FISHING

The offshore waters of Green Bay are ideal for boating, fishing, sailing, and kayaking. The park has a motorboat launch along Nicolet Bay and a small launch area for kayaks or canoes within the **Tennison Bay Campground.**

Shoreline fishing for smallmouth and rock bass is popular along the concrete pier at the **Weborg Point Picnic Area.** Charter boat fishing for salmon, steelhead, walleye, and brown trout can be booked at the nearby Fish Creek Harbor.

ACCESSIBLE FACILITIES AND TRAILS

The park's entrance station, nature center, concession stand, shelters, and most restrooms are accessible for mobility-impaired visitors. A 0.6 mile section of the **Sentinel Trail** has a near-level crushed gravel surface for wheelchair access. Nicolet Bay and Tennison Campgrounds offer wheelchair-accessible campsites that have electrical service and paved walks to a shower/restroom facility.

The **Eagle Bluff Observation Tower** has an 850-foot-long wheelchair ramp that ascends 60 feet to the top of the structure for a panoramic view of the park's forests and the waters of Green Bay below.

CAMPING

Peninsula has **472** campsites (**165** electric) located within four campgrounds. All campgrounds have restroom/shower facilities.

> **Ranger Note**
>
> The campsites at Peninsula are some of most sought after in the state throughout the summer season and weekends in spring and fall. Advance campsite reservations are recommended.

A tent-only group camp with three campsites is located east of the Nicolet Campground. Each site can accommodate up to 50 campers. None of the sites have electrical service but all have drinking water fountains and vault-type restrooms.

WINTER ACTIVITIES
Park staff grooms 16 miles of cross-county ski trails for both classic and skate skiing. The park also has six miles of snowshoe trails and 17 miles of snowmobile trails. Winter camping is available at the Tennison Bay Campground. A popular sledding and tubing hill is located above fairway 17 at the golf course.

AREA ATTRACTIONS
Door County and the village of Fish Creek have many fine restaurants, specialty shops and outdoor sites to explore.
www.visitfishcreek.com   www.doorcounty.com
The **Northern Sky Theater** is a 675-seat outdoor theater located within Peninsula State Park between Nicolet Bay and Tennison Bay campgrounds. Musical theater comedic performances are presented by professional actors throughout the summer season. Tickets can be purchased in advance at www.northernskytheater.com.

DIRECTIONS TO THE PARK
Peninsula is located along the Green Bay side of Door County. From Interstate I-43 near Green Bay, exit onto Highway 57 North. At the intersection of Highways 57 and 42 north of Sturgeon Bay, take Highway 42 to Fish Creek and the state park entrance.

Peninsula State Park
9462 Shore Road
Fish Creek, WI 54212
(920) 868-3258

21. PIKE LAKE UNIT—
KETTLE MORAINE STATE FOREST SOUTH
Scenic Gem—A Gift of the Glaciers

Observation tower view of Pike Lake

PARK SNAPSHOT

The Pike Lake Unit of the Kettle Moraine State Forest is perched above the eastern shoreline of Pike Lake, a scenic glacial lake located south of the city of Hartford in Washington County. Although modest in size, this **825-acre** property offers a wide range of outdoor activities including camping, hiking, swimming, picnicking, fishing, nature study, bicycling, cross-country skiing, boating, and much more.

GEOLOGY AND HISTORY

The rolling landscape of the Kettle Moraine is often referred to as the "Gift of the Glaciers." Over the past **2.5** million years, several massive glaciers have pushed through this region of southeastern Wisconsin from the Canadian Arctic. Near the end of that last Ice Age about **12,000** years ago, retreating glaciers left behind thousands of moraine hills and plains composed of sand, gravel, and other glacial debris. Buried beneath some of these glacial deposits were massive chunks of ice that had broken off larger ice sheets. As the underground ice began to melt, depressions known as kettles began to appear in the landscape. Some the kettle formations remained dry while others filled with water seasonally and evolved into wetlands or ponds. The largest and deepest ones filled with ground water, creating permanent kettle lakes such as Pike Lake.

The Pike Lake area was home to several Native American cultures over time, including the Effigy Mound Builders, a Late Woodland tribe that lived in southern Wisconsin from 700 to 1100 A.D. These ancient people built nine effigy mounds in the form of animals and geometric shapes along the shoreline of Pike Lake. Unfortunately, these mounds were destroyed due to farming activity and development practices prior to state ownership.

Beginning in the 17th century, Native American people such as the Ho Chunk, Sauk, Fox, and Potawatomi began to trade with the French and later with English and American fur companies. The fur trade brought indigenous people European goods such as iron axes, knives, pots, and wool blankets but disrupted their traditional way of life. When the first permanent white settlers arrived to farm and cut timber in the area, disputes with indigenous people regarding land ownership arose. Due to ongoing conflicts between white settlers and Native Americans, the U.S. government induced Potawatomi leaders to sign the Treaty of Chicago in 1833, which ended all tribal rights to land within southeastern Wisconsin.

Swimming beach on Pike Lake

In 1834, government surveyors John Theil and Nicholas Simon began to map the township lines of Washington County. The men were so impressed with the landscape of the Pike Lake area that both returned the following year to settle here themselves. Soon after, thousands of German and Irish immigrants arrived to clear the land for farming and work for lumber companies engaged in logging the vast hardwood forests of the Kettle Moraine. In the 1960s, the state began to purchase property along the eastern shoreline of Pike Lake to establish a new state park, which opened in 1971. Pike Lake State Park was later reclassified as a unit of the Kettle Moraine Forest in 1998.

BEACHES, PICNIC AREAS, AND SHELTERS

The **Beach Picnic Area** is well shaded with a playground and two shelters that can be reserved in advance. The **North Shelter** is an open-air facility with a capacity of up to 100 people. The **South Shelter** is an enclosed six-sided building with a limit of 30 people. Also located here is a 500-foot sand beach and swimming area and a bathhouse/restroom facility. The water quality of Pike Lake is generally very good thanks to the many clear-running springs that feed into it.

HIKING, BIKING, AND NATURE TRAILS

Pike Lake has ten miles of trails to explore, including a 2.6-mile section of the **Ice Age National Scenic Trail**. Most trails are gently rolling and easy to hike except for a few sections that are a bit more challenging due to the hilly landscape of the property.

The **Black Forest Nature Trail** (0.8 miles) loops through a hardwood forest of sugar maple, beech and red oak trees. Along the trail are interpretive signs which describe the flora, fauna, and geologic history of the area. An observation platform along this trail provides a close-up view of a crystal-clear hillside spring that flows into nearby woodland ponds.

The **Pike Lake Bike Trail** (1.75 miles) begins in the Beach Picnic Area and leads to the northwest boundary of park with a side trail to the visitor entrance station. From here the trail follows a two-mile paved trail from Pike Lake to the nearby city of Hartford and the **Rubicon River Bike Trail**, which connects several scenic riverside parks.

Ranger Trail Pick

The **Powder Hill Trail** (0.5 miles) is a short trail that ascends to a 55-foot-tall wooden observation tower with outstanding views of Pike Lake and the rolling landscape of the Kettle Moraine below. On a clear day the iconic double steeples of the Basilica of Holy Hill National Shrine can be seen in the horizon.

WILDLIFE VIEWING

The hardwood forests, prairies, wetlands, and the shoreline of Pike Lake provide ideal habitat for a wide variety of wildlife including white-tailed deer, gray and fox squirrels, rabbits, chipmunks, coyotes, red fox, and several species of owls and hawks. Pike Lake is an important spring and fall migratory route for waterfowl, shorebirds, woodland songbirds, and raptors. Bald eagles and osprey are occasionally spotted here as well.

Attractive wildflowers such as trilliums, wood violets, bloodroot, and marsh marigolds can be seen in full bloom in spring. Black-eyed Susan, wild sunflowers, goldenrod, and a variety of colorful asters bloom in late summer and fall.

INTERPRETIVE PROGRAMS AND FACILITIES

Interpretive signs featuring the flora, fauna, and human history of property are located along the **Black Forest Nature Trail**, the shoreline of Pike Lake and atop the **Powder Hill Observation Tower**. Interpretive displays and animals mounts are located in the lobby area of the visitor entrance station. Nature hikes, evening programs and special events are scheduled throughout the use season.

The **Astronomy Trail** (0.5 miles) located near the **Sunrise Campground**, offers a unique and imaginative hike through the solar system. Interpretive panels along this trail highlight each planet and are spaced to represent their relative distance from the sun. One long step (about a yard) along the route is the equivalent of 3,600,000 miles.

> **Ranger Note**
>
> The observation tower at Pike Lake is located on a 1,350-foot high cone-shaped hill known to geologists as a kame formation. A kame is formed beneath glacial ice when a river of meltwater carrying glacial debris plunges down a large hole on top of glacier. Over time, deposits of sand, gravel and rocks form a cone-shaped hill similar to sand falling through an hourglass.

BOATING AND FISHING

Pike Lake is a 522-acre spring-fed lake with a maximum depth of 45 feet. The lake was named for its excellent walleye and northern pike fishing but anglers will also find bluegill, perch, crappie, and largemouth bass here. Fishing from the shoreline or fishing piers is available adjacent to Beach Picnic Area. Motor boating, sailing, canoeing, kayaking, and water skiing are very popular on Pike Lake. The property does not have boat launch but several public landings are located on the west side of the lake. A carry-in kayak/canoe launch area is located north of the swimming beach.

ACCESSIBLE FACILITIES AND TRAILS

Pike Lake's visitor entrance station, picnic shelters, and most restroom facilities are accessible for all mobility-impaired visitors. Campsite 21 is wheelchair-accessible with electric service and a paved walkway to a restroom/shower facility.

The **Boardwalk Trail** (0.5 miles) is a looped wheelchair-accessible path that begins on the north side of the Beach Picnic Area. A section of this trail follows a raised boardwalk that leads to an accessible fishing pier. The rest of the trail follows a combined asphalt-paved and crushed limestone trail. Due to the hilly topography of the property, most other hiking trails may be challenging for visitors with walking impairments.

CAMPING

The **Sunrise Campground** has 33 campsites (12 electric). Most campsites are well shaded within a mature forest area. The campground has a flush toilet/shower building and an RV water fill and sanitary station. The campground is open from late April through the third weekend in October. Pike Lake also has three backpack (walk-in) campsites located along the 2.6 mile segment of the Ice Age Trail.

WINTER ACTIVITIES

Property staff groom four miles of cross-country ski trails located east of Powder Hill Road. All trails west of this road are reserved for snowshoeing and hiking. The campground is closed in winter. Pike Lake is a popular ice fishing destination for many anglers in winter.

AREA ATTRACTIONS

Washington County is known for its many public parks, museums, hiking and biking trails, and other attractions. www.visitwashingtoncounty.com

The **Basilica of Holy Hill National Shrine** is perched 1,350 feet above sea level atop one of the highest hills in southeast Wisconsin. This awe-inspiring cathedral with its iconic two-towered steeples can be seen for miles around. The Basilica is on National Register of Historic Places and is visited by more than 500,000 people each year. Visitors can climb a staircase to the top of one of the steeples to enjoy a bird's eye view of the rolling Kettle Moraine forest landscape below. On a clear day, the Milwaukee skyline can be seen more than 30 miles away. Holy Hill is located five miles south of Pike Lake at 1525 Carmel Road, Hubertus, WI 53033. www.holyhill.com

DIRECTIONS TO PIKE LAKE

Pike Lake is located about 25 miles northwest of Milwaukee, about two miles southeast of the city of Hartford. From U.S. Highway 41 exit onto Highway 60 and travel west to the Pike Lake entrance off Kettle Moraine Road.

Kettle Moraine State Forest—Pike Lake Unit
3544 Kettle Moraine Road
Hartford, WI 53027
(262) 670-3400

22. POINT BEACH STATE FOREST

Wisconsin's Largest Beach

Point Beach State Forest shoreline

FOREST SNAPSHOT

Point Beach State Forest is located along the beautiful shoreline of Lake Michigan in Manitowoc County. The forest's wide, six-mile-long sandy beachfront is the largest and longest in Wisconsin. This popular 3,029-acre property offers camping, swimming, boating, nature study, biking, and miles of hiking trails. An historic U.S. Coast Guard lighthouse is also located here.

GEOLOGY AND HISTORY

The landscape of Point Beach was formed over time by the action of wind, waves, and glacial activity. Several glaciers have advanced through this area over the past two million years; each time excavating the Lake Michigan basin ever deeper. Near the end of the last Ice Age, about 12,000 years ago, fluctuating lake levels altered the position of forest's waterfront many times. Geologists have identified 11 extinct shorelines here in the form of sand ridges and wetland swales that run parallel to the current lakeshore. The oldest shoreline still visible in the forest was formed more than 7,500 years ago. It was created by Lake Nipissing, a large post-glacial lake that submerged most of the lakeshore area beneath 25 feet of water.

Today, the lake's ever-changing water levels, waves, and wind continue to reshape the beach and sand dunes of the property.

> **Ranger Note**
>
> Tens of thousands of hemlock trees were cut down in the forest area solely for their valuable bark, which was peeled off and sold commercially. Hemlock bark contained the chemical tannin, which was used in the leather tanning industry of the early 20th century.

Many Native American cultures have fished, hunted, and gathered wild plants along Lake Michigan or *Michi Gami* ("large lake"), as it was called by the Ojibwe. Archeologists have discovered artifacts from several different tribes that once lived along these shores in the Manitowoc County including the Ho-Chunk, Meskwaki, Menominee, Ojibwa, and Potawatomi people.

European immigrants, mostly from Germany and Ireland, arrived in the Manitowoc/Two Rivers area in the mid- to late-1800s to farm the rich soils found inland from Lake Michigan. The sandy soils of the forest area were too dry and unfertile to raise crops, but the pine and hemlock forests that grew here were valuable to the timber industry. By the early 1920s, most of the trees along the lakeshore had already been cut for lumber, fence posts, shingles, and firewood, or for use in the tanning industry.

> **Ranger Note**
>
> **Rawley Point** was named after Peter Rawley, an early fur trader who established a trading post near this location in 1835. Several lighthouses have been erected at this site since the early 1860s to guide sailing vessels through the shallow waters offshore. The current 111-foot-tall structure was built in 1894 and is one of the oldest lighthouses along Wisconsin's shoreline, as well as the tallest.

In 1937, the federal government transferred 70 acres of surplus land from its **Rawley Point Lighthouse** property to the State of Wisconsin to develop a state park along Lake Michigan. Additional lands were donated by Manitowoc County and the city of Two Rivers for the park. Due to available government funding at the time, Point Beach State Park was designated a state forest when opened to the public in 1938.

BEACHES, PICNIC AREAS, AND SHELTERS

The cool, refreshing winds and wide sandy beaches of Lake Michigan have attracted visitors to this shoreline for generations. The **Lodge Picnic Area** is the largest day-use area in the forest. This attractive log and stone structure houses the forest's nature center and a concession stand that offers grilled food, refreshments, ice cream, camping supplies, and bicycle rentals. The lodge was built in 1939 by craftsmen of the WPA (Works Projects Administration), a federal Depression-era employment program.

Rawley Point Lighthouse

The **Lakeshore** and **Lighthouse Picnic Areas** are located along the beach east of the campground. South of these picnic areas is a scenic overlook of the historic Rawley Point Lighthouse.

HIKING, BIKING, AND NATURE TRAILS

Point Beach has 17 miles of hiking trails to explore including a nine-mile section of the 1,200-mile-long Ice Age National Scenic Trail. The **Swales Nature Trail** (0.5 miles), located west of the Lodge Picnic Area, has interpretive panels that describe the flora and fauna of the forest and explain how the sand ridge and swale wetlands were formed here.

The **Ridges Trail** (7.5 miles) begins near the Lodge Picnic Area and is the longest trail in the forest. This route has three connecting loops that run parallel

---

Ranger Trail Pick

Rawley Point Trail (5.0 miles) is a crushed limestone biking/hiking trail that leads through pine and hardwood forests, open sand dune areas, and crosses several wooden bridges over wetlands and streams. The trail begins at the Lodge Picnic Area and ends in the southwest corner of the forest, where it connects to the Mariners Trail (12.0 miles). This community trail leads through the nearby city of Two Rivers and then travels along an asphalt-paved route along Lake Michigan to the north side of Manitowoc.

to the lakeshore through forested sand ridges and wetland swales. The **Red Loop**, along with the **Blue Loop**, is a popular hike for a combined total of 5.5 miles.

The **Red Pine Trail** (3.1 miles) is a hiking, skiing, and off-road biking route that loops through a mature red pine plantation. Nearby is the **Legacy Equestrian Trail** (2.5 miles), a looped trail for horseback riding. Both of these trails can be accessed west of County Highway O across the road from the forest entrance.

WILDLIFE VIEWING

The forests, lakeshore, sand dunes, and wetlands of Point Beach provide ideal habitat for many species of birds, mammals, reptiles, and amphibians. The forest has three designated state natural areas, which protect several threatened animal and plant species. White-tailed deer, rabbits, chipmunks, and four different species of squirrels can are often spotted throughout the forest. More secretive animals such as coyote, fox, and occasionally otters also live here.

Birdwatchers have recorded more than a 150 bird species in the forest, including chickadees, woodpeckers, nuthatches, cardinals, blue jays, ruffed grouse, wild turkeys, and several kinds of woodland songbirds, owls, and hawks. Sandpipers, terns, gulls, geese, and flocks of diving ducks can be spotted along the Lake Michigan throughout the warm season. The shoreline of Lake Michigan is an important migratory route for thousands of birds during the spring and fall migration periods.

INTERPRETIVE PROGRAMS AND FACILITIES

A small nature center located inside the lodge/concession building has several interpretive displays. Nature hikes and evening programs are offered on weekends throughout the summer season. The nearby **Woodland Dunes Nature Preserve** is 1,200-acre wildlife refuge that has miles of hiking trails, elevated boardwalks, a wildlife garden, and an impressive interpretive center. Woodland Dunes is located west of the city of Two Rivers at 3000 Hawthorn Avenue (Highway 310). www.woodlanddunes.org

BOATING AND FISHING

There are no boat launch facilities at Point Beach but several public ramps are located within the harbor areas of Two Rivers and Manitowoc. Shoreline fishing is popular along the breakwater piers within these harbors. Anglers can book charter boat expeditions for trout and salmon fishing on Lake Michigan here as well.

Carry-in canoes and kayaks can be launched from beach areas in the forest. Paddling along the Lake Michigan shoreline on calm wave days is a great way to view the beautiful dunes and shoreline of the forest. Kayakers can also paddle upstream along **Molarsh Creek**, a scenic tributary of Lake Michigan in southern section of the forest.

ACCESIBLE FACILITIES AND TRAILS

The visitor entrance station, shelters, lodge/concession, and most restrooms are accessible for mobility-impaired visitors. A wheelchair-accessible ramp to the beach is located adjacent to the Lodge Picnic Area parking area. Campsite **31** is wheelchair-accessible and has electrical service. This site is located across the road from a restroom/shower facility.

CAMPING

Point Beach has 129 campsites (71 electric). Most sites are well shaded within a mature forest. The campground has two flush toilet/shower facilities and an RV water fill and sanitary station located adjacent to the **Lakeshore Picnic Area.**

> **Ranger Tip**
>
> Point Beach offers 27 campsites with water connection hook-ups, the only campground in the state park and forest system to offer this service.

The forest has two rustic log cabins that can accommodate up to 16 campers each. The cabins have bunkbeds and wood stoves for heat. Drinking water, vault-type restrooms, and an outdoor shelter are located nearby. The forest also has a group campground (tent only) with a capacity of 60 people located west of County Highway O in the north section of the forest.

WINTER ACTIVITIES

Forest staff groom 7.5 miles of classic (in-line) and 3.5 miles of skate-skiing trails. Snowshoeing and hiking are allowed on all other trails in the forest. Several campsites are plowed open for winter campers. A snowmobile trail travels through a portion of the forest with connections to the Manitowoc County snowmobile trail network and the city of Two Rivers.

AREA ATTRACTIONS

The Manitowoc and Two Rivers area has many parks, historic sites, and other attractions to visit. visitmanitowoc@manitowoc.info (920) 686-3070

> **Ranger Trivia**
>
> The county and nearby city of Manitowoc were both named after the Ojibwa term *Munedoowk*, which means "Place of the Manitou" or "Home of the Good Spirit."

The **Wisconsin Maritime Museum** showcases rare Great Lakes shipping artifacts and tells the story of Wisconsin's long marine history. The site has many interesting displays, including a depiction of a 19th-century sailing vessel port town. The museum also offers a unique onboard tour of the USS *Cobia*, a fully restored WWII submarine. The museum is located at 75 Maritime Drive in Manitowoc. www.schoonercoast.org.

DIRECTIONS TO THE FOREST
Point Beach State Forest is located north of the city of Two Rivers in Manitowoc County. From Interstate I-43 take exit 154 (Hwy 310) and travel east towards Two Rivers. Take Highway 42 North through Two Rivers and then exit onto County Highway O about four miles north to the state forest entrance.

Point Beach State Forest
9400 County Highway O
Two Rivers, WI 54241
(920) 794-7480

23. POTAWATOMI STATE PARK

Gateway to Door County

Sawyer Harbor at Potawatomi

PARK SNAPSHOT

Potawatomi State Park is located along the scenic shoreline of Sturgeon Bay just northwest of the city of Sturgeon Bay, the gateway community to the Door County Peninsula. From atop the park's 150-foot-high bluff, visitors can get a bird's eye view of Green Bay to the west and Lake Michigan to the east. This popular 1,225-acre park has a campground and miles of hiking trails plus biking, boating, fishing, nature study, and many other outdoor activities.

GEOLOGY AND HISTORY

The park's landscape is composed of forested hills, limestone bluffs, and a rocky shoreline along Sturgeon Bay. Geologists have found evidence that Sturgeon Bay may have started as a wide river valley that formed beneath an immense glacier during the last Ice Age. When the last of the glaciers retreated north out of Wisconsin about 10,000 years ago, water levels of the Great Lakes fluctuated greatly over time. About 8,000 years ago, post-glacial Lake Algonquin and later Lake Nipissing, caused high water levels to further deepen the river valley and erode the limestone bluffs of the park. Extinct shorelines and wave-sculptured rock outcrops from that period can still be seen in the park high above the current water level of the lake.

Some of the earliest Native American cultures to inhabit the park area were the Late Woodland people who lived here from about 400 to 1100 AD. They hunted big game, planted crops of corn, beans, and melons and fished for lake trout, whitefish, and lake sturgeon. Later tribes to occupy the area included the Sauk, Ojibwa, Potawatomi, Ho-Chunk, and the Menominee.

In 1830, the U.S. War Department acquired land here to build a military fort due to the threat of another war with Great Britain through Canada. The fort was never built but the property known as Government Bluff became an important source for dolomite limestone rock for decades. Most of the rock quarried here was used to build breakwaters at city ports throughout the Great Lakes region.

> **Ranger Trivia**
>
> The Menominee people referred to Sturgeon Bay as *Namaew-Wihkit*, meaning "The Bay of the Sturgeon."

In 1836, Menominee tribal leaders were induced to sell four million acres to the United States, including all of Door County, through the Treaty of Cedars. Soon after Norwegian, German, and Irish immigrants arrived to farm the land. Many found it difficult to grow crops on the thin soils of the local area and decided to sell timber or find work in local limestone quarries instead. The nearby community of Sturgeon Bay remained a small town until 1881 when a group of private investors came up with the idea of creating a shortcut shipping route between the waters of Green Bay and Lake Michigan. This massive construction project involved dredging out the natural water basin of Sturgeon Bay and excavating a 1.3-mile canal east to Lake Michigan. When completed, the new shipping lane elevated Sturgeon Bay into a major Great Lakes port city and important ship-building center.

In 1928, a group of biologists petitioned state representatives to acquire the Government Bluff property to preserve limestone bluffs and rare plant life of the site. The following year, the state established **Nicolet State Park** and opened it to the public. The park's name was later changed to **Potawatomi State Park** in honor of the Potawatomi people who once lived here.

BEACHES, PICNIC AREAS, AND SHELTERS

The park has several picnic areas located along the shoreline of the Sturgeon Bay. A combination indoor/outdoor shelter building in the picnic area has two stone fireplaces and can be reserved in advance. The shoreline of Potawatomi is composed of limestone shelves and fragmented rock, making it unsuitable for swimming. A sand-lined swimming area is available at **Otumba Beach Park** located along West Locust Court in the nearby city of Sturgeon Bay.

HIKING, BIKING, AND NATURE TRAILS

The park has 9.5 miles of trails to explore, including a 2.8-mile section of the Ice Age National Scenic Trail. Potawatomi serves as the eastern terminus of the Ice Age Trail, which travels 1,200 miles across the state to **Interstate State Park** near St. Croix Falls. (See Chapter 70)

The **Hemlock Trail** (2.6 miles) begins along the Sturgeon Bay shoreline and gradually ascends above the campground area over limestone outcroppings through hardwood and hemlock forest. **Tower Trail** (3.6 miles) begins at the **Old Ski Hill Overlook** in the northwest corner of the park. The trail heads downhill to the north campground and then loops back uphill to the highest bluff in the park. A 75-foot observation tower along the trail offers scenic views of Sawyer Harbor and Green Bay.

> **Ranger Note**
>
> The **Potawatomi Observation Tower** was built through the efforts of local citizens in 1931. Due to the age of this wooden structure, the tower is currently closed but efforts are underway to replace it in the future.

Bike riders have eight miles of paved roadways and eight miles off-road trails to explore at Potawatomi. Bike rentals are available at the park's concession area. A hookup bike route from the park leads through Sturgeon Bay along the **Ahnapee State Trail** (48 miles). This biking/hiking trail follows the Ahnapee and Kewaunee rivers south to Algoma, Luxemburg, and Kewaunee. www.ahnapeestatetrail.com

> **Ranger Trail Pick**
>
> The **Ancient Shorelines Nature Trail** (0.5 miles) starts across the road from the concession area. Interpretive signs along the trail explain how the forces of water, wind, and glacial activity shaped the limestone bluffs and Sturgeon Bay itself. The route leads past extinct shorelines and rock outcrops left by post-glacial Lake Nipissing 5,500 years ago and the even older shoreline of Lake Algonquin more than 8,000 years ago.

WILDLIFE VIEWING

The park's forests, bluffs, grassy fields, and shoreline areas attract a wide array of wildlife species. White-tailed deer, gray squirrels, chipmunks, rabbits, raccoon, fox, and coyote are commonly seen here. Over 220 species of nesting and migratory birds have been recorded in the park, including blue jays, cardinals, chickadees, pileated woodpecker and turkey vultures, plus several kinds of hawks, owls, and woodland songbirds. Herring gulls, ring-billed gulls, common terns, and many species of ducks and geese can be observed along the Sturgeon Bay shoreline.

INTERPRETIVE PROGRAMS AND FACILITIES
Nature hikes and other interpretive programs are offered from Memorial Day weekend through Labor Day. A small nature center is located adjacent to the park concession building. Evening programs are presented at the parks amphitheater located near the start of the **Ancient Shores Nature Trail.**

BOATING AND FISHING
Anglers travel to Door County from throughout the Midwest to fish for walleye, perch, northern pike, and bass. Sturgeon Bay is widely regarded as having one of the best smallmouth bass fisheries in North America. Charter boat fishing for Great Lakes salmon, brown trout, and steelhead can be reserved at the Sturgeon Bay Marina. The **Sawyer Harbor Boat Launch and Fish Cleaning Station,** located within the park, is one of the finest of its kind in the state. The launch provides access to Green Bay and Sturgeon Bay for boating, sailing, and fishing. Canoe and kayak rentals are available through the park's concessions stand for paddlers who wish to explore the park's two miles of scenic shoreline.

ACCESSIBLE FACILITIES AND TRAILS
Potawatomi's **Cabin by the Bay** has an accessible kitchen, living room, bedroom, and wheel-in restroom/shower. The cabin can be reserved in advance by physically disabled campers. The park also has two wheelchair-accessible campsites with electrical service and paved pathways to the restroom/shower facilities. A specially-adapted kayak equipped with floatation outriggers is available for use for mobility-impaired paddlers. The kayak can be reserved at the park's concession.

The park's visitor entrance station, nature center, concession, and most restrooms are accessible for mobility-impaired visitors. A paved walkway through the picnic area provides wheelchair access to restrooms, picnic tables, shelters, and the parks wheelchair-accessible fishing pier along Sturgeon Bay. Potawatomi's steep topography and rocky bluff areas make access to most other trails challenging for visitors with walking impairments.

CAMPING
Potawatomi has 127 campsites (40 electric). Each site has a fire ring, picnic table, and a Leopold-style bench. All electric service sites are located in the south campground. A restroom/shower building is located between the two campgrounds. An RV dump and water fill station are located adjacent to the park entrance station.

Campsite at Potawatomi

A well-stocked concession stand located near the entrance to the campground offers firewood, camping supplies, grocery items, souvenirs, snacks, and refreshments. Bicycle, canoes, and kayaks can be rented here as well.

The park's group campground has four large campsites. The campground is located near the south end of the property and has vault-type restrooms and a solar-powered drinking water tap.

WINTER ACTIVITIES
Park staff groom nine miles of cross-country ski trails for both classic and skate skiing. All the trails begin near the picnic area shelter facility. Hiking and snowshoeing are allowed on all other trails. Wildlife watching and winter camping are also popular offseason activities at Potawatomi. The park has eight miles of snowmobile trails that connect to local county trails. Snowmobile parking is available at the park entrance visitor station.

AREA ATTRACTIONS
The Sturgeon Bay area has many fine restaurants, shops, parks, museums, and other attractions for visitors. www.sturgeonbay.net

The **Door County Maritime Museum** is a large, modern museum with marine artifacts and historical displays featuring local shipbuilding, shipwrecks, and legendary sailors of the Great Lakes. Visitors to the site can board a 1919-era tug boat called the *John Purves*. The museum is located at 120 N. Madison Avenue in Sturgeon Bay. www.dcmm.org

DIRECTIONS TO THE PARK

Potawatomi State Park is located on the south shore of Sturgeon Bay, adjacent to the city of Sturgeon Bay in Door County. From Interstate I-43 near Green Bay, take Highway 57 north 37 miles to County Highway PD. Turn left (north) on Hwy PD and travel 2.4 miles to the park entrance.

Potawatomi State Park
3740 County Highway PD
Sturgeon Bay, WI 54235
(920) 746-289

24. RICHARD BONG STATE RECREATION AREA

Woods and Wetland Wonderland

Bong wetland area

RECREATION AREA SNAPSHOT

When Wisconsin's first state parks were established more than 100 years ago, the general consensus was that these public areas should be reserved for "quiet" recreational activities such as hiking, swimming, camping, picnicking, and nature study. This long-held policy was challenged during the planning stages for the Richard Bong Recreation Area located in rural Kenosha County in 1974. The new vision for this large 4,515-acre property was to allow more non-traditional recreational pursuits.

Today, visitors to Bong may be surprised to see model rockets being launched or hang gliders and model airplanes flying above the trees. Dirt-track trails are often humming with all-terrain vehicles and off-road motorbikes. Visitors might even spot someone training a falcon to hunt or witness a hot air balloon being launched. All of these out-of-the-ordinary activities at Bong occur within a **Special Use Zone** of the property that co-exists with other more traditional park activities. Bong may not be all things to all people but this unique recreation area does seem to have something for everyone.

GEOLOGY AND HISTORY

The original landscape of the recreation area was formed about 10,000 years ago near the end of the last Ice Age. As the last of the glaciers receded north out of Wisconsin, they left behind a mix of sand and gravel moraine hills, tall-grass prairies, streams, and expansive wetlands here. Native American cultures such as the Ottawa, Ojibwa, and Potawatomi lived in this region of the state for generations before pioneer settlers from the East Coast arrived here early in the 19th century. These newcomers built permanent cabins, raised cattle, cut timber, and dammed local streams, which led to conflicts with native people, especially the Potawatomi, who claimed ownership of this region at that time.

European immigrants from England and Germany arrived here later in the 19th century primarily to raise crops on the rich soils of the local area. In 1954, after farming these lands for several generations, they too would be forced to leave. Unknown to them, the federal government had decided to build a U.S. Air Force strategic air defense base here to defend the nearby urban areas of Milwaukee and Chicago against possible Soviet air attacks. Within a year, government agents purchased 5,540 acres of land from 59 local landowners. When completed, this massive air base would provide housing for 1,500 airmen and shelter fifty F-28 Sabre jets.

> **Ranger Note**
>
> During the Black Hawk War of 1832, the Potawatomi fought on the side of the United States Army against the Sauk and Fox. They had hoped their allegiance to the U.S. government would result in their people being allowed to remain in their traditional homeland. It did not. In 1833 the Potawatomi were induced to cede all their land in northern Illinois and southeast Wisconsin to the U.S. government. In return, they were to receive annual cash payments and agree to move west of the Mississippi River to reservation land in the Iowa Territory.

In 1959, after years of bulldozing and leveling hundreds of acres of land, the project came to an abrupt end just three days before concrete was to be laid for a 12,550-foot runway. The construction of the air base was cancelled due to the development of long-range intercontinental missiles which would have made the Bong Airfield obsolete before its completion. The federal government began selling off large tracts of the abandoned air base property in the 1960s. Local sportsmen's clubs, conservationists, and legislators petitioned the state to acquire 4,515 acres of this

> **Ranger Trivia**
>
> The new air base was to be named in honor of Major Richard Bong, a native Wisconsin fighter pilot and Medal of Honor recipient who rose to fame by downing 40 enemy planes in the Pacific Theater during World War II.

land for public recreation. As a result, Richard Bong Recreation Area opened to the public in 1974.

BEACHES, PICNIC AREAS, AND SHELTERS

The **Wolf Lake Picnic Area** has a popular 400-foot sandy swimming beach with changing stalls, showers, and restrooms. The picnic area also has a playground and two open-air shelters that can be reserved in advance.

HIKING, BIKING, NATURE TRAILS, AND SPECIAL USE ZONES

Bong has over 41 miles of hiking trails marked with colored emblems. The northern section of the property was not altered by the construction of the air force base, so the landscape here has remained close to its original state. The **Red Trail** (8.3 miles) leads through hardwood forests, prairie areas, and around wetlands and ponds to a small picnic area that is located along County Highway BB. The **Vista Nature Trail** (1.0 mile) is located along County Highway B north of Highway 142. The trail leads through a woodland area with interpretive panels and has a scenic overlook.

In the southern region, the **Blue Trail** (4.2 miles) loops around Wolf Lake through woodlands, prairies, and crosses over the dam that impounds the lake. An equestrian trail (13.0 miles) encircles the entire southern section of the property. There are no horse rental stables in the area but riders can trailer in their own horses.

Bong has 12 miles of off-road bicycle trails located north of Highway 142. The **White River State Biking/Hiking Trail** (16.0 miles) can be accessed from the nearby city of Burlington along Spring Valley Road.

Most of the **Special Use Zone** is located along Suz Road in the center of the property. Non-traditional use activities allowed here including all-terrain vehicle and dirt-bike riding, model rocket and hot air balloon launching, model airplane flying, dog and falconry training areas, and many others.

Ranger Trail Pick

The **Green Trail** (1.8 miles) begins near the **Molinaro Visitor and Nature Center**. The trail leads through a variety of landscapes including hardwood forests, restored prairies and wetlands north of Wolf Lake. This trail also provides some of the best views of Wolf Lake.

WILDLIFE VIEWING

Bong's varied landscape of oak woodlands, grasslands, savannas, prairies, wetlands, ponds, and Wolf Lake provide ideal habitat for wildlife. White-tailed

Wildlife pond along the Green Hiking Trail

Ranger Tip

Bong's largest wildlife refuge is located between Suz Road and Highway 142. The refuge is part of a controlled flowage system that maintains open water marshes, savannas, and grasslands that attract many species of birds. An observation tower along Suz Road allows visitors to enjoy a panoramic view of the wildlife refuge.

deer, fox, coyote, raccoons, rabbits, squirrels, and chipmunks are commonly seen here. Muskrats and beaver are occasionally seen swimming inWolf Lake or wetland areas.

Birdwatching is a popular activity at Bong all year but especially during the spring and fall migration periods. Several species of ducks, geese, swans, shorebirds, sandpipers, egrets, and cranes can be spotted in wetland areas. Raptors such as red-tailed hawks and northern harriers are also commonly seen at Bong.

Bong Recreation Area has the largest managed prairie in the state. This area has been designated a **Wisconsin Important Land Legacy Site**, a distinction bestowed on sites with exceptional ecological, wildlife, and recreational significance. Throughout the summer, prairie areas bloom with colorful coneflowers, butterfly weed, blazing star, prairie smoke, and compass plants.

INTERPRETIVE PROGRAMS AND FACILITIES
Interpretive hikes and evening programs are offered throughout the year. Nature programs are presented at the outdoor amphitheater and at the Molinaro Visitor and Nature Center. The visitor center has several hands-on interactive displays, live animal exhibits, a gift shop, and interpretive panels describing the life of early Native American inhabitants. Adjacent to the nature center is the **Nature Explorer Classroom Trail** (0.7 miles), which has several outdoor learning stations linked by walking paths through a restored prairie, a small oak woodlot, and a boardwalk with a scenic view of Wolf Lake.

BOATING, FISHING AND HUNTING
Wolf Lake (150 acres) has a boat launch but watercrafts are limited to canoes, kayaks, and boats with electric motors. Northern pike and yellow perch are regularly stocked to supplement the largemouth bass, bluegill, and other panfish that thrive in this lake. Rainbow trout are stocked in the property's **Urban Fishing Pond** located east of the Molinaro Visitor and Nature Center.

Most of the undeveloped areas of the property are open for hunting for deer, small game, upland gamebirds, and waterfowl. Hunters are required to register and purchase a hunting permit at the entrance station. During the pheasant season, wildlife personnel stock birds nearly every day.

ACCESSBILE FACILITIES AND TRAILS
The Molinaro Visitor Center, Nature Explorer Classroom Trail, and most picnic areas, shelters, and restrooms are accessible for mobility-impaired visitors. Wheelchair-accessible campsites with electrical service are available within each campground. An accessible cabin located within the **Sunset Campground** can be reserved for campers with physical disabilities. The cabin features an accessible kitchen, living room, bedroom, and wheel-in shower/restroom.

Accessible fishing piers and playground equipment are available at the Wolf Lake Picnic Area. Accessible fishing piers are also located at the Urban Fishing Pond east the Molinaro Visitor and Nature Center.

CAMPING
There are 221 campsites (54 electric) located within **Sunrise** and **Sunset campgrounds**. Each campground has restroom/shower facilities and offer both shaded and open campsites. An RV water fill and sanitary station is located near the park entrance road contact station. The property has six group campsites that can accommodate up to 225 campers.

WINTER ACTIVITIES

Property staff groom 16 miles of inline cross-country ski trails in winter. Other off-season activities include snowshoeing, hiking, and ice fishing on Wolf Lake. There are 12 miles of snowmobile trails at Bong, all of which connect to the Kenosha County trail system.

AREA ATTRACTIONS

Kenosha County offers many attractions for visitors, including many fine restaurants, shops, parks, and historical sites. www.visitkenosha.com

The **Hawthorn Hollow Nature Sanctuary and Arboretum** is a 40-acre wildlife preserve dedicated to the study and appreciation of nature. The site has a large arboretum, hiking trails, a nature center, and a farmstead with historic buildings. Adjacent to the sanctuary is **Petrifying Springs County Park**, which offers picnic areas, hiking trails, and a public golf course. Both parks are located about 15 miles east of Bong at 880 Green Bay Road, Kenosha, WI 53144. (262) 552-8196. www.hawthornhollow.org

DIRECTIONS TO THE RECREATION AREA

Richard Bong State Recreation Area is located south of Kansasville in Kenosha County. From Interstate I-41 near Kenosha, exit onto Highway 142 and travel west about 17 miles to the entrance station. From Interstate I-43, take the Highway 11 exit until it merges with Highway 142 near Burlington. Continue east on Hwy 142 to the property entrance station.

Richard Bong State Recreation Area
26313 Burlington Road
Kansasville, WI 53139
(262) 878-5600

25. ROCK ISLAND STATE PARK
Door County's Emerald Island

Rock Island State Park and Viking Hall Boathouse

PARK SNAPSHOT

Rock Island State Park may be the most challenging property in the state to visit but it's definitely worth the effort to get there. This park is located as far out into Lake Michigan as possible and still be in Wisconsin. It can only be reached by personal watercraft or by taking commercial ferry boat rides to the island. The entire 912-acre island is owned by the State of Wisconsin except for a small parcel held by the U.S. Coast Guard.

There are very few places left in Wisconsin where one can experience the soothing sights and sounds of a wilderness area without the noise of cars and trucks in the background. Rock Island is one of these rare places. Motor vehicles and even bicycles are not allowed on the island. In addition to the quiet solitude found here, Rock Island offers hiking, picnicking, swimming, and walk-in camping. Park volunteers offer guided tours of island's historic **Pottawatomie Lighthouse** and the **Viking Hall Boathouse**, one of the most elaborate stone structures of its kind ever built.

GEOLOGY AND HISTORY

Rock Island, as its name implies, has lots of rocks. Much of its shoreline is lined with wave-smoothed stones except on the west end of the island, where 120-foot limestone bluffs tower above the shoreline. The bases of these massive bluffs plunge more than 140 feet into the depths of Lake Michigan. On the opposite side of the island, low-lying bluffs slope gently toward a shoreline of cobblestone and small sand beaches.

The bedrock of Rock Island is composed of dolomite, a type of limestone created over 400 million years ago from sand, calcium carbonite, and sea life deposited beneath the warm, shallow ocean that once covered this area. This hard rock resisted the crushing effect of glaciers that gouged out the Lake Michigan and Green Bay basins several times in the past two million years.

Although Rock Island is isolated from the mainland of Door County by nearly 12 miles, it was inhabited by several Native American cultures over time. The Menominee, Ottawa, Chippewa, Huron, and Potawatomi people all spent time on the island prior to the arrival of French explorers in the 1600s. Rock Island served as an important fur trading post for both France and England until after the War of 1812, when the United States took possession of the island.

Throughout the 19th century, Rock Island became an important stop along the shipping route from southeastern ports to Green Bay and Upper Michigan. In 1836, the U.S. government commissioned a lighthouse to be built on the island to provide safe passage for sailing vessels through the treacherous waters adjacent to the island. Within a decade, more than 200 immigrant fishermen and their families arrived on the island and built the first village in Door County. This new settlement was short-lived, however. By 1895, nearly all of the immigrants had moved to the mainland, leaving the island deserted except for the lighthouse keeper and his family.

> **Ranger Trivia**
>
> Rock Island almost became a "Michigan" state park. A border dispute regarding the location of the state line between Wisconsin and Michigan was settled in 1936 when the U.S. Supreme Court ruled that Rock Island and three other nearby islands were legally within the state boundary of Wisconsin—not Michigan.

In 1910, Chester Thordarson, a Chicago businessman, purchased nearly all the property of Rock Island. Thordarson made his fortune by designing and marketing the world's first high-voltage electrical transformer. He said he was attracted to the island because its rocky shoreline, forested hills, and the deep blue waters of Lake Michigan reminded him of his ancestral homeland of Iceland. Thordarson built several buildings in the Icelandic style on the island using local limestone and rocks. His most impressive structure was the **Viking Hall Boathouse**, which he built as a gathering place for his family and friends. After Thordarson's death in 1945, his family decided to honor his long-held belief in preserving the wilderness aspect of the island by selling it to the state of Wisconsin. Twenty years later, in 1965, Rock Island State Park was opened to the public.

BEACHES, PICNIC AREAS, AND SHELTERS

A picnic area is located near the Viking Hall Boathouse. An historic stone and timber open-air shelter facility located here once served as a greenhouse for the Thordarson family. Nearby at the south side of the island is a small sand dune area and sand beach swimming area.

HIKING, BIKING, AND NATURE TRAILS

Rock Island offers 11 miles of hiking trails to explore. **Michigan Avenue** (.25 miles) is a short walking trail that provides access to the swimming beach from the picnic area. The **Algonquin Nature Trail** (1.0 miles) is a looped trail that begins near the camping area. Interpretive signs along this trail highlight the flora, fauna, local history, and geology of the island.

The **Fernwood Trail** (1.2 miles) leads across the park from east to west through a mature hardwood forest to the highest point of the Rock Island. The **Havamal Trail** (1.0 miles) and **Blueberry Trail** (.5 mile) are both located at the

Pottawatomie Lighthouse

southern end of the park near the site of the once a thriving fishing village of the mid-1800s.

The **Pottawatomie Lighthouse Trail** (1.2 miles) is a gravel-surfaced walking trail that may be challenging for visitors with walking impairments. The trail ascends 122 feet to the lighthouse perched above Lake Michigan.

The Pottawatomie Lighthouse was built in 1836, making it the first lighthouse in Wisconsin. The original structure was replaced by the current lighthouse in 1858. It was manned by an onsite lightkeeper until 1946, when an automated signal light was installed. This historic lighthouse was authentically restored by the Friends of Rock Island, a not-for-profit support group for the park. Lighthouse tours are given by park volunteers from 10 a.m. to 4 p.m. daily during the use season.

Ranger Trail Pick

The **Thordarson Trail** (5.2 miles) encircles the entire perimeter of Rock Island. The trail has several overlook areas with views of the blue-green waters of Lake Michigan. Historical sites along this trail include the **Pottawatomie Lighthouse**, the ruins of the island's 19th-century fishing village, and a one-of-a-kind water tower/ fireplace stone structure designed by Chester Thordarson.

WILDLIFE VIEWING

The undeveloped shoreline, bluffs, and pristine forests of Rock Island are a true oasis for wildlife. White-tailed deer, fox, coyote, rabbits, squirrels, and several species of non-poisonous snakes and amphibians thrive here. Over 200 species of birds are known to either nest on the island or migrate through the area and include a few uncommon birds, such as the spotted sandpiper, American redstart, and several kinds of warblers.

Ranger Trivia

Due to the island's remote location, there are no raccoons, skunks, or bears on Rock Island. Wood ticks are also rare and often non-existent on the island.

The **Rock Island Woods State Natural Area** (558 acres) protects several types of ecosystems on the island, including wet mesic forests, limestone rock outcrops, woodland seeps, and upland forests of beech and sugar maple. These areas provide undisturbed habitat for many of the islands rare and endangered plants and animals.

INTEPRETIVE PROGRAMS AND FACIILITIES

The impressive Viking Hall and boathouse built by Chester Thordarson in the 1920s now serves as the park's interpretive center. Interpretive displays highlight the flora and fauna of the park and Native American artifacts uncovered during archeological surveys conducted on the island. The Viking Hall dining room still has the original hand-carved tables, chairs, and wooden chandeliers commissioned by Thordarson. Beneath Viking Hall are stone-arched boat mooring slips, which are still in use today. Viking Hall is listed on the National Register of Historic Places. Park volunteers offer guided tours of Viking Hall from time to time.

BOATING AND FISHING

Private watercrafts are allowed to dock at Rock Island. Overnight mooring fees are based on the length of the watercraft. There are no streams or lakes on the island so fishing is limited to Lake Michigan. Charter boat rentals are available on Washington Island for anglers interested in fishing for Great Lakes trout and salmon.

ACCESSIBLE FACILITIES AND TRAILS

Due to the rugged terrain and steep hills of the island mobility-impaired visitors may require companion assistance.

CAMPING

Rock Island has 40 walk-in campsites and two group campsites. All sites are located in the southwestern corner of the island with easy access to the beach. Firewood can be purchased on the island but all other camping supplies must be brought along by campers themselves. Vault-type restrooms are located within the camping area and drinking water is available at the picnic area.

WINTER ACTIVITIES

Rock Island is closed during the winter season. The passenger ferry to Rock Island from Jackson Harbor on Washington Island operates from late May through mid-October.

AREA ATTRACTIONS

Washington Island is one of the top tourist destinations in Door County and offers many unique shops, parks, bicycle tours, and other attractions. www.washingtonisland.com

The **Fragrant Isle Farm and Shop** is the largest lavender farm in the Midwest with more than 20,000 lavender plants under cultivation. This unique site has a gift shop with lavender products for sale and a café that serves lavender-

infused coffee, tea, wine, and bakery. The lavender farm is located at 1350 Airport Road on Washington Island.
www.fragrantisle.com

Ranger Tip

Park visitors can purchase round-trip tickets for both the Washington Island Ferry and the Karfi Passenger Ferry at the Northport Pier Ferry Landing.

DIRECTIONS TO THE PARK

Rock Island is located off the tip of Door County. From Sturgeon Bay, take Highway 57/ 42 north to Gills Rock and the Northport Pier. The Washington Island Car Ferry here crosses the channel to the Detroit Harbor on Washington Island. Follow County Hwy W (Main Road) north past the airport and take Jackson Harbor Road to the Karfi Ferry landing, where you'll board passenger-only ferry to Rock Island State Park. Passengers can park their vehicles in the State Park lot near the ferry free of charge.

Rock Island is open from Memorial Day weekend through mid-October. Ferry schedules and prices can be found at www.wisferry.com/rockisland

Rock Island State Park
1924 Indian Point Road
Washington Island, WI 54246
(920) 847-223

26. WHITEFISH DUNES STATE PARK

Home of Wisconsin's Tallest Sand Dunes

Whitefish Dunes beach

PARK SNAPSHOT

Whitefish Dunes State Park is located along the beautiful blue-green waters of Lake Michigan in Door County. This popular 867-acre day-use park is renowned for its wide sandy beaches and towering sand dune formations, the tallest in the state. The property offers swimming, sunbathing, picnicking, fishing, nature study, and biking plus miles of hiking trails and beachfront to explore.

GEOLOGY AND HISTORY

Whitefish Dunes State Park is located on a forested isthmus situated between Clark Lake and Lake Michigan's Whitefish Bay. The landscape of the park area has been reshaped many times by the action of waves, wind, and glaciers over the past two million years. Near the end of the last Ice Age about 10,000 years ago, retreating glacial ice excavated the Lake Michigan basin and deposited enormous amounts of sand along the park's lakeshore area.

About 7,500 years ago, Whitefish Dunes and Clark Lake would have been submerged under a single open water bay of Lake Nipissing, an expansive post-glacial lake 25 feet higher than what Lake Michigan is today. As this ancient lake receded, water currents deposited deep layers of sand within Whitefish Bay and formed a wide isthmus that permanently isolated Clark Lake from Lake Michigan. Today, wave action and wind continue to reshape the park's sand dune formations and relocate the shoreline of Whitefish Dunes as lake levels rise and fall over time.

The waters of Lake Michigan and Clark Lake and the forests of Whitefish Dunes have long attracted indigenous people to this area. Archaeological excavations at the property have shown that at least eight different Native American cultures spent time along these shores over the past 3,000 years. Many built seasonal villages along the Lake Michigan shoreline in spring and summer to spear or net lake trout, walleye, suckers, sturgeon, and other fish, including whitefish, the park's namesake. The Oneota people lived here from about 1300 to 1400 A.D. and were probably the first to establish a permanent settlement in the park area. In addition to hunting and fishing, the Oneota planted corn and squash gardens here.

Most European immigrants arriving here in the early 19th century were engaged in farming or the timber industry but a few began commercial fishing operations on Lake Michigan. Two of the first settlers, James Clark and his brother Isaac, set up the **Whitefish Bay Fishing Camp** here in the early 1840s. The Clarks employed local Native American workers to net, clean, and salt fish, which were then packed into barrels for shipment to city ports as far away as Cleveland and Detroit. The Clark brothers also shipped lumber cut from Door County forests until the early 20th century.

Beginning in the late 1930s, local citizens and conservation groups petitioned state legislators to preserve the rare sand dune formations and Lake Michigan shoreline along Whitefish Bay. Their grassroots efforts led to the establishment of Whitefish Dunes State Park, which opened in 1967.

BEACHES, PICNIC AREAS, AND SHELTERS

The park's two-mile-long sandy beach is one of the most popular in the state. In summer, thousands of visitors enjoy swimming in the refreshing waters of the Lake Michigan or just relaxing along the shoreline, enjoying a cool breeze off the lake. The park's beach area can be accessed from a trail near the visitor welcome center or from several side trails along dune area hiking trails.

All picnic areas are well shaded and are located along the Lake Michigan shoreline. An attractive combination indoor/outdoor shelter

> **Ranger Note**
>
> Lake Michigan wave action can create dangerous rip currents at times, especially between the first and second beach access points at Whitefish Dunes. Warning signs are posted and buoy markers help identify this area.

with a fieldstone fireplace and a restroom with changing stalls for swimmers is located in the main picnic area.

HIKING, BIKING, AND NATURE TRAILS

Whitefish Dunes has 16 miles of hiking trails to explore. Trails routes are identified by color-coded emblems. Off-trail hiking is not allowed within the state natural area to protect the fragile dune formations and the rare plants that grow in this unique ecosystem. The **Black Trail** (2.5 miles) leads to the northeast section of the park through woodland areas and along limestone outcroppings adjacent to Lake Michigan. This trail connects to **Cave Point County Park**, a popular Door County attraction featuring scenic wave-sculptured Lake Michigan sea caves. (See Area Attractions)

The **Red Trail** (2.8 miles) begins near the visitor center and leads south through forested areas and open sand dunes. A 2.25-mile section of the Red Trail is open for bike riding from the nature center to South Cave Point Drive along the park's south boundary. About halfway along the Red Trail, a side bike trail leads west and connects to Clark Lake Road, which loops back north to the entrance of the park. Another side trail near the southern end of the Red Trail leads up a series of wooden steps to the top of **Old Baldy**, a massive dune that rises 93 feet above the lake level. Old Baldy is considered the tallest sand dune in the state. An overlook platform here provides a panoramic view of the forests, dunes, Lake Michigan and Clark Lake in the distance.

Ranger Trail Pick

The **Brachiopod Nature Trail** (1.5miles) leads through seven different ecosystems of the park, including the Lake Michigan shoreline, mature hardwood forests, conifer swamps, wetlands, and sand dunes. A trail guide booklet available at the park office lists 16 interpretive stops along this trail. One of them is a limestone ledge imbedded with dozens of 425- million-year-old fossils, including brachiopod mussels.

WILDLIFE VIEWING

Whitefish Dunes has a rich diversity of habitat for wildlife, including both hardwood and conifer forests, wetlands, small creeks, sand dunes, Clark Lake, and the Lake Michigan shoreline. White-tailed deer, chipmunks, raccoons, squirrels, and porcupines are commonly seen in the park. More secretive animals such as red fox, coyote, and black bear are occasionally spotted here as well.

Over 200 species of nesting and migratory birds have been recorded at Whitefish Dunes. A bird list is available at the park office/ visitor station. Other brochures highlight the park's dune plants, fungi, and plant life. A publication entitled *Watchable Wildlife of Whitefish Dunes State Park* is also available.

INTERPRETIVE PROGRAMS AND FACILITIES

The park's nature center is located inside the park office/visitor station. The center has several interactive displays highlighting the flora, fauna, and sand dune geology of the park. Native American artifacts recovered from archeological studies in the park are also on display. Adjacent to the nature center is a re-creation of a Native American fishing encampment. Nature hikes and programs are offered throughout the season by park naturalists and guest speakers.

BOATING AND FISHING

There is no Lake Michigan-access boat launch at Whitefish Dunes. Small watercraft such as canoes and kayaks can use the **Schauer Road Boat Launch** located a few miles north of the park. Larger watercraft can access the lake from the **Lily Bay County Boat Launch** located along County Highway T south of the park. Charter boats can be booked for Great Lakes trout and salmon fishing at the **Sturgeon Bay Marina**, located about 15 miles southwest of the park.

Clark Lake (865 acres), located on the west side of the park, is the second largest lake in Door County. Anglers can expect to catch bluegill, northern pike, walleye, brook trout, and bass in the lake. A public boat launch is located on the west side of the lake. Whitefish Bay Creek, located on the southern end of the park, has a small population of trout. The creek can be accessed by vehicle off Clark Lake Road or via the **Whitefish Creek Spur Trail** (1.0 miles) off the **Yellow Trail**.

ACCESSIBLE FACILITIES AND TRAILS

The park office visitor station, nature center, picnic area, shelter, and most restrooms are accessible for mobility-impaired visitors. A plastic grid walkway from the visitor station to the Lake Michigan shoreline is available for wheelchair access to the beach. A specially designed beach wheelchair with large balloon tires is available free of charge and can be reserved at the park office/nature center.

> **Ranger Note**
>
> The beach access trail and wheelchair grid route from the visitor station may be closed at times depending on the water level of Lake Michigan. Beach access is available from side trails off the Red Trail. These routes are not considered wheelchair accessible, however.

CAMPING

Whitefish Dunes is a day-use-only property so there are no camping facilities here. Overnight camping is available at **Peninsula State Park** near Fish Creek about 17 miles north of the park. **Potawatomi State Park**, located 19 miles southwest of Whitefish Dunes near Sturgeon Bay, also has camping facilities. (See Chapters 20 and 23)

WINTER ACTIVITIES

Whitefish Dunes is open in winter for hiking and wildlife watching. Park staff groom nine miles of cross-country ski trails for inline skiing. Snowshoeing is popular along the 2.5-mile **Black Trail**. Sledding on the dunes and snowmobiles are not permitted in the park.

AREA ATTRACTIONS

Lake Michigan's eastern shoreline is often referred to as the "quiet side" of the Door County peninsula. The lakeshore communities here have great restaurants, unique gift shops, parks, and many other attractions for visitors. www.doorcounty.com

Cave Point County Park is one of Door County's most scenic natural wonders. Limestone bluffs perched above Lake Michigan here feature underwater caves that were carved by relentless wave action over time. This small but popular

Cave Point County Park

19-acre county park is located along Schauer Road within the northern section of Whitefish Dunes State Park.

DIRECTIONS TO THE PARK

Whitefish Dunes is located south of the community of Jacksonport along the eastern shoreline of Door County. From Interstate I-43 near Green Bay, take Highway 57 north to Sturgeon Bay. After crossing the Sturgeon Bay Bridge, travel another 10 miles north on Highway 57 and then turn right (east) onto Clark Lake Road (County Hwy. WD). Follow the signs to the park entrance. The park is open daily from 6 a.m. to 8 p.m. year round.

Whitefish Dunes State Park
3275 Clark Lake Road (Co. WD)
Sturgeon Bay, WI 54235
(920) 823-2400

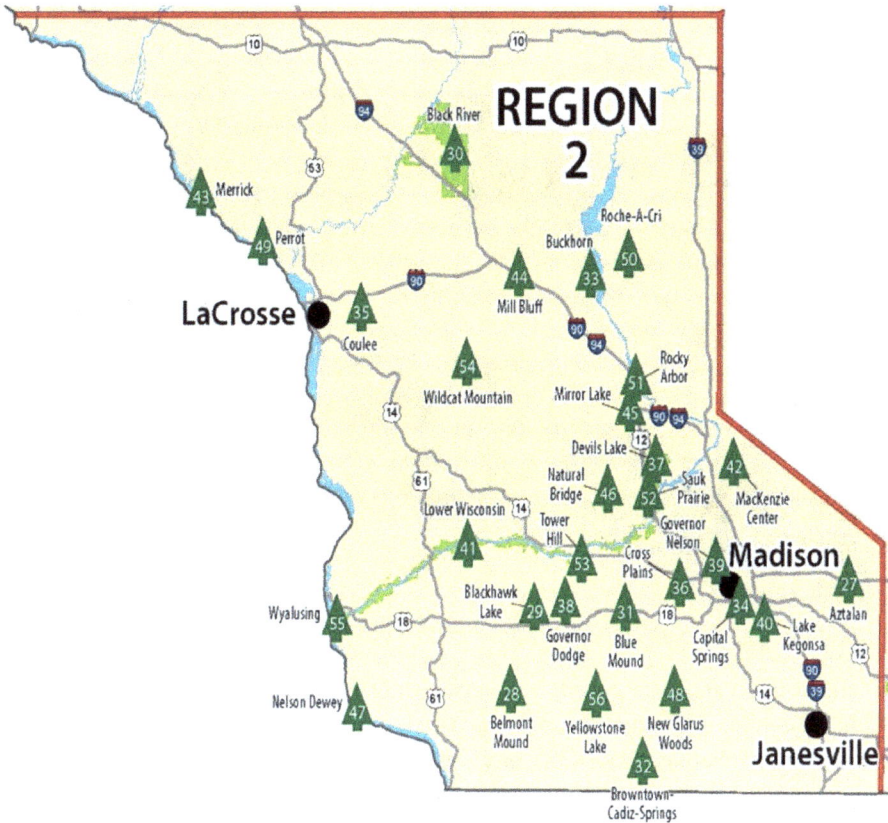

Region Two—South Central and Southwestern Wisconsin: Mississippi, Kickapoo, Black and Lower Wisconsin rivers, Castle Rock Flowage, Central Sands Region, Blue Mounds, New Glarus, Baraboo Hills, Wisconsin Dells, and Madison area.

27. AZTALAN STATE PARK
Wisconsin's Archaeological Treasure

Restored Temple Mound

PARK SNAPSHOT

Aztalan State Park is the largest and most significant archeological site in the state. This 172-acre property located in rural Jefferson County was once an extraordinary Native American "city" with hundreds of houses, a flat-topped temple mound, and an expansive ceremonial plaza, all surrounded by a towering log stockade. This mysterious site has been excavated and studied by numerous archeologists for nearly a century.

The park also offers hiking, picnicking, nature study, birdwatching, fishing, canoeing, and kayaking but its primary purpose is to preserve and interpret the rare archeological features found here. Aztalan is a designated National Landmark and part of the National Register of Historic Places.

GEOLOGY AND HISTORY

Aztalan is located along the Crayfish River in an area geologists refer to as a glacial plain or flat area, between glacial hills called drumlins. These landforms were formed by retreating glaciers near the end of the last Ice Age about 10,000 years ago.

The rich prairie soils, natural springs, streams, lakes, and forests of this area have attracted several Native American cultures to this area for thousands of years. Sometime after 1050 AD, a mysterious group of people, now known as the Aztalans, migrated to this area from western Illinois and built what could only be described as "city" along the banks of the Crayfish River. At the time, Woodland Indians were already living in bark wigwams in this area. One can only imagine what they thought as the Aztalans arrived and built massive pyramid-like temple mounds, ceremonial plazas, and hundreds of plaster and pole houses, granaries, and other structures. The Aztalans also installed towering defensive log stockades around the entire city to the banks of the Crawfish River.

The Aztalans hunted deer, elk, small game, waterfowl, turkeys, and other upland gamebirds. Nearby rivers and lakes provided fish, clams, and crawfish (crayfish). The Aztalans also excelled in growing crops of corn, squash, beans, sunflowers, and gourds to feed their growing community. Archeologists have estimated that at least 500 people lived in or near this Native American community. Around 1200 AD, after nearly 150 years of living here, the Aztalan people mysteriously vanished. Why the Aztalan city was abandoned is unknown. Some speculate it may have been due to climate change, crop failure, disease, or perhaps even war with neighboring tribes. Other Native American people eventually settled in the same region over the next 600 years including the Menominee, Ho-Chunk, and the Santee Sioux.

The first Yankee settlers arrived here from New York in the early 19th century. They questioned native people still living in the area about the mysterious mounds located along the Crawfish River but none had any knowledge of who built them. Over the next century, several archeologists explored and mapped what became known as the "Aztalan" site.

By 1922, most of the original 40 mounds recorded earlier at the Aztalan site had been leveled to make way for cropland by local farmers. In 1928, a local historical organization purchased several acres of the site and donated it to the Wisconsin Archaeological Society to create "Mounds Park." In 1947, the

Ranger Trivia

In 1836, Nathaniel Hyer, a Milwaukee Judge and amateur archeologist, heard reports a mysterious abandoned ancient "city" in Jefferson County. Judge Hyer traveled on horseback to investigate the site for himself. He came to the conclusion that the settlement must have been built by descendants of the Aztec people from the city of Aztalan, Mexico. Although Judge Hyer's theory was later discounted his Aztalan description of these people remains to this day.

State of Wisconsin purchased additional property along the Crawfish River to develop Aztalan State Park, which opened in 1952.

State- and university-sponsored archeological excavations of the site have revealed many rare artifacts, including beautifully painted pottery, stone tools, human burials, ornaments, necklaces, and thousands of other items. More recent discoveries have provided evidence to trace the Aztalan people to their ancestral home at Cahokia, located in western Illinois.

BEACHES, PICNIC AREAS, AND SHELTERS

Aztalan State Park has several picnic spots located near the Rock River and an open-air shelter facility. The park does not have a swimming area but public beaches are located along the east shoreline of Rock Lake in the community of Lake Mills about two miles west of the park.

> **Ranger Note**
>
> Cahokia was a Native American "city" built around 600 A.D. This large urban community may have had more than 20,000 inhabitants, making it the largest indigenous settlement ever built north of Mexico. The city had a central urban/trading center, several flat-topped pyramid temples, ceremonial plazas, thousands of houses and other structures, all nearly identical to those found at Aztalan.

HIKING, BIKING, AND NATURE TRAILS

The park has four miles of hiking trails that loop through upland hardwood forests, oak savannas, and restored prairies or follow the shoreline of the Crawfish River.

A few miles south of the park is the **Glacial Drumlin State Trail** (52.0 miles), a crushed gravel biking/hiking route that follows an abandoned railroad corridor between Waukesha and Cottage Grove just east of Madison. A 13-mile asphalt-paved section of this trail connects cities of Waukesha and Dousman.

> **Ranger Trail Pick**
>
> A short self-guided interpretive trail begins at a kiosk in the lower picnic area. A 14-page brochure available at kiosk follows numbered stops along the trail. Each stop reveals what is known of the mysterious Aztalan people who lived here nearly a thousand years ago. The trail leads to the site's largest ceremonial mound surrounded by wooden poles, which simulate the defensive stockades that once lined the perimeter of the city.

WILDLIFE VIEWING

The park's forests attract several species woodland birds, hawks, and owls. Open grasslands provide habitat for bluebirds, tree swallows, meadowlarks, and upland game birds. Native prairie plants and grasses bloom in early summer and attract a colorful array of butterflies, moths and many other insects.

The Crawfish River and wetland areas provide essential habitat for cranes, herons, shorebirds, waterfowl, muskrats, mink, and occasionally otter. White-tailed deer, fox, cottontail rabbits, squirrels, raccoon and opossum are commonly seen in the park

INTERPRETIVE PROGRAMS AND FACILITIES

Interpretive panels and self-guiding brochures describe the site's most important archeological features. Interpretive programs and public tours are presented by volunteer naturalists most Sunday afternoons from June through August. Upcoming programs are listed on the Friends of Aztalan website at www.aztalanfriends.org. Friends group members are also engaged in raising funds to build a visitor/interpretive center for the park.

> **Ranger Note**
>
> Aztalan State Park preserves the burial sites of many indigenous people who once lived here. Native American people consider this site to be sacred so please treat the area with respect. State and federal laws prohibit digging or artifact collecting at this site.

> **Ranger Trivia**
>
> The Crawfish River is named after a small crustacean known as crawfish, also called a "crayfish." Crayfish thrive in nearly every river, lake, and pond in Wisconsin. In the southern states they are known as crawdads and are considered a delicacy when boiled and seasoned similar to lobsters.

BOATING AND FISHING

The Crawfish River is known for its walleye, catfish, and white bass fishing, but other fish inhabit these water as well, including northern pike, carp, suckers, dogfish, bullheads, and freshwater drum. Many paddlers enjoy canoeing and kayaking the Crawfish River. A carry-in launch is located along the river east of the lower picnic area. The Crawfish is a very shallow stream so the use of motorboats is not recommended.

ACCESSIBLE FACILITIES AND TRAILS

The picnic areas, restrooms, and shelter facility at Aztalan are accessible for mobility-impaired visitors. The self-guided interpretive trail is surfaced with crushed limestone but visitors who require wheelchairs may need assistance along this route. Most other hiking trails in the park have mowed grass surfaces.

CAMPING

Aztalan is day-use state park so there are no camping facilities here. Lake Kegonsa State Park, located about 28 miles from the park near Madison, has a campground with restrooms and showers. (See Chapter 40)

WINTER ACTIVITIES

Aztalan is open in winter for hiking, wildlife viewing, fishing, and cross-country skiing. None of trails are groomed. Sledding is not allowed in the park to protect the ancient mounds of the property.

AREA ATTRACTIONS

Jefferson County has many parks, museums, bike trails, and small towns and villages for visitors to explore. www.enjoyjeffersoncounty.co

The **Aztalan Museum** is located adjacent to the state park. The site has several 19th-century pioneer buildings, including an 1852 Baptist church, which houses most of the museum's collection. The property also has an Aztalan-era burial site known as the "Princess Mound." When excavated in 1919, the remains of a young woman believed to a high-ranking member of an Aztalan ruling family was discovered. At the time of her death, she was interned with three large belts decorated with nearly 2,000 circular clam shells, some of which originated from as far away as the Gulf Coast. www.lakemillsaztalanhistory.com

DIRECTIONS TO THE PARK

Aztalan is located about two miles east of Lake Mills in Jefferson County. From Interstate I-94, take the County Highway B exit at either the Highway 89 or Highway 26 interchange. The park is located at the corner of County B and County Highway Q.

Aztalan State Park
N6200 Hwy. Q
Lake Mills, WI 53551
920-648-8774

28. BELMONT MOUND STATE PARK AND FIRST CAPITOL HISTORICAL SITE

Wisconsin's "Beautiful Mountain"

View from Belmont Mound

PARK SNAPSHOT

Belmont Mound State Park is located atop a 400-foot forested hill in Lafayette County. This scenic 274-acre park offers hiking, picnicking, nature study, off-road biking, and many other outdoor activities. Just down the road from the park is **First Capitol Historical Site**; the site of Wisconsin's first legislative session in 1836.

GEOLOGY AND HISTORY

The crest of Belmont Mound is 1,400 feet above sea level, but 400 million years ago, the entire park area would have been submerged beneath a shallow ancient sea. During that period, hundreds of feet of sand, silt, and sea life accumulated on the bottom of this ocean. As the sea receded, these deposits were transformed into layers of limestone and sandstone. Today, exposed outcrops of these ancient rock formations can still be seen along the park's bluffs and ledges.

Belmont Mound is located within the driftless area of southwestern Wisconsin. This region lacks the glacial deposits or "drift" found in other areas of the state that were covered by glaciers during the last Ice Age. Since the crushing effect of glaciers didn't occur here, most of the area's sandstone has been eroded into small hills or flat plains by rain, wind, and frost over millions of years. A few towering remnant sandstone hills, such as Belmont Mound, stand today thanks to the hard dolomite limestone "caps" that have shielded them from erosion.

Archeological excavations at Belmont Mound have revealed that Native American people occupied this area as far back as the late Archaic Period 1,000- 500 B.C. By the time the first French explorers reached Wisconsin in 1634, the Ho-Chunk (Winnebago) people held claim to this part of the state.

> **Ranger Note**
>
> Belmont Mound was named *Belle Monte* or "Beautiful Mountain" by early French explorers.

The first European settlers to the local area came from the East Coast about the same time lead miners arrived from Illinois. Conflicts with Ho-Chunk people over land use, especially lead mining, led to the Winnebago War of 1827. Attacks against settlers in the nearby Prairie du Chien area by Ho-Chunk warriors were met with military action by the territory's volunteer militia. As a result, defeated Ho-Chunk leaders were induced to cede all their lands south of the Wisconsin River to the U.S. government.

> **Ranger Trivia**
>
> One year before the first meeting of Wisconsin's territorial government, John Atchison from Galena, Illinois, purchased land near Belmont Mound and laid out a village plat for a new town. He then commissioned four buildings to be built in Pittsburg, Pennsylvania and had them dismantled, shipped, and reassembled in his new town of Belmont.

In 1836, the U.S. Congress created the Wisconsin Territory, which included Minnesota, Iowa, and the eastern Dakotas. Henry Dodge, a well-known, prominent citizen and militia colonel during the Black Hawk War of 1832, was chosen to be the Governor of the Wisconsin Territory. In 1836, Dodge called for the first territorial legislative session to be held in the tiny village of Belmont, located a half mile west of Belmont Mound.

John Atchison's vision to have Belmont become the permanent capitol of Wisconsin Territory nearly succeeded. He did convince Henry Dodge to hold his first legislative meeting in Belmont, but ultimately the representatives voted to move the capitol to a new, undeveloped town called Madison. The fate of this tiny hamlet was doomed when the railroad chose to lay tracks three miles to the south of town. As a result, most people chose to relocate to the "new" village of Belmont along the railroad line.

The restoration of the original first capitol buildings and site began in 1910 by the Wisconsin Federation of Women's Clubs and later purchased by the state. In 1961, Wisconsin legislators approved a plan to develop a new state park adjacent to the First Capitol Historic Site, which led to the establishment of Belmont Mound State Park. The park is currently co-managed between the state and the local Belmont Lions Club.

BEACHES, PICNIC AREAS, AND SHELTERS

The park's picnic area has tables, grills, a small playground, and an open-air shelter facility. Adjacent to the picnic area are remnants of a 19th-century wood-fired lime kiln. The kiln was used to heat limestone rock to make quicklime, a compound used to make plaster and cement and also in the iron, steel, and paper industries.

There are no lakes or rivers at Belmont Mound. A swimming beach is available at Lake Joy Campground located about three miles northeast of the village of Belmont.

HIKING, BIKING, AND NATURE TRAILS

The park has three miles of hiking trails. The **Belmont Mound Woods State Natural Area Trail** (0.60 miles) follows a service road along the western border of the park to the state natural area in the northwest corner of Belmont Mound. The **Belmont Mound Extended Trail** (1.9 miles) circles the perimeter of the mound with a side trails to the northeast section of the property. Off-road biking is allowed on these trails.

The **Belmont Mound Trail** (0.50 miles) begins at the picnic area and leads uphill past scenic rock outcrops to the summit of Belmont Mound 400 feet above the surrounding landscape. A 64-foot-tall wooden observation tower is located here but is currently closed. Hopefully, it will be repaired or replaced in the future.

The nearby **Pecatonica State Trail** (10.0 miles) begins near the village of Belmont and follows a scenic section of the Pecatonica

Ranger Trail Pick

Near the observation tower are several rustic side trails that lead to some amazing rock formations, including **Devil's Dining Table**. This massive horizontal limestone slab is 40 feet across and balanced atop narrow pedestal of rock giving it the appearance of a gigantic stone table. A similar but smaller rock outcrop called the **Devil's Chair** is located nearby. Another rock formation, **The Cave**, is a dark passage beneath several massive boulders.

River along a former railroad corridor to the village of Calamine. The trail connects to the **Cheese Country Trail** (47 miles), which winds through three counties.

Mound View State Trail (7.0 miles) is an asphalt-paved trail that leads from the village of Belmont to the city of Platteville, where it joins the **Roundtree Branch Bike Trail**. Adjacent to Mound View trail is the **Belmont Prairie State Natural Area**, a remnant dry prairie site that supports over 80 types of native grasses and forbs. Both Belmont Mound and the towering **Platte Mound** can be viewed from this trail.

> **Ranger Trivia**
>
> Platte Mound is best known for the massive 241-foot-tall letter "M" on its side that can be seen for miles around. This iconic landmark is made up of thousands of white-washed rocks placed there by college students in 1936 as a memorial to the lead miners of the region. The letter also commemorates the dedication of the first mining school in the United States at UW-Platteville.

WILDLIFE VIEWING

The park's hardwood forests provide habitat for white-tailed deer, fox, raccoon, chipmunks, squirrels, woodland songbirds, owls, and wild turkey. During the spring and fall migration period, bird watchers from throughout the region enjoy observing many species birds not usually seen here the rest of the year.

The **Belmont Mound Woods State Natural Area** (80 acres) is located in the northwest corner of the park along an exposed dolomite limestone bluff area. This remote section of the park preserves old-age sugar maple, red oak, white ash, and black oak trees. The site also has an exceptional diversity of plant life, including spring beauties, hepatica, wild leek, May apples, and hundreds of other native flowers, forbs, and grasses.

INTERPRETIVE PROGRAMS AND FACILITIES

There are no interpretive programs or self-guided nature trails at Belmont Mound State Park but the nearby First Capitol Historic Site, operated by the Wisconsin Historical Society, is well worth a visit. Two of the original buildings used for Wisconsin's first legislative session in 1836 are open for visitors to tour. The buildings house antique furniture of the era and have several interesting historical displays and artifacts. www.firstcapitol.wisconsinhistory.org

BOATING AND FISHING

Boat rentals and fishing opportunities are available nearby at Joy Lake (56 acres), which has good populations of northern pike, bass, walleye, catfish, and bluegill.

First Capitol historic site

ACCESSIBLE FACILITIES AND TRAILS
The picnic area has an accessible picnic table, grill, and open-air shelter facility. Due to the steep topography of the park, hiking trails may be difficult to access for visitors with walking disabilities.

CAMPING
Belmont Mound is a day-use property so there are no campsites here. **Yellowstone Lake State Park**, located about **25** miles west of Belmont, has **128** (**38** electric) campsites. (See Chapter **56**) **Lake Joy Campground**, a private resort northeast of the village of Belmont, has **240** campsites along the south shore of Lake Joy. www.lakejoy@mhtc.net

WINTER ACTIVITIES
Belmont Mound is open in winter for wildlife watching, birding, snowshoeing, hiking, and cross-country skiing. None of the trails are groomed for skiing.

AREA ATTRACTIONS
Lafayette County has many parks, bike trails, historic sites, and small villages to explore. www.lafayettecounty.org/explore

The **Mining and Rollo Jamison Museum** houses artifacts from Wisconsin's early 19th-century lead and zinc mining period. Museum guests can take a unique underground guided tour of an actual lead mine dug by miner Lorenzo Bevens in 1845. Visitors can also ride an above-ground 1931 mining train. The Rollo Jamison Museum is located nine miles southwest of Belmont Mound State Park along Main Street in Platteville. www.mining.jamison.museum

DIRECTIONS TO THE PARK

Belmont Mound State Park and First Capitol Historic Site are both located along County Highway G, north of the village of Belmont in the northwestern corner of Lafayette County.

Belmont Mound State Park
222 S. Mound Road (County Highway G),
Belmont, WI 53510
(608) 762-5142 (Yellowstone Lake State Park)

29. BLACKHAWK LAKE PARK AND RECREATION AREA

A Man-made "Wilderness" Park

Blackhawk Lake

RECREATION AREA SNAPSHOT

Blackhawk Lake Park and Recreation Area is located in Iowa County about 17 miles west of Dodgeville. This popular park offers camping, picnic areas, hiking, swimming, fishing, and other amenities found at most state parks—except Blackhawk Lake is "not" a state park.

The 2,050-acre property was purchased by the Wisconsin Department of Natural Resources in the early 1970s to create a 220-acre man-made lake. Instead of developing a state park here, state officials decided to designate most of property as a state wildlife recreation area and agreed to lease 220 acres of the state-owned land to a local citizen commission to develop a park.

GEOLOGY AND HISTORY

The landscape of Blackhawk Lake is a mix of upland fields, steep-sided hills, oak forests, small creeks, and rocky bluffs. The property is located within the Driftless Area of southwestern Wisconsin, a region that remained untouched by the advance of glaciers during the last Ice Age. As a result, the Driftless Area does not have glacial deposits (drift), moraine hills, or the numerous kettle lakes found elsewhere in Wisconsin.

Several Native American cultures are known to have hunted, fished, and gathered wild plants in this region, some even before the end of the Ice Age about 10,000 years ago. When French and English explorers reached this area in the late 17th century, they found several established Ho-Chunk, Sauk, and Fox villages scattered through-out southern Wisconsin.

Beginning in the 1820s, an influx of pioneers from the East Coast, along with lead miners from Illinois, arrived to settle in this region. Conflicts with indigenous people over land use and mineral rights eventually led to the Winnebago War of 1827. At the end of the conflict, tribal leaders were forced to sign treaties with the U.S. government, forfeiting all their lands in southern Wisconsin and agreeing to move onto reservations west of the Mississippi River in the Iowa Territory.

The first permanent settlers in the Black-hawk Lake area were miners who founded the nearby village of Highland in 1842. Unlike most other miners in southern Wisconsin, the local miners here excavated deposits of "zinc," not lead. Zinc was in high demand in the 19th century for use in the production of brass. In the 1870s, many German, Irish, and other European immigrants arrived to farm the rich soils of the area and establish new communities.

> **Ranger Trivia**
>
> Blackhawk Lake was named after Chief Black Hawk, one of the most recognized Native American leaders in U.S. history. In 1832, Black Hawk led a group of 1,000 Sauk, Kickapoo, and Fox people back across the Mississippi River into Illinois and southern Wisconsin in defiance of treaty agree-ments. Black Hawk hoped to regain the tribe's former homeland by driving out white settlers from the area. His actions led to the Black Hawk War of 1832, an unfortunate conflict that ended tragically for Black Hawk's followers. Most were killed or captured by U.S. Army and Territorial Militia during the Battle (Massacre) of Bad Axe along the Mississippi River while attempting to escape back to Iowa.

In the early 1960s, a group of local businessmen petitioned state legislators to establish a new state park and create a fishing lake in the area in hopes of improving the region's economy. State officials declined their request to build another state park since Governor Dodge State Park in nearby Dodgeville had just been opened a few years earlier. As an alternative, the state acquired 2,050 acres of land for a new state wildlife recreation area and built earthen dams across Otter Creek and Cave Hollow Creek to create Blackhawk Lake. A long-term lease of 220 acres of state land was awarded to the Cobb-Highland Commission with permission to build Blackhawk Lake Park, which opened on July 4, 1972.

BEACHES, PICNIC AREAS, AND SHELTERS

A wide, sandy 300-foot-long swimming beach is located along Blackhawk Lake, along with changing stalls, restrooms, showers, and the park's concession stand. A large picnic area with volleyball courts and an open-air shelter is located adjacent to the beach. A boat dock with canoe and other watercraft rentals is located nearby. A second picnic area is located north of the park's public boat launch.

HIKING, BIKING, AND NATURE TRAILS

Several hiking trails link Blackhawk Lake Park to the **State Recreation Wildlife Area**. All are mowed but none have directional signs or trail markers. Hikers are advised to stop at the park office and obtain a trail map before venturing out on these trails.

The **Around the Lake Trail** (8.0 miles) generally follows the shoreline as it circles Blackhawk Lake. The **Beaver Run Trail** and the **White Oak Trail** are short trails located in the southeastern section of the recreation area. Both trails connect to the Around the Lake Trail with side trails that lead to the park's group camp.

The nearby **Military Ridge State Trail** (40 miles) is a popular biking/hiking trail that follows a former railroad corridor from Dodgeville to the city of Fitchburg near Madison. The trail can be accessed at Governor Dodge State Park about 17 miles east of the Blackhawk Lake.

Swimming beach

Ranger Trail Pick

A section of the **Around the Lake Trail** can be accessed directly from a parking area located along Plank Road in the northeastern side of the lake. A mowed trail north of the parking area leads uphill past an old cemetery and connects to a section of the trail that passes over the crest of both earthen dams with scenic views of Blackhawk Lake below and forested hills that surround it.

WILDLIFE VIEWING

The oak and pine forests, wetlands, streams, prairies, and rocky wooded bluffs of the Blackhawk Lake Wildlife Recreation Area provide habitat for white-tailed deer, raccoon, cottontail rabbits, opossum, squirrels, chipmunks, fox, beaver, mink, and otter. Birdwatching can be exceptional here with occasional sightings of uncommon birds such as bobolinks, meadowlarks, red-headed woodpeckers, and several rare species of warblers. Bald eagles are often seen snatching fish from the waters of Blackhawk Lake. Herons, shorebirds, and many kinds of waterfowl can be viewed along the lake and wetland areas.

INTEPRETIVE PROGRAMS AND FACILITIES

Interpretive programs and nature hikes are offered in the park from time to time. A nature center room at the park office has several wildlife mounts and interpretive displays. The **Red Pine Nature Trail** (0.2 miles) leads from the upper campground area through oak and red pine forests and down a rocky bluff area to the park's beach and picnic area.

BOATING AND FISHING

Blackhawk Lake (220 acres) is a man-made lake with a maximum depth of 42 feet. The lake has underwater rock outcrops and artificial fish cribs that provide habitat for a variety of fish including bluegill, crappie, large-mouth bass, northern pike, and walleye. Contour lake maps and fishing bait are available at the park office. A fish-cleaning station is located adjacent to the RV sanitation station near the campground area.

Ranger Note

Except for the beach and picnic area the shoreline of Blackhawk Lake is undeveloped. Slow-no-wake is enforced for motorboats to maintain a quiet, near-wilderness experience of the lake.

The park has a paved boat launch and overnight mooring slips that can be rented for long-term or seasonal boat storage. Motorboats, pontoons, canoes, and kayaks can be rented through a private concessioner.

ACCESSIBLE FACILITIES AND TRAILS

The park office, picnic areas, and all restroom and shower facilities are accessible for mobility-impaired visitors. Several campsites in the main campground are wheelchair accessible. A log cabin in the upper campground can be reserved for campers with physical disabilities for overnight stays.

A wheelchair-accessible fishing pier is located adjacent to the park's swimming beach. An accessible fishing platform can be accessed along a paved walkway near the park's boat launch. The steep topography of the property can be challenging for visitors with walking difficulties.

CAMPING

Blackhawk Lake has 150 campsites (92 electric) located in the upper and lower campgrounds. Most campsites have mowed grass and shade trees. A restroom/shower building is centrally located within both campgrounds. An RV water fill and sanitary station is located near the entrance of the lower campground area. The park has 14 group campsites (5 electric) located in forested areas around the perimeter of a central picnic area. The group campground has a shared open-air shelter and shower/restroom facility. The park also has five small log cabins overlooking Blackhawk Lake for overnight rental. Each cabin has electric service and air conditioning with access to shared vault-type restrooms.

> **Ranger Note**
>
> Camping fees at Blackhawk are comparable to those charged at Wisconsin state parks. Daily or annual park admission stickers are required for both campers and day users but Wisconsin State Park stickers are not honored at Blackhawk Lake.

WINTER ACTIVITIES

Several campsites and cabins are available in the off season at Blackhawk. Reduced rates area offered from October 1st through April 15th. Wildlife watching, hiking, snowshoeing, and skiing are favorite winter activities. Trails are not groomed for skiing. Ice fishing on Blackhawk Lake is a popular winter activity for anglers.

AREA ATTRACTIONS

Iowa County has several well-known tourist attractions to visit including the **House on the Rocks**, Frank Lloyd Wright's **Taliesin,** and the **Pendarvis State Historic Site.** www.iowacounty.org

DIRECTIONS TO THE RECREATION AREA

Blackhawk Lake Recreation Area is located in Iowa County about 60 miles west of Madison. From State Highway 18 west of Dodgeville, turn north onto Highway 80 in the Village of Cobb. After about six miles, turn right (east) onto to County Highway BH, which leads to the park entrance.

Blackhawk Lake Recreation Area
2025 County Highway BH
Highland, WI 53543
(608) 623-2997
www.blackhawklake.com

30. BLACK RIVER STATE FOREST

The "Northwoods" of Western Wisconsin

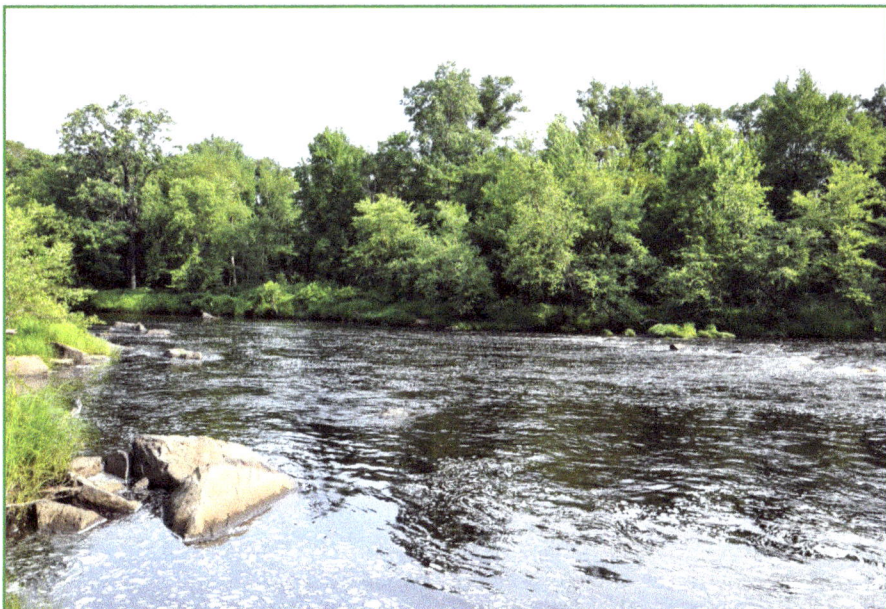

East Fork of the Black River

FOREST SNAPSHOT

The Black River State Forest is often described as a Northwoods wilderness area despite the fact that the property located far to the south in west-central Wisconsin. This 68,000-acre property does seem like a northern forest with its vast woodlands, flowage lakes, pristine creeks, and the wild, fast-moving Black River.

The forest offers camping, hiking, picnicking, swimming, nature study, biking, canoeing, kayaking, fishing, hunting, and many other outdoor pursuits. The property is a "working" state forest as well, with several active timber harvest operations and wildlife habitat enhancement projects underway throughout the property.

GEOLOGY AND HISTORY

The landscape of the Black River Forest can be traced back 600 million years when a warm, shallow sea covered most of Wisconsin. Over time, deep deposits of sand settled to the bottom of this sea and were transformed into Cambrian sandstone. Exposed sandstone can still be seen today within the rocky bluffs of the forest. During last Ice Age, these towering bluffs would have overlooked massive ice fields and Glacial Lake Wisconsin to the east and the unglaciated hills and valleys of the Driftless Area to the west. When the last of the glaciers retreated north about 10,000 years ago, new grasslands and forests attracted big game animals and several Native American cultures who hunted them. Sometime between 800-1200 A.D., the Ho-Chunk (Winnebago) people arrived here from the Ohio River Valley to hunt, fish, and establish permanent villages.

The Black River Forest area attracted French fur traders to the region in the late 18th century. Jean (Joseph) Rolette, a French Canadian fur trader from Prairie du Chien, is believed to have been the first European to live here. In 1818 Rolette built a dam and a sawmill on the Black River but his stay was short. Local indigenous people objected to his presence on their land and burned his mill to the ground.

> **Ranger Note**
>
> The Ho-Chunk people consider the Black River region as their ancestral home. The national Ho-Chunk government headquarters is located in the nearby community of Black River Falls.

> **Ranger Trivia**
>
> The Black River was named by early French fur traders due to the dark reddish-brown color of the water. The color is caused by tannic acid leaching from decaying oak leaves and tamarack trees and also from the high iron content in the soil.

Ho-Chunk warriors fought against the U.S. government during the Winnebago War of 1827 and some participated in raids against white settlements during the Black Hawk War of 1832. After the war, the Ho-Chunk were forced to sell their lands to the U.S. government. Soon after, lumber companies purchased nearly all the forest land in the local area. By 1844, there were eight sawmills along the Black River alone.

Near the end of the 19th century, most of the timber in the Black River area had been leveled. Immigrant farmers primarily from Western Europe attempted to farm cutover lands but the dry, sandy soil of this area was not suitable for growing crops. The federal government purchased many of these abandoned farms and deeded them to the State of Wisconsin under the condition that land be used for timber production and for public recreation. This action led to the establishment of the Black River State Forest in 1957.

BEACHES, PICNIC AREAS, AND SHELTERS

The **Pigeon Creek Flowage Picnic Area** on the south end of the forest has a wide, sandy swimming beach, tables, and vault-type restrooms. The **Castle Mound Recreation Area**, located south of the city of Black River Falls, has a picnic area, an observation tower, and an indoor shelter facility. The **Oxbow Pond Picnic Area**, located in the northern part of the forest, has several picnic tables situated along the shoreline of the pond.

HIKING, BIKING, AND NATURE TRAILS

The Black River Forest has nearly 30 miles hiking trails. Most of them (24 miles) are located within the **Wildcat** and **Smrekar Trail System**. These trails are also open for off-road biking during the warm season and cross-country skiing in winter. The forest has 33 miles of ATV/ UTV trails with links to both Jackson and Clark counties' all-terrain trail system.

> **Ranger Bike Trail Tip**
>
> The **Foundation Trail** (4.1 miles) in the nearby city of Black River Falls is a paved biking/hiking trail that circles the outer perimeter of the city adjacent to scenic sections Town Creek and the Black River. Access is available at the Black River Falls Chamber Visitor Center along North Water Street.

The **Castle Mound Nature and Overlook Trail** (2.0 miles) features some of the most impressive sandstone bluff formations in the area, including Castle Mound, which towers 180 feet above the surrounding forest. **Overlook Trail** ascends to the top of Castle Mound with panoramic views of the surrounding forest landscape. The nature trail encircles the lower perimeter of Castle Mound. Interpretive signs highlight the geology and plant life of the area. Both trails are located near the forest headquarters.

> **Ranger Trail Pick**
>
> The **Perry Creek Nature Trail** is a beautiful but often overlooked trail that follows the banks of Perry Creek. This clear-running stream flows through a scenic sandstone gorge with several small waterfalls. Perry Creek is ideal for wading and viewing the fern-covered canyon walls. Perry Creek is located along the Black River southwest of the forest headquarters.

Perry Creek Nature Trail

WILDLIFE VIEWING

The forest's 68,000 acres of undeveloped woodlands and waterways provide habitat for a wide range of wildlife including white-tailed deer, squirrels, rabbits, beavers, muskrats, waterfowl, wild turkeys, woodland songbirds, hawks and owls. More reclusive animals such as coyote, bobcat, fisher and black bear inhabit the forest as well.

In recent years, elk from Kentucky have been transported and released into the forest. These large members of the deer family had been absent from Wisconsin for more than 125 years. Thanks to these reintroduction efforts, the Black River State Forest now has more than 100 elk roaming the property.

> **Ranger Tip**
>
> The **North Settlement Road** on the west side of the forest is one of the best routes to watch for elk especially early in the morning or near sunset.

Dike 17 Wildlife Area, located along North Settlement Road, is a great place to see flocks of waterfowl, especially during the spring and fall migration period. This former 1,300-acre commercial cranberry bog is also home to mink, beaver, and herons.

INTERPRETIVE PROGRAMS AND FACILITIES

The forest headquarters has interpretive displays, property maps, and information regarding the forest's elk herd and the endangered Karner blue butterfly. Special events and evening programs are occasionally offered at the forest but there are no regularly scheduled interpretive programs or hikes.

BOATING AND FISHING

The Black River is known for its muskie, northern pike, smallmouth bass, and walleye fishing. Flowage lakes within the forest have good populations of blue gill, perch, largemouth bass, and crappie. A boat launch for small watercraft (electric motors only) is located on the **Pigeon Creek Flowage**. Brook and brown trout fishing is available at several streams, including Halls, Robinson, and Hay creeks.

> **Ranger Tip**
>
> The **Oxbow Ponds Day Use Area,** located in the northern section of the forest, offers shoreline fishing primarily for bluegill and other sunfish. Rainbow trout are occasionally stocked in some of the ponds as well.

The East Fork of Black River and the main channel of the Black River offer outstanding canoeing and kayaking opportunities. A brochure entitled, *Paddling the Black River through Jackson County* is available at the forest headquarters.

ACCESSIBLE FACILITIES AND TRAILS

All the restrooms and shower facilities at the forest headquarters are wheelchair-accessible. Accessible campsites are located at both Castle Mound and Pigeon Creek campgrounds. The indoor group camp cabin near the **East Fork Campground** has an accessible restroom and shower. Due to the hilly topography and sandy soils of the forest, many hiking trails may be challenging for visitors with walking difficulties.

CAMPING

The **Castle Mound Campground** has 35 campsites (14 electric). Restrooms, shower facilities, and an RV water fill/ sanitary station are available at the nearby forest headquarters. The **Pigeon Creek Campground** has 38 campsites (no electric). The campground has vault-type restrooms and hand water pumps. The **Pigeon Creek Flowage** swimming beach is located nearby.

The **East Fork Camping Area** has 24 rustic campsites (no electric). The campground is located above the east branch of the Black River in the far north end of the forest. Nearby is the **Outdoor Group Camp**, which can accommodate up to 50 people and eight wheeled camping units. Both the East Fork and Outdoor Group campgrounds have vault-type restrooms and hand water pumps.

The forest's **Indoor Group Camp**, located south of the East Fork camping area has bunk beds and can accommodate 12 people. The facility has a stove, refrigerator, gas fireplace, restrooms, and showers. Backpack camping is allowed in most areas of the forest except in the Hawk Island area along the Black River. A backpack camping permit is required and can be obtained online or at the forest headquarters. Two first come-first serve canoe campsites are located along the Black River south of Black River Falls.

WINTER ACTIVITIES
Forest staff groom 24 miles of cross-country ski trails for both traditional and skate skiing within the Wildcat and Smrekar trail system. A marked 1.6-mile snowshoe trail is also available here. The forest has 48 miles of snowmobile trails with access to the Jackson County trail system.

AREA ATTRACTIONS
Jackson County and Black River Falls have many unique shops, restaurants, parks, and other tourist attractions to explore. www.blackrivercountry.net or www.co.jackson.wi.us

The **Sandhill State Wildlife Area** (9,150 acres) is a popular wildlife viewing area. Many species of waterfowl, shorebirds, and songbirds can be seen here along with sandhill cranes, herons, osprey, and bald eagles. The property has an education center, marsh boardwalks, three observation towers, and a 260-acre bison enclosure. Sandhill Wildlife Area is located 38 miles northeast of the Black River Forest at 1715 County Road X near Babcock. www.dnr.wi.gov.wildifeareas/sandhill

DIRECTIONS TO THE FOREST
The Black River Forest is located in Jackson County in west central Wisconsin. From Interstate I-94, take Highway 54 exit west towards the city of Black River Falls. Exit south onto Highway 12/27 and then left (east) onto Highway 12 south. The forest headquarters is a mile south of Black River Falls near the Castle Mound Campground area.

Black River State Forest
W10325 Highway 12
Black River Falls, WI 54615
(715) 284-4103

31. BLUE MOUND STATE PARK

Highest Point in Southern Wisconsin

Sunrise observation tower

PARK SNAPSHOT

Blue Mound State Park is located atop the highest hill in the Southern Wisconsin more than 1,719 feet above sea level. The park has two observation towers that provide panoramic views of the scenic forests and rural countryside below.

This popular 1,153-acre state park offers camping, picnicking, off-road biking, nature study, wildlife viewing, and miles of hiking trails to explore. Blue Mound also has the only aquatics center and swimming pool in the state park system.

GEOLOGY AND HISTORY

Blue Mound is located within the Driftless Area of the state, a region of southwest Wisconsin that escaped the crushing effect of continental glaciers that covered most of the rest of the state 30,000 years ago. Although not altered by glacial activity, this area's landscape was shaped by the erosional forces of water, wind, and frost action over millions of years. The sandstone and limestone bedrock of this area was formed from deposits beneath an ancient sea that once covered most of the state more than 600 million years ago. Over time, most sandstone formations in southern Wisconsin have eroded into rolling plains except for a few tall remnant hills such as the park's 500-foot-high West Blue Mound and the nearby 300-foot-high East Blue Mound. These two towering geologic relics owe their existence to the hard dolomite limestone cap and an even-harder covering of a quartz-based rock called chert or flint, which sheltered them from erosional forces.

> **Ranger Trivia**
>
> Why are these hills called "blue mounds"? In 1766, Jonathan Carver, an early English explorer, was traveling down the Wisconsin River by canoe about 20 miles north of here when he noticed and recorded what he called "bluish mountains" rising above the rolling hills to the south. Carver later climbed to the top of one of the mounds and wrote about the fantastic views of the surrounding prairies.

Native American people have lived and hunted throughout the Blue Mounds area for more than 7,000 years. They frequently set fire to the surrounding prairies and oak savannas to maintain the lush grasslands of the region, critical in attracting big game browsers such as deer, elk, and bison.

The first white settler to the Blue Mounds area was Ebenezer Brigham, who arrived in 1829. Although Brigham moved here to mine lead, he was soon engaged as a colonel in the militia during the Black Hawk War of 1832. This was during a period of much unrest and fear of Native American uprisings among local citizens. The U.S. Army built a fort at the base of the East Blue Mound to protect local settlers. Although Blue Mound Fort was never directly attacked, the fort's first commander and two other soldiers were killed by Indian attacks in the local area.

After the Black Hawk War in 1835, the U.S. Army built a military road through the area. Ebenezer Brigham erected an inn to serve as a rest stop for settlers and militia traveling along this road. Within a few years, thousands of German and Norwegian immigrant families arrived to settle in the area. The rich prairie soils of the area were ideal for farming and the forested hillsides of Blue Mounds provided lumber and firewood for the growing population of the area.

The cool, breezy hilltop of Blue Mound became a favorite spot for local people to have a picnic or take a stroll through the shady forest. With onset of automobile travel, Blue Mound was transformed into a statewide tourist destination. It was developed into a privately-owned park in 1933. When this property became available for sale, the State purchased it and established Blue Mounds State Park in 1959.

Blue Mounds swimming pool

BEACHES, PICNIC AREAS, AND SHELTERS

Blue Mounds has a modern aquatic center that features a 1,950 square-foot heated swimming pool and a splash pad. The center has restrooms, showers, lockers, and changing stalls. Adjacent to the aquatic center is an indoor shelter facility with a large gathering room used for evening programs and special events. The shelter can be rented for private use as well.

A large picnic area is located on top of Blue Mound. The area has a playground area and an open-air shelter that can accommodate up to 100 people. On both the east and west end of the picnic area are 40-foot-tall observation towers that offer spectacular views of the fields, forests, villages, and other points of interest. On clear days, the Lower Wisconsin River Valley and the Baraboo Bluffs can be seen on the horizon.

HIKING, BIKING, AND NATURE TRAILS

Blue Mound has 25 miles of trails to explore. The **Indian Tree Marker Trail** (0.5 miles) loops through an oak forest along the north slope of the park. The trail was named after an oak tree that was bent at an angle as a young sapling by indigenous people more than a hundred years ago. This "guide tree" was believed to point towards a fresh water spring to refresh those who climbed to the top of Blue Mound many years ago.

Ranger Trail Pick

The **Flint Rock Nature Trail** (1.3 miles) has interpretive signs highlighting the unique geology of the park. Along the trail are several unique limestone rock formations and colorful chert-encrusted boulders. Chert, or flint, is a form of quartz rock used by indigenous people for thousands of years to fashion spear and arrow heads, axes, knives, and ornaments. It was also used to strike sparks in fire-making.

The **Ridgeview Trail** (0.9 miles) and the **Walnut Hollow Trail** (0.5 miles) are located along the south side of the park. Both are perched above the **Military Ridge State Biking/Hiking Trail** and offer scenic views of the surrounding countryside. The **Pleasure Valley Trail** (1.0 mile) is a gently rolling hike that leads through an attractive sugar maple/oak forest and a restored prairie. The **Weeping Rock Trail** (1.0 mile) follows **Ryan Creek** through a beautiful rocky glen where natural springs seep through sandstone walls and also provides access to a large restored prairie region of the park.

Ranger Trivia

The **Military Ridge Trail** was named after a military road built by the U.S. Army in 1835. This rustic wilderness road was laid out on the crest of upland ridges, often following the foot trails used by generations of Native American people. The Military Road provided a vital connection between Fort Howard in Green Bay, Fort Winnebago in Portage, and Fort Crawford along the Mississippi River in Prairie du Chien.

Blue Mounds is a popular off-road bicycle destination for many. The **CORP Single Track Biking Trail** (15.5 miles) is a challenging looped trail system that winds through the hilly terrain of the property. The trail was developed through the efforts of the **Capital Off-Road Pathfinders** or "CORP," a Madison-based mountain bike group.

The **Military Ridge Access Trail** (.25 miles) is a short hookup route to the **Military Ridge State Trail**, a popular 40-mile-long biking/hiking trail that follows a former railroad grade between the cities of Dodgeville and Fitchburg near Madison.

WILDLIFE VIEWING

The oak and maple forests, open fields, prairies, and rocky ravine areas of the park provide ideal habitat for raccoon, opossum, 13-lined ground squirrels, white-tailed deer, rabbits, gray and fox squirrels. More secretive animals, including flying squirrels, fox, and coyote, are also found here. More than 150 species of birds have been recorded in the park, including dozens of kinds of woodland and grassland songbirds, thrushes, ovenbirds, chickadees, blue jays, woodpeckers, and many others. Turkey vultures, owls, red-tailed hawks, and bald eagles are often spotted here as well.

INTERPRETIVE PROGRAMS AND FACILITIES

A small nature center is located in the picnic area and has interpretive displays that highlight the natural features and history of the park. Nature hikes and evening programs are presented throughout the summer season at the indoor shelter and the park's outdoor amphitheater.

BOATING AND FISHING

Blue Mound does not have any major water resources. **Mount Vernon Creek State Fishery Area**, located southeast of the park, is one of the state's most popular trout streams. The nearby **Steward Lake** (6.0 acres) offers shoreline fishing for large-mouth bass, bluegills, and other panfish. A public launch facility (electric motors only) is available at **Steward Lake County Park** along County Road JG in Mount Horeb.

ACCESSIBLE FACILITIES AND TRAILS

The park's visitor entrance station, playground, picnic area, and restroom/shower facilities are accessible for mobility-impaired visitors. Both indoor and outdoor shelter buildings are also wheelchair accessible. The park's swimming pool has a lift chair that turns 180 degrees for easy wheelchair loading and unloading.

Campsites **28** and **32** are wheelchair-accessible sites. Both have electrical service and walkways to the restroom/shower facility. A rustic wheelchair-accessible cabin is also available campers with physical disabilities. The cabin has electrical service and an accessible fire ring and picnic table. The park's steep and rocky topography makes access to hiking trails challenging for visitors with walking disabilities.

CAMPING

Blue Mound has 90 campsites (25 electric), all located within a mature oak/maple forest. The campground has a central restroom/shower facility and RV water fill and sanitation station. The park also has 12 "Bike-Hike" Campsites located near the Military Ridge State Trail. The sites are reserved for bike-in and backpack campers.

WINTER ACTIVITIES

Park staff groom nine miles of cross-county trails for both inline and skate skiing. All ungroomed trails are open for snowshoeing and hiking. A popular sledding hill is located adjacent to the swimming pool parking lot. Walk-in or ski-in winter camping is allowed throughout the off season.

AREA ATTRACTIONS

The nearby village of Mount Horeb is known as the "Troll Capital of the World." This popular Norwegian community has whimsical wood-carved trolls in its downtown area plus art galleries, restaurants, museums, and gift shops. www.trollway.com

The **Cave of the Mounds** is a National Natural Landmark and has been a popular Wisconsin attraction since it was discovered in 1939. Staff-guided tours reveal the cave's spectacular crystallized rock formations and clear underground pools. The cave is located about two miles southeast of the park along Cave of the Mounds Road. www.caveofthemounds.com

DIRECTIONS TO THE PARK

Blue Mound State Park is located in Dane County about 28 miles west of Madison. From U.S. Highway 18/151, exit onto North County Highway F. Turn left (west) onto County Highway ID and travel about a half mile to the village of Blue Mounds. Turn right (north) on Mounds Road to the park entrance.

Blue Mounds State Park
4350 Mounds Park Road
Blue Mounds, WI 53517
(608) 437-5711

32. BROWNTOWN-CADIZ SPRINGS STATE RECREATION AREA

Spring-fed Lakes—A Scenic Wonder

Zander Lake

RECREATION AREA SNAPSHOT

Browntown-Cadiz Springs Recreation Area is a quiet, day-use property tucked in the southwest corner of Green County. This scenic 645-acre gem has two spring-fed lakes and offers swimming, hiking, picnicking, nature study, fishing, canoeing, kayaking, and much more.

The property protects several freshwater springs, clear-running creeks, old oak forests, sandstone outcrops, and rare ecosystems that support several uncommon plants and animals that are found here.

GEOLOGY AND HISTORY

Browntown-Cadiz Spring is located within a region of Wisconsin geologists refer to as the Driftless Area. The continental glaciers that deposited glacial drift (silt, sand, and rocks) throughout most of Wisconsin near the end of the Ice Age never reached this area. The hills, valleys, and lowlands here were shaped by wind, water and frost erosion over millions of years.

Several different Native American tribes were drawn to the clear water springs, forests, and long-grass prairies of this area to hunt, fish, and gather wild plants over time. Archeological evidence has shown that early Woodland Indians occupied this region of Wisconsin as early as 500 B.C. When the first European explorers arrived here in the early 18th century, the Meskwaki, Sauk, and Ho-Chunk people had already lived here for generations. Indigenous claim to the region came to an abrupt end after Winnebago War of 1927, however. The defeated Ho-Chunk people were induced to cede all their lands in southern Wisconsin to the U.S. government and forced to move to reservation land west of the Mississippi. Within a few years, federal land surveys were underway setting township boundaries within Green County.

> **Ranger Trivia**
>
> Green County was created in 1837, more than ten years before Wisconsin became a state. The county was named in honor of Nathanael Greene, a major general of the Continental Army and hero of the American Revolutionary War.

The rolling hills and fertile soils of the area were ideal for small-scale farming. European immigrants, primarily from Switzerland, arrived to set up farmsteads throughout Green County in the mid-1800s. They brought with them exceptional knowledge of both dairying and cheese-making. Within a few years, Swiss milk cows could be seen grazing on most every hillside in the county and dozens of cheese factories sprang up throughout the area. By 1870, Green County was already well known as the "Swiss Cheese Capital" of the United States.

One of the more eccentric citizens of the local area was Hennie Zander, who lived in the nearby village of Browntown. Zander, with the help of several other landowners, hauled rock and soil from the surrounding farm fields to build a U-shaped earthen dike blocking the outflow of Cadiz Springs. The dike flooded the marsh around the spring to create a nine-acre lake, now called Zander Lake.

Zander's lake soon became a favorite recreation spot for local people to picnic, swim, or take a stroll along its scenic dike. Natural lakes are rare in southwestern Wisconsin, so local citizens often petitioned their legislators to fund the development of man-made lakes not only for fishing and swimming but to attract tourists to their communities.

In 1965, the Green County Conservation League raised funds and provided volunteer labor to create **Beckman Lake**, a 63-acre lake adjacent to Zander Lake. The new lake was named after the organization's long time president, Ray Beckman, of the nearby community of Monroe. The property was transferred to State of Wisconsin in 1970 to develop Cadiz Springs State Park. Additional lands were purchased by the State adjacent to the park to create **Browntown State Wildlife Area**. Today both properties are managed jointly as Browntown-Cadiz Springs Recreation Area.

BEACHES, PICNIC AREAS, AND SHELTERS

Beckman Lake Picnic Area has a 150-foot-long sandy swimming beach, an open-air shelter facility, and a modern playground. The nearby **Zander Lake Picnic Area** has an outdoor shelter, a small playground, vault-type restrooms, and drinking water. Both shelter facilities have electrical service and can be reserved in advance.

> **Ranger Trivia**
>
> Hennie Zander is said to have created his lake to commercially raise and sell bullfrogs. Frog legs were a popular delicacy at many big-city restaurants at the time. It's unclear how successful Zander's frog-raising enterprise became but it was certainly a novel, if unusual, business venture.

HIKING, BIKING, AND NATURE TRAILS

There are nine miles of hiking trails to explore at Cadiz Springs. A looped trail leads through the **Browntown Oak Forest Natural Area**, located north of Cadiz Springs Road in the northern section of the property. Several un-named trails lead through upland forests and lower wooded areas adjacent to Zander Creek in the eastern section of the park.

Zander Lake Nature Trail (1.0 miles) begins at the picnic area and follows the dike between Zander and Beckman lakes to a forested area on the north side of Zander Lake. Interpretive signs along this trail highlight the flora, fauna, and history of the property.

The **Devil's Tea Table** is an interesting geologic rock formation located along the northwest side of Beckman Lake. The rock formation is somewhat hidden by vegetation so watch for an unmarked but well-used trail through a grassy area which leads towards a grove of trees.

The **Badger State Trail** (40 miles) is a biking/hiking trail that follows a former railroad grade from the Illinois state line to the city of Madison. The trail leads through rolling hills, scenic farmsteads, and over 40 bridges. A highlight of the trail is traveling through 1,200-foot-long railroad tunnel. The Badger Trail is located about 7 miles east of Cadiz Springs with parking at **Twinning Park** in the city of Monroe. www.friendsofbadgerstatetrail.org

> **Ranger Trail Pick**
>
> A popular hiking route begins by taking **Zander Lake Nature Trail** between Zander and Beckman lakes. At the junction of the nature trail and Zander Creek, turn right (south), cross the bridge, and follow the hiking trail along the creek. From here, hikers can opt to explore a series of looped paths through upland woods in the eastern part of the property. If not, turn west at the bridge area and follow the open grassy trail along the south and west side of Beckman Lake back to the picnic area.

WILDLIFE VIEWING

Browntown-Cadiz Springs is well known for its exceptional birdwatching opportunities. The **Great Wisconsin Birds and Nature Trails Organization** lists sandhill cranes, northern shoveler ducks, horned grebes, Bell's vireos, and willow flycatchers as regularly seen here, plus several species of hawks and owls. The property has several bluebird nest boxes and a nesting platform for osprey.

The **Browntown Oak Forest State Natural Area** features sandstone outcrops and upland forests of white and black oak, sugar maple, hickory, and butternut. Common wildlife seen here and elsewhere at Cadiz Springs include white-tailed deer, raccoon, fox, coyote, chipmunks, rabbits, opossum, squirrels, and several species of woodland songbirds such as blue-gray gnatcatchers, orchard orioles, and tufted titmouse. Waterfowl, cranes, and herons, along with several species of turtles and frogs (including bull frogs), can often be seen at Zander and Beckman Lakes

INTERPRETIVE PROGRAMS AND FACILITIES

Interpretive panels are posted at various locations throughout the property and describe the flora, fauna, and history of the area. Nature hikes are offered by property volunteers from time to time throughout the use season.

BOATING AND FISHING

Beckman and Zander lakes provide excellent fishing opportunities for large-mouthed bass, bullheads, northern pike, catfish, bluegill, and other panfish. State fishery crews occasionally stock rainbow and brown trout in the lakes as well. Canoeing and kayaking are very popular on both lakes. A boat launch (electric motors only) is located on Beckman Lake.

ACCESSIBLE FACILITIES AND TRAILS

Both picnic areas have paved walkways to picnic tables, grills, and shelters. A wheelchair-accessible boat launch and fishing pier is located along Beckman Lake. The dike section of the **Zander Lake Nature Trail** is nearly level and is partially paved with asphalt and crushed gravel. Visitors with walking disabilities should be able to access this scenic area with some assistance.

CAMPING

Brownstone-Cadiz Springs Recreation Area is a day-use property so there are no campsites here. **Yellowstone Lake State Park**, located about 24 miles northwest of Browntown-Cadiz Springs near Blanchardville, has 128 sites (38 electric) with restroom and showers facilities. (See Chapter 56)

WINTER ACTIVITIES

The recreation area is open during the winter months for hunting, hiking, snowshoeing, wildlife watching, and cross-country skiing. None of the trails are groomed. Anglers enjoy ice fishing, especially for panfish, on both Beckman and Zander lakes.

AREA ATTRACTIONS

Green County has long been a popular tourist destination thanks to its Swiss heritage, scenic rolling hills, small farmsteads, museums, and cheese factories. The nearby city of Monroe is known as Green County's "Gateway to Cheese Country." www.greencounty.org

The **National Historic Cheese-Making Center** and **Imobesteg Farmstead Factory** preserve many artifacts relating to dairying and cheese-making. Guided tours are offered at the center's **Cheese Factory**, which was built in 1890 by Imobesteg family members who immigrated to this area from Switzerland. The **National Cheese-Making Center** is located about 7 miles east of Cadiz Springs along 6th Avenue in Monroe. www.nationalhistoriccheesemakingcenter.org

DIRECTIONS TO THE RECREATION AREA

Browntown-Cadiz Springs Recreation Area is located southwest of the city of Monroe. From State Highway 11, exit south on Cadiz Springs Road and travel about a half mile to the park entrance.

Browntown-Cadiz Springs State Recreation Area
N 2241 Cadiz Springs Road, Highway 11
Browntown, WI 53522
(608) 523-4427

33. BUCKHORN STATE PARK

A Wisconsin Water and Wilderness Wonderland

Swimming beach

PARK SNAPSHOT

Buckhorn State Park has some of the finest outdoor recreation opportunities in the state. This scenic 8,190-acre park offers camping, backpacking, hiking, swimming, nature study, fishing, boating, kayaking, canoeing, hunting, cross-country skiing, and much more.

The park is located on the southern tip of the Buckhorn Peninsula within the **Castle Rock Flowage** of the Wisconsin River. The property is bordered by ten miles of undeveloped shoreline, which gives it the impression of a wilderness island.

GEOLOGY AND HISTORY

The landscape of the park is a mix of wetlands, flowages, upland forests, dry prairies, and rare oak and pine sand barrens. Near the end of Wisconsin's last Ice Age about 14,000 years ago, a massive ice and earthen dam holding back the waters of Glacial Lake Wisconsin to the north collapsed. Almost immediately, torrents of flood water thundering down the Wisconsin River reshaped the landscape and deposited sand as it roared downstream toward the Mississippi River. After the floods subsided and the Wisconsin River settled into its current basin, Archaic people, some as early as 5,000 B.C., entered the river valley to hunt, fish, and collect wild plants. When French fur traders arrived in the mid-1600s to explore the lower Wisconsin River, they found the region inhabited by the Fox, Sauk, and the Ho-Chunk people.

The extensive forests that grew adjacent to the Wisconsin River lured lumbermen during the early years of the lumber boom in the 1840s. Nearby sawmill towns such as Necedah and Germantown literally sprang up overnight, but the timber industry here was short-lived. By 1877, most of the timber in southern and central Wisconsin had already been cut, forcing timber crews to move further north.

European immigrants attempted to farm cutover forest land but crops were difficult to grow in the dry, sandy soil of the local area. Many farmers simply abandoned their properties and moved on. In the late 1920s, the Wisconsin River Power Company purchased thousands of acres of idle land all along the Wisconsin River in anticipation of building hydroelectric dams. The nearby **Castle Rock Dam** and **Petenwell Dam** were built across the Wisconsin River in 1951. The immense flowage lakes that formed behind these dams submerged thousands of acres of forests, agricultural land, and in some cases, entire communities. The flowage lakes also created new and valuable shoreline property in upland areas, including the newly formed **Buckhorn Peninsula**. The State purchased 3,970 acres along the southern tip of the Buckhorn Peninsula to develop Buckhorn State Park in 1971, which opened as a day-use-only property. Additional park development including a swimming beach, picnic areas, and walk-in campsites led to a second opening of Buckhorn State Park in 1984.

> **Ranger Trivia**
>
> The park's name, Buck-horn, is often misinter-preted as referring to a buck's (male deer) ant-ler (horn). Actually, buckhorn was derived from the Ho-Chunk word *Pur-kane* meaning "unbroken wilderness." The Ho-Chunk also called this area "land of yellow waters" in refer-ence to the water color of the nearby Yellow River.

BEACHES, PICNIC AREAS, AND SHELTERS

Buckhorn has a 330-foot-long swimming beach with restrooms, changing stalls, and an outdoor rinse station. A picnic area adjacent to the beach has picnic tables and two open air shelters that can be reserved in advance.

HIKING, BIKING, AND NATURE TRAILS

There are nine miles of hiking trails to explore at Buckhorn. The **Turkey Hollow Trail** (1.5 miles) and the **Partridge Trail** (1.2 miles) lead through scenic oak woodlands in the west side of the peninsula. Side paths along these trails provide access to several walk-in shoreline campsites. The **Central Sands Nature Trail** (1.4 miles) has interpretive panels that describe native plants and the significance of the sandy regions of the park.

The **Island Canoe Trail** (1.8 miles round trip) is a water trail that loops through a sheltered inlet of the Castle Rock Flowage. The trail has 10 numbered interpretive stops that highlight various animals and plants found along this waterway. A self-guided trail map is available at the park's canoe launch area. Canoe and kayak rentals can be made at the park office.

Bicycling on park and local township roads is a favorite activity for many visitors. The area's roadways are relatively flat and most lead to the village of Necedah, about 10 miles north of the park.

Ranger Trail Pick

The **Oak Barrens Trail** (1.5 miles) is a self-guided nature trail located west of the campground area. The trail loops through a dry, scarcely wooded area of the park known as a "sand barren." Pine and oak sand barrens once covered over four million acres of southern and central Wisconsin. Today only about 10,000 acres of these rare ecosystems remain. A 20-foot observation tower along this trail overlooks this unique landscape.

WILDLIFE VIEWING

Buckhorn's pine and oak woodlands, prairies, wetlands, and the Castle Rock Flowage provide ideal habitat for many wildlife species. Game animals such as deer, squirrels, rabbits, raccoon, wild turkey, grouse, pheasants, and many kinds of ducks and geese abound both in the park and the adjoining state wildlife areas.

Birdwatching at Buckhorn can be excellent anytime of the year but is especially good during spring and fall migration periods. Bald eagles, red-tailed hawks, and other raptors are often seen flying over the park. Several pairs of osprey make use of the nesting platforms installed along shoreline areas of the park. Dozens of species of woodland and grassland birds can be spotted here, as well as and shorebirds including sandpipers, plovers, egrets, and herons.

INTERPRETIVE PROGRAMS AND FACILITIES

A small nature center room with interpretive displays and children's activities is located inside the park entrance station. Interpretive programs, hikes, and special events are offered throughout the use season. Evening programs are often presented at the park's amphitheater located south of the picnic area.

The **Sand Blow Vista** located north of the Oak Barrens Nature Trail, has a ground-level viewing platform. Interpretive panels here explain vegetation and the significance of this dry, desert-like, sand blow area.

BOATING AND FISHING

Castle Rock Lake (16,640 acres) is the fourth largest man-made flowage of its kind in Wisconsin. Boating, waterskiing, canoeing, kayaking, and fishing are favorite activities on this lake. Anglers can expect to catch a variety of fish in the flowage including panfish, bullheads, crappie, perch, musky, northern pike, walleye, sturgeon, and catfish.

Shore fishing is also popular at Buckhorn, especially near the County Highway G Bridge west of the park entrance. A **Kid's Fishing Pond** is located along **Turtle Trail** (0.4 miles) north of the park entrance. This attractive pond has a fishing pier and is occasionally stocked with panfish and large-mouth bass.

A boat launch is located along the west shoreline of the park. A separate kayak and canoe launch is available on the far eastern shoreline of the park. Canoe and kayak rentals can be obtained at the park office.

Kid's fishing pond

ACCESSIBLE FACILITIES AND TRAILS

The park's visitor entrance station, showers, shelters, and most restrooms are accessible for mobility-impaired visitors. The picnic area and several walk-in campsites have paved walkways for wheelchair access. A special lightweight PVC beach wheelchair with balloon-type tires is available for use free of charge.

Campsites 61 and 63 are wheelchair-accessible sites. Both offer electric service and paved walkways to nearby shower/restroom facilities. Accessible watercraft-boarding piers are available at both launch areas. A wheelchair-accessible fishing pier is located at the north end of the picnic area

The **Cabin of the Setting Sun** has a fully accessible kitchen, bedroom, living room, and a wheel-in restroom/shower. The cabin is located along the Castle Rock Flowage and can be reserved for campers with physical disabilities.

CAMPING

Buckhorn has 60 vehicle-access campsites (20 electric). The campground has a central restroom/shower facility and an RV water fill/sanitary station. The park also offers 42 walk-in campsites along the shoreline of the Castle Rock Flowage located on both the east and west side of the peninsula. Campers can backpack or use a wheeled cart to haul their gear to these campsites. Most of these sites are also accessible by watercraft. Walk-in sites are served by vault-type or portable restrooms.

The park's group campground has three large sites, each of which can accommodate up to 40 people. Drinking water and vault-type restrooms are located nearby. Wheeled camping units are allowed at the group campground but must be kept on paved parking areas. There is no electrical service here.

WINTER ACTIVITIES

Park staff groom five miles of cross-country ski trails for inline skiing. Other off-season activities include hunting, hiking, wildlife watching, winter camping, and ice fishing. Snowshoes can be borrowed free of charge at the park office.

AREA ATTRACTIONS

Juneau County is a well-known vacation destination and has many fine lakeside resorts, restaurants, gift shops, parks, and museums. www.juneaucounty.com

The **Necedah National Wildlife Refuge** is an internationally recognized federal Wildlife Area acclaimed for its groundbreaking efforts to enhance trumpeter swan and whooping crane populations. The 45,000-acre refuge has several miles of roadways, nature trails, boardwalks, and observation platforms for birdwatching. The Necedah Wildlife Refuge office is located about 12 miles northwest of Buckhorn off Headquarters Road near Necedah. www.fws.gov/refuge/necedah

DIRECTIONS TO THE PARK
Buckhorn State Park is located in eastern Juneau County along the Castle Rock Flowage. From Interstate I-90/94 or Highway 82, take Highway 58 north about 8 miles. Exit onto County Highway G and travel east about 3.5 miles to the park entrance.

Buckhorn State Park
W 8450 Buckhorn Park Avenue
Necedah, WI 54646
(608) 565-2789

34. CAPITAL SPRINGS
STATE RECREATION AREA
Madison's Green Oasis

Lake Waubesa

RECREATION AREA SNAPSHOT

Capital Springs State Recreation Area is a 2,500-acre property located on the western shore of Lake Waubesa just minutes away from downtown Madison. The recreation area is cooperatively managed by the State of Wisconsin, Dane County, and local non-profit organizations. Capital Springs has several hiking trails, paved bike paths, and offers camping, fishing, picnicking, nature study, and other outdoor activities.

The recreation area is the hub of a seven-mile-long environmental corridor known as the **Nine Springs E-Way**, which follows Nine Springs Creek to Lake Waubesa. The property is also the trailhead for both the **Capital City State Trail** and the **Lower Yahara River Trail**.

GEOLOGY AND HISTORY

The landscape of the Capital Springs Recreation Area has been altered many times over thousands of years. Prior to the last Ice Age, the Yahara River ran through a very deep and wide valley in the Madison area. As glaciers advanced through this area about 30,000 years ago, they deposited hundreds of feet of sand, gravel, and rocks, filling the entire river valley. When they retreated north again about 18,000 years ago, an earthen dam was formed, which submerged the entire Madison area under melt water. Retreating glaciers also excavated four large depressions in the local landscape. Over time, the Yahara River eventually breached the glacial dam and filled these deep depressions with flood water. Today they are known as Madison's Four Lakes Region: Mendota, Monona, Waubesa, and Kegonsa.

Several Native American cultures lived, hunted, and fished within the Four Lakes Region for thousands of years. Archaeological excavations conducted at Capital Springs revealed that indigenous people were using this site as early as 8,500 B.C. The Late Woodland culture known as the Mound Builders lived here from about 1,000 B.C. to 1,600 A.D. Evidence of their presence can still be seen today in several effigy burial mounds located at Capital Springs. More recent Native American cultures to occupy this area were the Ojibwe and the Ho-Chunk people.

The first white settlers to the area were Yankees from the East Coast who arrived in the early 1800s. When the U.S. Congress created the Wisconsin Territory in 1836, many European immigrants, primarily Germans, Irish, and Norwegians, moved here to farm the land and build settlements. Treaties signed in 1837 between the U.S. government and the Ho-Chunk nation required that all tribe members move onto reservations west of the Mississippi River, but many decided to stay. Several groups of Native American people continued to camp in the Four Lakes Region as late as the 1940s.

Madison-area citizens have always treasured their community parks. In 1969, local support to preserve undeveloped areas led to the establishment of the **Nine Springs E-Way Natural Area** on the south side of Madison.

> **Ranger Note**
>
> The original plan for the E-Way was the brain-child of Professor Philip Lewis Jr. of UW-Madison's Department of Landscape Architecture. The "E" in the E-Way stands for Education, Ecology, Esthetics, and the Environment.

Capital Springs State Recreation Area was created in 2000 to help celebrate the 100th anniversary of the Wisconsin State Parks and Forest system. The state, along with Dane County, the city of Madison, and other local government agencies, joined forces to expand the existing Nine Springs E-Way into a state recreation area. The property is made up of the **Centennial State Park, Lake Farm County Park**, the **Monona Conservancy, MMSD Wildlife**

Observation Area, Nevin Springs Wildlife Area, and the Jenni and Kyle Nature Preserve.

BEACHES, PICNIC AREAS, AND SHELTERS

Capital Springs has a large picnic area along Lake Waubesa. The area has three open-air shelters with restrooms and electrical service that can be reserved in advance. An observation tower here offers a panoramic view of the lake and the surrounding countryside. There are no swimming areas at Capital Springs but an excellent swimming beach is available at **Goodland County Park**, located along the southwest shore of Lake Waubesa.

HIKING, BIKING, AND NATURE TRAILS

Capital Springs has eight miles of hiking and nature trails. The **Lussier Moraine Nature Trail** (1.6 miles) begins near the **Lussier Family Heritage Nature Center**. The trail loops through gently rolling hills before crossing Lake Farm Road, where it leads to a glacial moraine hilltop with a great view of Nine Springs Creek area.

    Lake Farm Heritage Trail (2.1 miles) and **Whitetail** Trail (1.3 miles) both follow the Lake Waubesa shoreline to a restored prairie in the northeast part of the property. Interpretive signs along the trail highlight the historical and cultural significance of the area. The **MMSD Wildlife Observation Trail** (3.3 miles) leads over several dikes adjacent to wetlands, ponds, and open-water lagoons. Birdwatching can be excellent along these trails.

    The **Lower Yahara River Trail** (2.5 miles) is a biking/hiking route that begins at Capital Springs and travels south along Lake Waubesa to the Village of McFarland. The trail crosses a fun, one-mile-long bridge and boardwalk, the longest of its kind in Wisconsin.

    The **Capital City State Trail** (9.5 miles) leads north from Capital Springs to access the **Capital City Path** in the heart of downtown Madison. This state biking/hiking trail also leads west through the beautiful Nine Springs E-way natural area with connections to the **Military Ridge State Trail** (40 miles) and the **Badger State Trail** (40 miles).

Ranger Trail Pick

The **Nine Springs E-Way Trail** (2.7 miles) begins at its trailhead along Lake Farm Road. The trail heads west along the banks of Nine Springs Creek into the E-Way nature preserve. The streams, springs, woodlands, and wetlands along this trail provide ideal habitat for hundreds of species of plants and animals. The trail crosses over the creek and then travels east along the Capital City State Trail back to the trailhead.

WILDLIFE VIEWING

Capital Spring's diverse landscape of woodlands, wetlands, creeks, springs, prairies, and Lake Waubesa provides ideal habitat for many wildlife species including white-tailed deer, fox, raccoon, muskrats, squirrels, chipmunks, rabbits, painted turtles, and several types of frogs.

More than 200 different species of birds have been recorded at Capital Springs. A popular birding site is the 600-acre **Madison Metropolitan Sewerage District (MMSD)** Wildlife Observation Area located on the north side of the recreation area. Lagoons that once held treated waste from a water treatment plant have been cleaned up and repurposed into wetland habitat for waterfowl and other marshland birds.

INTERPRETIVE PROGRAMS AND FACILITIES

The **Lussier Family Heritage Nature Center** is an education and interpretive center located near the campground area. This facility has interpretive displays and an outdoor observation deck with great views of the recreation area. Volunteer naturalists offer nature hikes and interpretive programs throughout the year at Capital Springs. Upcoming hikes and programs are posted on the Friends of Capital Springs website: www.friendsofcapitalsprings.org.

A rare grouping of Late Woodland Indian effigy mounds can be viewed in the southwestern section of **Lake Farm County Park**. Some of mounds are

Lussier Family Heritage Nature Center

geometric in shape while others form the outline of animals. Interpretive signs explain how the mounds were built and what they may have represented to these early inhabitants.

BOATING AND FISHING

Lake Waubesa (2,080 acres) is a popular boating and fishing lake. Anglers can expect to catch bluegill, crappie, bass, northern pike, walleye, and muskie. A boat launch and fish cleaning station are located along Libby Road just north of Capital Springs State Park. A pier for launching canoes and kayaks is also located here.

> **Ranger Trivia**
>
> Lake Waubesa was named after an Ojibwa term meaning "Lake of the White Bird" (swan). Early settlers to the Madison merely called it "Lake Two." It seems immigrants named each of Madison's four lakes by number as they encountered them while boating upstream along the Yahara River. In 1858, state legislators passed a bill renaming each lake back to its original Native American designation.

ACCESSIBLE FACILITIES AND TRAILS

The Capital Springs office, Lussier Nature Center, and most restrooms, picnic areas, and shelter facilities are accessible for mobility-impaired visitors. A wheelchair-accessible campsite with electric service and a paved walkway to a restroom/shower building are located in the Lake Farm County Park campground. A wheelchair-accessible fishing pier is located adjacent to the boat launch area.

> **Ranger Pick**
>
> The **Jenni and Kyle Preserve** is a fully-accessible natural area that was designed specifically to accommodate children and adults with physical disabilities. The site has asphalt-paved walkways that provide access to shelter buildings, interpretive signs, and a fully accessible playground that features a wheelchair swing. The preserve has two fishing ponds that are stocked with panfish and trout. Fishing is limited to children 14 years or younger and disabled adult anglers. The preserve is located on the far western side of Capital Springs at 925 Post Road, Madison.

CAMPING

The **Dane County Lake Farm Park** has 54 campsites (39 electric). The campground has a toilet/shower facility and an RV water fill/sanitary station. A group campground is located north of the picnic areas near Lake Waubesa. Advance campsite reservations can be made online through the Dane County Park System. www.danecountyparks.com

WINTER ACTIVITIES

The recreation area has six miles of cross-country ski trails that are groomed for both classic and skate skiers. Snowshoeing, wildlife watching, hiking, and ice fishing are also popular winter activities. The campground is closed in winter.

AREA ATTRACTIONS

The Madison area is one of the top tourist destinations in Wisconsin and offers many parks, museums, cultural centers, sports arenas, restaurants, bike trails, and a popular free zoo. www.visitmadison.com

The **Nevin State Fish Hatchery**, built in 1876, is the oldest state fish hatchery in Wisconsin. Thousands of cold-water fish such as brook, brown, and rainbow trout are raised here annually. The hatchery makes use of natural artesian springs that release over two million gallons of water daily. Self-guided interpretive panels at the site explain how the hatchery operates. The Nevin Fish Hatchery is located on the far west side of Capital Springs Recreation Area at 3911 Fish Hatchery Road in Fitchburg.

DIRECTIONS TO THE RECREATION AREA

Capital Springs Recreation Area is located southeast of the city Madison. From Interstate I-90 take the Highway 12/18 west exit (West Beltline). Exit onto South Towne Drive and travel south for about a mile and then turn onto Moorland Road to Lake Farm Road. (608) 275-3246

Capital Springs State Recreation Area
3101 Lake Farm Road
Madison, WI 53711
(608) 224-3606

35. COULEE EXPERIMENTAL STATE FOREST

A Remote Wilderness Gem

Coulee Forest view

FOREST SNAPSHOT

The Coulee Experimental State Forest may be the least-visited state forest in Wisconsin, even though its only 15 miles from La Crosse, one of the largest cities in the state.

Unlike most other state forests, the property doesn't have a visitor entrance station, campground, picnic areas, or even a restroom. The lack of visitor amenities here is by design, however, since the purpose of this state forest is very different from other properties. As its name suggests, this "experimental" state forest was developed primarily to serve as an outdoor research lab to test various erosion control and forest management methods.

The 3,000-acre state forest "is" open for public recreation, however. The property offers hiking, cross-country skiing, horseback riding, wildlife viewing, birding, hunting, and most other activities that conform to the forest's low-impact policy. To many visitors, the lack of development at this state forest makes it ideal for those who prefer a quiet, peaceful, uncrowded wilderness experience.

GEOLOGY AND HISTORY

The Coulee State Forest is located within the Driftless Area of the state that escaped the crushing effect of glaciers. As a result, this area has very little glacial drift (sand, gravel, and rocks) that were deposited by glaciers in many other places in the state. The rugged landscape here of deep valleys, steep hills, high flat ridges, and rocky bluffs was formed over millions of years by wind, water, and frost erosion. The unique topography of this region of Wisconsin is often referred to as "Coulee Country."

> **Ranger Note**
>
> A coulee is a geologic description of a deep ravine or steep-sided valley that collects and drains water by runoff and small streams. The term "coulee" was derived from the French-Canadian word *couler,* which means "to flow."

The forested hills, valleys, creeks, and prairies of this area have attracted indigenous people to hunt, fish, and gather wild plants here for thousands of years. Archeologists have found evidence that the prehistoric Oneota people roamed these hills as far back as 1200 B.C. In more recent times, both the Dakota (Sioux) and the Ho-Chunk people laid claim to this region.

The first Europeans to reach this area were French fur traders and missionaries who paddled down the Mississippi River in the 17th century. These early explorers came ashore and made contact with Native American people living in a prairie area of what is now the city of La Crosse.

> **Ranger Note**
>
> While visiting Native American villages along the Mississippi River in the 1600s, French missionaries recorded seeing young tribesmen playing a rough, competitive ball game using a cupped racquet on the end of long-handled stick. They described the game stick as resembling a Catholic bishop's cross staff or *crosier*. Thereafter, the name "La Crosse" was used to describe this ball game. It was also selected to name the La Crosse River, La Crosse County, and the city of La Crosse.

Many early 20th-century immigrants to the local area worked in lumber camps, limestone quarries, or cut firewood to fuel Mississippi River steamboats. The nearby community of La Crosse grew quickly to become a major port of call for trade and steam-powered vessels along the river. Farmers found the deep, rich soils of the area ideal for growing crops. Lumber companies found high-grade timber trees along the hills and valleys throughout the coulee region. Unfortunately, cutting trees on the steep-sided ridges led to serious erosion problems. The loss of topsoil on the hilly terrain of the area grew even worse when used for cattle grazing or plowed for cropland.

Better farming techniques, such as strip-cropping, were introduced in the 1940s and helped to reduce the loss of soil in many agricultural areas. More research was needed to control erosion on steep hills and valleys of the coulee area. As a result, the U.S. Forest Service, in cooperation with the State of Wisconsin, purchased thousands of acres of land here in 1960 to create the Coulee Experimental State Forest. Ongoing research to control erosion caused by water runoff, cattle grazing, and harvesting timber continues today in the forest. Experimental plantings of different species of trees and shrubs for erosion control or for timber production are also conducted here.

BEACHES, PICNIC AREAS, AND SHELTERS
Picnic areas, restrooms, and a swimming beach are available at **Neshonoc Swarthout Beach County Park**, located a few miles north of the forest along State Road 16 near West Salem.

HIKING, BIKING, AND NATURE TRAILS
There are 12 miles of marked hiking trails and dozens of logging roads to explore throughout the forest. Horseback riding is allowed on most logging roads and trails except on marked hiking/ski trails. Off-road bike riding, ATVs, or other motorized vehicles are not allowed in the forest to protect the thin soils and steep topography of the property.

The **La Crosse River State Trail** (22 miles) is a popular biking/hiking trail that links the cities of La Crosse/Onalaska to the community of Sparta. The trail also provides access to the **Great River State Trail** (24 miles). The La Crosse River Trail can be accessed a few miles north of the Coulee State Forest along Interstate I-90 in the village of Bangor. www.lacrosseriverstatetrail.org.

> **Ranger Note**
>
> Most of the forest's hiking trails follow ridgetop cross-country ski routes. These trails are sparsely signed and are often not mowed until late summer or early fall. A property map and compass are recommended when hiking in the forest.

WILDLIFE VIEWING
Coulee State Forest has thousands of acres of remote, old-growth forests, cropland, and prairie areas that provide ideal habitat for a wide variety of wildlife, especially for white-tailed deer. Other animals found here include rabbits, squirrels, ruffed grouse, wild turkey, fox, coyote, and several species of hawks and owls. The property is also home to several uncommon species, such as black and yellow-billed cuckoos, hooded warblers, Acadian flycatchers, bobwhite, and even box turtles.

The **Northeast Coulee Oak Woods State Natural Area** (92 acres) is located along a sandstone ridge in the northern section of the forest. This area preserves old-growth red and white oak forests and other trees typically found in dry mesic woodlands.

The **Berg Prairie** and **Billy Goat Ridge State Natural Areas** (377 acres) protect small remnant goat prairies that have native grassland plants, such as little bluestem, Indian grass, wild bergamot, asters, lead plants, and pasque flowers. The Berg Prairie is located in the southern section of the forest with access from a parking lot off Antony Road. Billy Goat Ridge prairie area is in the northern area of the property off of Russian Coulee Road.

> **Ranger Note**
>
> A "goat prairie" is a term used to describe a specific kind of dry prairie that grows on steep, south-facing bluffs. This type of prairie has very thin soil that discourages tree growth but supports native prairie grasses, forbs and flowering plants.

INTERPRETIVE PROGRAMS AND FACILITIES

There are no interpretive programs or self-guided nature trails in the forest.

BOATING AND FISHING

There are several trout streams in La Crosse County. **Dutch Creek** east of the forest and **Bostwich Creek** to the south are Class II Trout streams. Both have brown trout and other species of fish. **Neshonoc** Lake (606 acres) is located about 11 miles north of the forest near West Salem. The lake has a public boat launch and good populations of panfish, large-mouth bass, northern pike, and catfish.

ACCESSIBLE FACILITIES AND TRAILS

Due to the extreme topography of the forest, access for visitors with walking difficulties may be difficult without some assistance.

CAMPING

The Coulee State Forest is a day-use property so there is no camping here. **Veteran Memorial Campground** offers both tent and RV camping along the La Crosse River. The campground is about nine miles northwest of the forest along County Road VP near West Salem. **Goose Island Park and Campground** has 250 tent and RV campsites along the backwaters of the Mississippi River. This park is located about 19 miles southwest of Coulee State Forest off County Road GI south of La Crosse. www.lacrossecounty.org/facilities

WINTER ACTIVITIES
The forest has 12 miles of cross-country ski trails located along bluff ridge tops within the central area of the property. The trails are groomed for both classic and skate skiing by volunteer members of the La Crosse Ski Club. Other off-season activities in the forest include hiking, wildlife observation, snowshoeing, and especially deer hunting.

AREA ATTRACTIONS
The La Crosse area is a popular year-round tourist destination with many outstanding attractions to explore. www.explorelacrosse.com

Grandad Bluff Park, located in the city of La Crosse, has one of the most outstanding vistas in Wisconsin. This towering 600-foot-tall bluff offers incredible views of the Mississippi River Valley below. On clear days, the states of Minnesota and Iowa can be seen on the horizon. The park has a picnic area, shelter, and a viewing platform with spotting scopes. **Hixon Forest** (800 acres) is located adjacent to Granddad Bluff. This city-run forest also has scenic bluff views and ten miles of hiking and mountain bike trails. www.citiyoflacrosse.org/parks

DIRECTIONS TO THE FOREST
From Interstate I-90 east of La Crosse exit onto Highway 162 south from Bangor about two miles. Exit onto Russian Coulee Road from highway 162 to the parking areas for the Northern Coulee Woods and Billy Goat Ridge Natural Areas. To access the Berg Prairie Natural Area, take the Highway11 exit off of Highway 162 to the south parking areas. Access to the ski and hiking trails is along the west side of the forest. Take the Russian Coulee Road off County Road M just east of the village of Barre Mills.

Coulee Experimental State Forest
Wisconsin Dept. of Natural Resources
3550 Mormon Coulee Road SOB
La Crosse, WI 54601
(608) 785-9007

36. Cross Plains State Park

Ice Age Crossroad

Wilke Gorge

PARK SNAPSHOT

Cross Plains State Park is part of a 1,700-acre preserve known as the **Ice Age Complex at Cross Plains**. This unique property, located just west of Madison, is one of the few places in Wisconsin where both glacial moraine and non-glacial landforms can be seen at the same site. The park has several miles of hiking trails that lead through old-growth hardwood forests, deep ravines, moraine hills, wetlands, and a restored prairie and oak savanna. The Ice Age Complex property has wildlife ponds, an historic farmstead, and miles of walking trails including a section of the **Ice Age National Scenic Trail**.

GEOLOGY AND HISTORY

Most of the Cross Plains landscape was shaped by the Green Bay Lobe of the Wisconsin Glacier more than 18,000 years ago. As the earth's climate began to warm near the end of the Ice Age, a massive ice field located here came to a near standstill. The stalled glacier deposited large amounts of sand, gravel, rocks, and other glacial debris, or drift, at its base, creating a large hill geologists refer to as an end moraine. Beyond the edge of this glacier, the land remained free of glacial ice and became part of southwest Wisconsin's Driftless Area. When the ice fields began to melt, a large, temporary lake was formed and eventually drained through turbulent rivers beneath the glacier. Massive amounts of meltwater and glacial debris surged down 600-foot-deep canyons here, filling them with hundreds of feet of sand and gravel. Other bursts of melt water cut deep valleys down to the bedrock in some areas such as the 100-foot-deep **Wilke Gorge** found at Cross Plains State Park.

> **Ranger Note**
>
> An impressive group of effigy mounds can be seen at nearby Governor Nelson State Park. (See Chapter 39)

Early indigenous people who moved to this part of Wisconsin included a Woodland Period tribe known as the Mound Builders who lived in this region from about 500 B.C. to 1650 A.D. These early inhabitants are best known for the elevated geometric and animal-shaped effigy burial mounds they built throughout southern and eastern Wisconsin.

The Ho-Chunk (Winnebago), Sauk, and Meskwaki (Fox) people lived in villages throughout southern Wisconsin at the beginning of the 19th century. Conflicts regarding land ownership arose when white settlers arrived to mine lead in the region in the 1820s. This led to the Winnebago War of 1827, after which tribal leaders were forced to cede their lands to the U.S. government and agree to move to reservations west of the Mississippi River.

One of the early pioneers to the local area was Berry Haney who moved here from Tennessee in 1838. Haney opened the first post office and helped built the first town in the area, which he named after his home town in Cross Plains, Tennessee. Haney's "new" Cross Plains became a bustling town for immigrant farmers from Germany, Norway, and other western European nations. The town also served as crossroad stop for two U.S. military roads. One road led from Prairie du Chien to Fort Howard in Green Bay and the second one connected Galena, Illinois and Fort Winnebago near Portage.

Cross Plains State Park (168 acres) was established in 1975 adjacent to other public lands owned by Dane County and the U.S. Fish and Wildlife Service. In 2002, the National Park Service purchased an historic 157-acre farmstead south of the state park. Today, all of the properties are managed jointly within the Ice Age Complex at Cross Plains.

BEACHES, PICNIC AREAS, AND SHELTERS

There is no swimming beach or a picnic area at Cross Plains State Park. **Governor Nelson State Park**, located along the northwest shoreline of Lake Mendota, has one of the finest beaches and picnic areas in the Madison area. The park is located about 17 miles west of Cross Plains along County Highway M near Waunakee. (See Chapter 39)

Prairie hiking trail

HIKING, BIKING, AND NATURE TRAILS

There are several hiking trails at Cross Plains but most do not have trail markers. Property maps indicating walking routes are available at the information kiosk near the **Wilke Farmstead** entrance.

A 2.6-mile section of the Ice Age National Scenic Hiking Trail begins at the parking lot along the Old Sauk Pass Road. The trail crosses a wooden bridge to the historic Wilke farmstead. From there it loops south and east through open fields and woodlands perched atop a massive glacial terminal moraine. On a clear day the bluish outline of the Blue Mound State Park (See Chapter 31) can be seen nearly 20 miles away in the horizon.

Ranger Trail Pick

The **Wilke Gorge Trail** (2.3 miles) is a bluff-top trail that begins at the parking area along Old Sauk Pass Road. The trail leads through a restored prairie and hardwood forest perched above a 100-foot-deep ravine known as Wilke's Gorge. Rocky outcrops of native sandstone here afford scenic vistas of the gorge below. Large round volcanic (basalt) boulders can be seen along this trail. These enormous rocks were picked up by glaciers in northern Wisconsin and Upper Michigan and transported here near the end of the Ice Age. The trail loops back to the parking area through a restored oak savanna with scenic views of the expansive valley below and Katherine Lake in the distance.

There are no off-road bicycle trails within the Cross Plains Ice Age Complex. Many bikers enjoy peddling the lightly used local township and county roads through this area.

WILDLIFE VIEWING

Cross Plain's diverse landscape of mature oak and maple woodlands, forested ravines, prairies, and oak savannas provide habitat for a wide range of wildlife species. Common animals seen here include white-tailed deer, raccoon, squirrels, rabbits, chipmunks, and wild turkey. More secretive animals such as fox and coyote also make their home here.

Many people travel here just to enjoy the park's outstanding birdwatching opportunities. More than 200 species of birds are known to either nest or migrate through this area. Forest areas attract woodpeckers, chickadees, blue jays, nuthatches, barred owls, and several species of hawks. Bluebirds, tree swallows, and gold finches can be spotted in the park's open grassland and prairie areas. Waterfowl, herons, cranes, and shorebirds can observed at **Shoveler's Sink**, a shallow 174-acre glacial lake located on U.S. Fish and Wildlife Service lands in the southeast part of the Ice Age Complex.

INTERPRETIVE PROGRAMS AND FACILITIES

The historic 1850s Wilke farmstead house, located along Old Sauk Pass Road, is being remodeled to serve as the property's visitor center by the National Park Service. When completed, the center will have an Ice Age Trail information desk, a small picnic area, public restrooms, and interpretive displays explaining the unique geology and the human history of the Cross Plains area. The National Park Service has recently opened its administrative office at this location and is open Monday through Friday. Office hours 8 a.m. to 4:30 p.m. Closed Saturday and Sunday.

BOATING AND FISHING

The **Black Earth Creek Fishery Area** located just north of Cross Plains State Park, is a nationally-recognized Class 1 cold water trout stream with good populations of rainbow, brook, and brown trout. A boat launch is available at Governor Nelson State Park for access to **Lake Mendota** (9,781 acres). Anglers can expect to catch most species of gamefish and panfish in this lake. Governor Nelson Park is located about 17 miles west of the Cross Plains.

ACCESSIBLE FACILITIES AND TRAILS

The path to the Wilke Farmstead National Park Visitor Center office and its restrooms are wheelchair-accessible. Hiking trails within the Ice Age Complex are not surfaced, making them challenging for visitors with walking disabilities. A wheelchair-accessible route to Wilkie Gorge at Cross Plains State Park is planned to be developed soon.

CAMPING

Cross Plains State Park is a day-use property, so there are no campsites here. **Mendota County Park**, located about 13 miles west of the park, has 30 well-shaded campsites with electric service and a restroom/shower facility. The park is located along County Highway M near Middleton. www.countyofdane.com/parks/mendota

WINTER ACTIVTIES

The Cross Plains Ice Age Complex is open year round. Off-season activities include cross-country skiing, snowshoeing, hiking, hunting, and wildlife observation.

AREA ATTRACTIONS

The Madison area has many scenic parks, bike trails, historical sites, restaurants, fine arts centers, a free zoo, and many other attractions. www.visitmadison.com

Indian Lake County Park (450 acres) is one of Dane County's largest and most popular parks. The park has several outstanding glacial features, including a 64-acre kettle lake and miles of hiking trails that lead through old-growth forests, oak savanna, and prairies. A small historic chapel built in 1857 is located on the crest of a hill in the park and offers a beautiful view of the lake below. Indian Lake Park is 12 miles north of the park along State Highway 19 near Cross Plains. www.countyofdane.com/parks

DIRECTIONS TO THE PARK

Cross Plains State Park is located four miles west of Madison in Dane County. From Highway 12 in Middleton, take Old Sauk Road west to Timber Lane. Turn right onto Timber Lane and then left on Old Sauk Pass Road. Parking is available at the Ice Age Trail lot along this road.

Cross Plains State Park
8075 Old Sauk Pass Road
Cross Plains, WI 53520
(608) 831-3005
(Governor Nelson State Park)

37. DEVIL'S LAKE STATE PARK

Crown Jewel of Wisconsin State Parks

Devil's Doorway

PARK SNAPSHOT

Devil's Lake may be the most visited state park in the Midwest. Each year, more than three million visitors flock to this property to marvel at the towering 500-foot-high rock bluffs and the shimmering, crystal-clear waters of Devil's Lake.

Despite the park's popularity, visitors still find lots of room to explore this 10,200-acre property. Devil's Lake offers camping, swimming, boating, biking, fishing, picnicking, nature study, rock climbing, SCUBA diving, and much more.

GEOLOGY AND HISTORY

The powerful tectonic forces that uplifted the rocky bluffs of Devil's Lake and the entire Baraboo Hills Range have lured geologists to this area for generations. The towering quartzite rock bluffs of the park were shaped by ice and erosion from powerful rivers that once thundered through this gorge over millions of years.

The rugged bluffs of Devil's Lake resisted the advance of the Wisconsin Glacier, which reached this area about 15,000 years ago. As global temperatures rose, the advance of glacial ice came to a standstill along the eastern edge of the park. As a result, much of the western section of the park remained ice-free and became part of the Driftless (non-glaciated) Area of southwestern Wisconsin. Devil's Lake itself was formed when two separate lobes of the retreating glacier deposited layers of sand, gravel, and other debris on both the north and south ends of river valley between the bluff areas. These earthen dams blocked the river on both ends, transforming it into the beautiful lake seen today.

> **Ranger Note**
>
> Quartzite is a very hard metamorphic rock that was formed 1.6 billion years ago from quartz-rich sand deposited beneath an ancient ocean that once covered this area. Over time, the intense heat and pressure of the earth's crust chemically transformed the sandstone into quartzite. The reddish-color of quartzite rock is due to the presence of iron within the stone.

Several Native American cultures lived, hunted, fished, and gathered plants in the Devil's Lake area for thousands of years. Late Woodland people known as the Effigy Mound Builders lived here between 800 to 1,400 years ago. These ancient people are best known for the earthen effigy burial mounds they built throughout southern Wisconsin. Several of these mounds are located within the park. In more recent times, the Sauk, Fox and Ho-Chunk people called this region home until 1829, when treaty concessions required tribal leaders to surrender all their land south of the Wisconsin River to the U.S. government.

Many European immigrants arrived to farm and establish the nearby community of Baraboo in the 1850s. The rocky bluffs, forests, and waters of Devil's Lake were favorite spots for local families to swim, hike, or have a picnic and over time, it became a popular summer retreat for people throughout the Midwest. Several large lakeside resorts, hotels, and dozens of cottages were built around the lake to serve summer vacationers from Madison, Milwaukee, Minneapolis, and Chicago. The Ringling Brothers of circus fame also built a summer home along Devil's Lake in the late 1800s. On occasion, they would bring elephants from their circus headquarters in nearby Baraboo to the lake for bathing.

To accommodate visitors, a railroad line was built to Devil's Lake in 1871. Hotel concessionaires offered their guests special amenities during their stay such as steamboat tours of the lake. Resort guests over the years included Mary Todd, Abraham Lincoln's wife, Ulysses S. Grant, and other national celebrities.

The State of Wisconsin purchased much of the property around the lake and opened Devil's Lake State Park to the public in 1911. The train routes to park hotels came to an end in the late 1920s due to the advent of automobile travel. At the same time, tent camping was becoming more popular and was much cheaper than staying fancy hotels, which brought about the end of the lodging at Devil's Lake.

During the Depression era of the 1930s, a C.C.C. (Civilian Conservation Corps) camp was established at Devil's Lake. More than 200 young men went to work building roads, trails, parking areas, and campgrounds in the park. They also assisted more skilled craftsmen in the construction of several stone buildings still in use today in the park.

> **Ranger Trivia**
>
> How did Devil's Lake get its name? To many Native American people, it was known as "Spirit" or "Sacred" Lake. To immigrant settlers of the 1850s, it was called "Lake of the Hills". It was renamed in the late 19th century by hotel owners and railroad promoters on a "very" loose translation of the Ho-Chunk term *Ta-wa-cun-chuk-day* or Devil's Lake.

BEACHES, PICNIC AREAS, AND SHELTERS

Devil's Lake has swimming beaches and picnic areas along both the north and south shores of the lake. Each has open-air and enclosed shelters facilities that can be reserved in advance. Private concessionaires offer prepared food, refreshments, clothing, camping supplies, and water craft rentals.

North Shore swimming beach

HIKING, BIKING, AND NATURE TRAILS

Devil's Lake has 31 miles of hiking trails to explore including a 14-mile segment of the **Ice Age National Scenic Trail**. Some trail segments can be challenging, especially those that require climbing steep stone staircases to the top of bluff areas.

The **West Bluff Trail** (1.4 miles) ascends to the crest of the west bluff. The trail offers outstanding scenic views of the east bluff across the lake and has several interesting rock formations, including **Cleopatra's Needle.**

The **Tumbled Rocks Trail** (1.0 miles) is one of the most popular hikes in the park. This nearly level asphalt-paved trail leads from the north shore picnic area between large quartzite boulders along the entire western shore of Devil's Lake to the south shore boat launch area.

The **Steinke Basin**, located east of the lake off of County Highway DL, has several trail loops that lead through open meadows and hardwood forests. Off-road biking is allowed on the **Upland Loop Trail** (3.8 miles) in this area. Many visitors also enjoy biking the two-mile asphalt-paved trail that leads from the park's north entrance to the nearby city of Baraboo.

Ranger Trail Pick

The **East Bluff Trail** (1.7 miles) is one of the best routes to see many of the park's unique rock formations and enjoy panoramic views Devils' Lake. The easiest route to the top of this bluff begins at the trailhead located within the north shore picnic area. Near the top of the bluff are the **Elephant Rock** and **Elephant Cave** rock formations. A short side trail leads to **Devil's Doorway**, the park's most iconic and photographed rock formation. For those who want to extend this hike into a 4.7 mile-loop hike around the lake; take either the **Balanced Rock Trail** (0.4 miles) or the **Pothole Trail** (0.3 miles) down the **East Bluff** to the **Grottos Trail** (0.7 miles) below. Walk east through the south shore picnic area and boardwalk towards the boat launch area and then take either **Tumbled Rocks Trail** shoreline route or the **West Bluff Trail** back to the north end of the lake.

WILDLIFE VIEWING

Devil's Lake's diverse landscape of bluffs, upland forests, wetlands, streams, and grasslands provides habitat for a wide array of wildlife. The park is home to over 100 species of nesting birds, 880 kinds of plants, and many types of mammals, amphibians, reptiles, and birds. White-tailed deer, squirrels, raccoons, fox, coyote, chipmunks, blue jays, crows, ravens, turkey vultures, woodland songbirds, and hawks and owls are often seen in the park. Occasionally, a bobcat, black bear, and (rarely) a timber rattlesnake may be spotted in remote areas of the park.

The **Baxter's Hollow State Natural Area** (5,914 acres) is a great place to experience the flora and fauna of the region. A somewhat challenging 2.5-mile one-way trail here follows **Otter Creek**, a crystal-clear stream that runs through a scenic, forested, rocky gorge. Baxter's Hollow is owned by the Nature Conservancy and is located west of Devil's Lake off Highway 12 and County Road C. www.nature.org.wisconsin. (262) 642-7276

INTERPRETIVE PROGRAMS AND FACILITIES

The park's interpretive center has live animal displays, historical artifacts, geology dioramas, and hands-on activities for children. Nature hikes and evening programs are presented by park naturalists throughout the year. The park has two self-guided nature trails and several interpretive panels that highlight the animal-shaped burial mounds built the Effigy Mound Builder Culture more than 600 years ago.

The **Parfrey's Glen State Natural Area** offers one of the most beautiful hikes in the park. A short interpretive trail (0.7 miles) follows a cold clear-running stream through a deep gorge lined with towering sandstone walls imbedded with quartzite stones and rare plant species. Parfrey's Glen is located in the far southeast corner of the park off County Road DL about two miles east of Highway 113.

BOATING AND FISHING

Devil's Lake (374 acres) has a maximum depth of 47 feet. Anglers will find good populations of northern pike, bass, panfish, brown trout, and walleye. Boat launch facilities (electric motors only) are located along both the north and south shoreline of the lake. SCUBA diving is very popular at Devil's Lake throughout the warm season.

ACCESSIBLE FACILITIES AND TRAILS

The park's entrance stations, nature center, shelters, concessions, and most restrooms are accessible for mobility-impaired visitors. Wheelchair-accessible campsites are available in each campground with access to restroom/shower facilities. Some sections of the Tumbled Rocks, Grottos trails and the Parfrey's Glen Trail are wheelchair-accessible with assistance. Due to the steep and rocky topography of the park, most other hiking trails may be challenging for visitors with walking difficulties.

37. Devil's Lake State Park

CAMPING

Devil's Lake has 423 campsites (154 electric) located within the **Quartzite, Northern Lights**, and the **Ice Age** campgrounds. Electric service is available at most campsites within the Quartzite and Northern Lights campgrounds but not in the Ice Age Campground. All camping areas have restrooms, showers, and a RV sanitary dump station.

A group campground is located in the south side of the park along South Lake Road. The campground has nine campsites with a total capacity of 240 campers.

WINTER ACTIVITIES

Hiking, birdwatching, snowshoeing, ice fishing, and winter camping are popular off-season activities in the park. Cross-country skiing is available along six miles of multi-use trails at **Steinke Basin** off of County Highway DL. None are groomed.

AREA ATTRACTIONS

The nearby city of Baraboo is a popular tourist destination and offers many historical sites, museums, fine restaurants, theaters, and other attractions. www.baraboo.com/visit-baraboo

The **Circus World Museum** houses thousands of rare circus-related artifacts, including the largest collection of hand-crafted 19th- and 20th-century circus parade wagons in the world. Live circus performances under a canvas big top tent are offered daily during the summer season. This unique 64-acre National Historic Landmark site is operated by the State Historical Society. The museum is located along Water Street in Baraboo. www.circusworldbaraboo.org

DIRECTIONS TO THE PARK

Devil's Lake is located between Highway 12 and Highway 113 in Sauk County. The park's main entrance is on Park Road at the end of Highway 123 about four miles south of Baraboo.

Devil's Lake State Park
S. 5975 Park Road
Baraboo, WI 53913
(608) 356-8301

38. Governor Dodge State Park

Outdoor Recreation Paradise

Cox Hollow Lake

PARK SNAPSHOT

Governor Dodge is the second-largest state park is Wisconsin. This popular 5,300-acre park is known for its scenic forested bluffs, rolling hills, expansive prairies, and picturesque lakes.

The park offers camping, swimming, picnic areas, nature study, boating, canoeing, kayaking, bicycling, hunting, and fishing and has miles of hiking and equestrian trails.

GEOLOGY AND HISTORY

Governor Dodge is located within the Driftless Area of southwestern Wisconsin. This region of the state remained ice-free during the last Ice Age and lacks the glacial "drift" (sand, gravel, and rocks) deposited by glaciers throughout most of the rest of the state 12,000 years ago. As a result, the hills, valleys, sandstone canyons and bluffs here were shaped by wind, frost, and water erosion over the course of millions of years.

The sandstone rock formations in the park were formed from sand that was deposited beneath a warm, shallow sea that once covered this area more than 470 million years ago. Some of this sandstone was softer and more easily eroded than the rock above it, which led to the creation of stone ledges, overhangs, and small caves found in some bluff areas of the park. Archeological excavations at these sites have revealed that many of these rock shelters were used by early nomadic hunters who roamed this area more than 8,000 years ago.

The Sauk, Fox, and Ho-Chunk people lived, hunted, and gathered plants in this area for generations before the first American miners entered the region in search of lead ore in the 1820s. One of them was General Henry Dodge, who opened the first commercial lead mine just south of the park in what is now the city of Dodgeville. As more miners arrived in the area, conflicts with local indigenous people over land ownership arose. Fortunately, Dodge had prior experience in negotiating similar conflicts with native people during his military career. He was able to bring peace to the local area without an outbreak of war for many years until the Blackhawk War of 1832 which led to the removal of all Native American people from southern Wisconsin.

> **Ranger Trivia**
>
> Governor Dodge State Park and the city of Dodgeville were named in tribute to Henry Dodge, who moved here with his wife and children from Missouri in 1827 to mine lead. Dodge served as a commander of the local militia during the Winnebago (Red Bird) war of 1827 and the Blackhawk War of 1832. General Dodge was appointed Governor of the Territory of Wisconsin in 1836 and later became a U.S. Senator when Wisconsin became a state in 1848.

Many European immigrants who settled in the Dodgeville area in the mid-1800s were farmers who found the rich soil of the prairies easy to plow and raise crops. Even today, agriculture continues to be main occupation for many local families. In 1948, Iowa County donated 160 acres of public land north of Dodgeville to the State of Wisconsin to develop a new state park. The State purchased thousands of additional acres in the area and built earthen dams to create **Twin Valley** and **Cox Hollow** lakes. Governor Dodge State Park was opened to the public in 1955.

> **Ranger Note**
>
> The **Deer Cove Picnic Area**, located near **Cox Hollow Lake**, has an interpretive panel highlighting the archeological findings in the park. A short trail from here leads to the base of an ancient sandstone bluff and rock ledge shelters.

BEACHES, PICNIC AREAS, AND SHELTERS

There are several picnic areas in the park. The most popular are adjacent to the swimming beaches at **Twin Valley Lake** and **Cox Hollow Lake**. Both areas have a bathhouse and restrooms. A concession stand at the Cox Hollow is open during the summer months and offers snacks, refreshments, watercraft rentals, and camping supplies. Open-air shelter facilities are located near both swimming beaches and at the **Trails End** and **Enee Point** picnic areas. All shelters have electric service and can accommodate between 30 to 75 people.

HIKING, BIKING, AND NATURE TRAILS

Governor Dodge has 30 miles of hiking trails to explore. The **Meadow Valley Trail (5.0 miles)** is the longest hiking trail in the park. It leads through several forests and restored prairie areas. **Cave Trail (1.0 miles)** loops off the Meadow Valley Trail and offers scenic views of Twin Valley Lake. A rustic side trail leads to colorful sandstone bluffs and **Thomas' Cave**. The cave is closed to public access to protect the roosting bats inside.

Pine Cliff Trail (4.6 miles) begins at the Enee Point Picnic Area, where visitors can view the park's iconic Enee Point, a dramatic 80-foot sandstone bluff. Pine Cliff Trail leads uphill and skirts the south side of Cox Hollow Lake with beautiful views of the valley below. A two-mile, self-guided nature trail along this route highlights the history, flora, and fauna of the area.

The **Meadow Valley Trail** and **Mill Creek Trail** offer eight miles of challenging off-road biking through the park. A paved bike trail from the park provides access to the Military Ridge State Trail (40-miles), a popular limestone-surfaced trail that travels east to city of Madison.

Stephens Falls

Governor Dodge has 22 miles of marked equestrian trails. Horseback riders can camp overnight at the **Trails End Horse Camp**, which has 11 non-electric campsites.

Ranger Trail Pick

The **Stephens Falls Trail** (0.5 miles) is a natural wonderland of waterfalls, rock gorges, and cold-water streams and springs. The trail begins just north of the park's entrance along a paved walkway that leads to an overlook above the 30-foot Stephens Falls. Rock steps and hand railings allows access to the base of the waterfall. From here, the trail follows the creek through a picturesque sandstone canyon filled with massive rocks and lush ferns. An uphill path near the end of the creek trail joins the **Lost Canyon Trail** (3.0 miles), a mostly upland looped trail that leads through forests and open fields back to the start of the trail.

WILDLIFE VIEWING

The park's diverse landscape of oak-hickory woodlands, wetlands, creeks, lakes, prairies, and sandstone bluffs provide ideal habitat for wildlife. The park is an oasis for many bird species, including red-tailed hawks, turkey vultures, bald eagles, osprey, owls, pileated woodpeckers, wild turkey, and many kinds of woodland and grassland songbirds. Bird watchers have recorded more than 150 species of birds in the park.

Mammals such as red and gray fox, coyote, beaver, woodchuck, squirrels, rabbits, and white-tailed deer are commonly seen in the park. The park also has hundreds of species of wild plants, including several types of rare ferns that only grow on dry sandstone bluffs or cool, spring-fed canyon areas.

INTERPRETIVE PROGRAMS AND FACILITIES

Wildlife and geology interpretive displays can be viewed inside the visitor entrance station. Nature hikes are scheduled throughout the summer season. Some evening programs are held at the park's amphitheater adjacent to the Cox Hollow picnic area.

The **Spring House Tour** is a self-guided hike that leads to several 19th-century spring house sites. The tour begins at the historic spring house located along the **Stephens Falls** hiking trail. Interpretive signs here explain how important these natural "pioneer refrigerators" were to local farm families.

BOATING AND FISHING

Cox Hollow Lake (81 acres) and **Twin Valley Lake** (136 acres) each have boat launch areas (electric motors only). A fishing pier is located near the Cox Hollow launch area. Anglers can expect to catch panfish, large-mouth bass, trout, walleye, and muskie in both lakes. Row boats, canoes, kayaks, and paddle boats can be rented at the **Cox Hollow Concession.**

ACCESSIBLE FACILITIES AND TRAILS

The visitor entrance station, picnic areas, shelter facilities, and most restrooms are accessible for mobility-impaired visitors. Cox Hollow and the Twin Valley campgrounds have wheelchair-accessible campsites with electrical service and access to nearby restrooms and shower facilities. An asphalt-paved accessible walkway leads from the Stephens Falls parking area to a shaded viewing area above the waterfall. Due to the hilly topography of the park, most hiking trails are not wheelchair-accessible.

CAMPING

Governor Dodge has **269** campsites (**80** electric) within Cox Hollow and Twin Valley campgrounds. Both campgrounds are located in upland forest areas high above the lakes. Each has restrooms, shower facilities, and an RV water fill and sanitary station.

The **Hickory Ridge Group Campground**, located in the far northern section of the property, has eight campsites that can accommodate between **15** to **40** people per site. The group camp has vault-type restrooms and a shower building. There are also six backpack (walk-in) campsites located about a half mile from the group camp.

> **Ranger Tip**
>
> Camping is very popular at Governor Dodge. Campgrounds often fill to capacity most days in summer and on weekends in late spring and early fall. Advance reservations are highly recommended.

> **Ranger Tip**
>
> Stephens Falls often freezes into a magical ice sculpture in winter. Water that seeps through the sandstone walls of the gorge also creates intricate icicles and fanciful frozen sculptures as well.

WINTER ACTIVITIES

Park staff groom 12.5 miles of cross-country ski trails. Snowshoeing, sledding, wildlife viewing, and hiking are also popular winter activities. Park roads are plowed from the entrance station to both Twin Valley and Cox Hollow lakes for ice fishing access for anglers. Winter camping is available at the Twin Valley Campground.

AREA ATTRACTIONS

The Dodgeville area has several fine museums, parks, restaurants, wineries, shops, and other attractions to explore. www.dodgeville.com/visit/attractions

The **House on the Rock** is one of the most popular tourist attractions in Wisconsin. This one-of-kind house was built on the very top of a 60-foot-tall sandstone pinnacle in the early 1950s by Alex Jordan Jr., an eccentric architect and antiquities collector. This site was opened to the public in 1959 and has since expanded into a world-class museum. The museum complex houses an enormous collection of exotic and unusual artifacts including the world's largest indoor carousel and dozens of mechanical music machines. A long glass-enclosed infinity room extends over the edge of a sandstone bluff. The House on the Rock is located six miles north of the park along Highway 23 near Spring Green. www.thehouseontherock.com

DIRECTIONS TO THE PARK

Governor Dodge is located in Iowa County about 47 miles west of Madison. From Highway 151, take the Highway 18 exit into the city of Dodgeville. Turn north onto Highway 23 and travel three miles to the park entrance.

Governor Dodge State Park
4175 Highway 23 N
Dodgeville, WI 53533
(608) 935-2315

39. GOVERNOR NELSON STATE PARK
Capital City's Best Kept Secret

Lake Mendota shoreline

PARK SNAPSHOT

Governor Nelson State Park may be small in size but this scenic 422-acre gem has much to offer. The park is located along shoreline of Lake Mendota and has one of the finest swimming beaches in the Madison area. This day-use property also has hiking trails and picnic areas and offers boating, birding, fishing, nature study, and many other outdoor opportunities.

Governor Nelson is located just north of Madison, the second largest city in Wisconsin. Despite being so close to a metro area, the park is rarely crowded and provides a tranquil green oasis for urban visitors. On a clear day, a stunning view of the state capitol dome can be seen across the shimmering waters of the lake.

GEOLOGY AND HISTORY

The park area and Lake Mendota are all part of Madison's iconic Four Lakes Region. Geologists believe that prior to Wisconsin's last Ice Age about 30,000 years ago, this region was a wide river valley more than 400 feet deep. As glaciers pushed through the Madison area, the river valley was filled with deposits of sand, gravel, rocks, and other glacial debris with the exception of four large depressions in the earth. As the glaciers began to retreat north again about 18,000 years ago, they created a moraine dam, which blocked the flow of the nearby Yahara River. The backed-up water submerged most of the Madison area beneath a massive glacial lake. Eventually, the river breached the dam, releasing a torrent of glacial melt water downstream, excavating a new channel for the Yahara River. When all the melt water finally subsided, the four water-filled depressions became Madison's Lakes Mendota, Waubesa, Menona, and Kegonsa. The landscape of park was also shaped by the glacial activity. Disintegrating glacial ice formed outwash plains, wetlands, kettle depressions, and moraine hills throughout the property.

> **Ranger Trivia**
>
> Lake Mendota was named after a Sioux term meaning "Mouth of the Rivers," referring to the Yarhara River, which connects all four Madison lakes. The Yarhara River itself was originally called the Gahara River, a Ho-Chunk word for "catfish."

Several different Native American cultures have lived in the Madison area for thousands of years. The Yahara River and each of the local lakes, including Lake Mendota, provided a vast source of edible and medicinal plants, game animals, fish, and fur.

When early European explorers arrived in the Madison area, they encountered the Sauk, Fox, and Ho-Chunk people who lived throughout what the local inhabitants referred to as *Taychopera*, "Land of Four Lakes." In the early 19th century, many lead miners and settlers flocked to southern Wisconsin, which inevitably led to conflicts with native people over land ownership. Native American raids against settlers resulted in Winnebago War of 1827 and the Black Hawk War of 1832. Treaty settlements after the wars forced tribal leaders to cede their lands to the U.S. government.

> **Ranger Trivia**
>
> The park was named as a tribute to Gaylord Nelson, who served as Wisconsin's 35th governor and was a U.S. senator for 20 years. Nelson was a strong advocate in protecting the environment and helped pass many laws to reverse decades of air and water pollution. He was also credited in establishing **Earth Day**, a world-wide annual event that highlights local, national, and worldwide environmental concerns.

In 1836, the U.S. Congress created the Wisconsin Territory. The first legislative session was held in tiny community of Belmont in La Fayette County. (See Chapter 28) Representatives at that meeting voted to move the territorial capital to Madison in Dane County. Soon after, a flood of immigrants from Germany, Italy, Russia, and Norway arrived to settle in the Four Lakes Region, which transformed Madison into a thriving city.

In 1905, a summer retreat for boys called Camp Indianola was established along the shoreline of Lake Mendota. Over its 62 years of existence, the camp had several notable campers, including a young Orson Wells. When Camp Indianola closed in 1967, the property was purchased by the State to develop **Governor Nelson State Park**, which opened in 1975.

BEACHES, PICNIC AREAS, AND SHELTERS

Governor Nelson has a 500-foot-wide swimming beach and a separate pet beach area. Adjacent to the beach is an attractive multi-purpose building, which has restrooms, changing stalls, and a shelter area with picnic tables.

The park has a large, well-shaded picnic area with two open-air shelters. The **Sandy Beach Shelter** adjacent to the beach can hold up to 30 people. The nearby **Dragonfly Shelter** has a capacity of 70 people. Both have electrical service and can be reserved in advance.

Swimming beach

HIKING, BIKING, AND NATURE TRAILS

There are eight miles of hiking trails to explore in the park. The **Morningside Trail** (2.4 miles) travels through restored prairies and wetland areas and leads to the shoreline of Lake Mendota. The north section of this trail crosses a small bridge and loops through a prairie area adjacent to **Six Mile Creek**.

The **Oak Savanna Trail** (1.8 miles) leads through a restored prairie and oak savanna area before ascending a hill with a scenic overlook of Lake Mendota. The trail connects to the **Redtail Hawk Trail** (1.0 mile) as well.

Ranger Trail Pick

The **Woodland Trail** (1.2 miles) is self-guided nature trail with numbered marker posts that correspond to a trail guide. The trail begins west of the boat launch area and leads through a restored prairie, an oak savanna, and a collection of rare prehistoric effigy burial mounds.

There are no bike trails in the park, but the nearby **Middleton Bike Park** has several off-road bike trails that range from an easy 1.4-mile route to more challenging trails with jumps. Middleton Bike Park is located along Pleasant View Road about eight miles southwest of Governor Nelson. www.madcitiydirt.com

WILDLIFE VIEWING

The park's diverse landscape of prairies, oak savannas, hardwood forests, wetlands, creeks, and the Lake Mendota shoreline provide habitat for white-tailed deer, red fox, squirrels, chipmunks, hawks, wild turkeys, woodland songbirds, and many other wildlife species. More nocturnal wildlife such as skunks, raccoon, coyotes, and owls also make their homes here.

Waterfowl, cranes, herons, and other marshland birds are often spotted in wetland areas and along the shoreline of Lake Mendota. The park's restored prairies and open fields attract grassland birds such as bobolinks, bluebirds, and tree swallows.

The **North Mendota Wildlife and Prairie Area**, located across the road from Governor Nelson along Highway M, is 63-acre Dane County wildlife preserve. The property has two miles of hiking trails, which lead through open prairies and woodland areas.

INTERPRETIVE PROGRAMS AND FACILITIES

The park's interpretive panels and self-guided nature trails describe the human history, geology, flora, and fauna of the area. A short path off the **Woodland Trail** loops around a group of ancient effigy burial mounds. Archeologists believe these earthen structures were built by early indigenous people known as Mound Builders between 500-1000 A.D. Some of the mounds are conical or oval, while others are in the shape of real or mythical animals.

BOATING AND FISHING

Governor Nelson State Park has one of the best boat launch and trailer parking areas on Lake Mendota. Restrooms and a fish cleaning station are available here as well.

Lake Mendota (9,781 acres) is the largest and the deepest (83 feet) of Madison's four lakes. The lake has good populations of panfish, northern pike, bass, walleye, perch, and catfish, plus a few musky and sturgeon. Boaters can access the adjacent Lake Monona and the other two lakes via the **Tenney Lock and Dam** located along the southeast shoreline of the lake.

ACCESSIBLE FACILITIES AND TRAILS

The park's entrance station, restrooms, and shelter facilities are accessible for mobility-impaired visitors. Asphalt-paved walkways provide wheelchair access to the picnic and beach area along Lake Mendota. A wheelchair-accessible fishing pier is adjacent to the park's boat launch area. Some of the park's hiking trails are fairly steep or have mowed grass tread, which may make them challenging for some visitors.

CAMPING

Governor Nelson is a day-use park so there are no camping facilities here. **Mendota County Park** has 30 wooded campsites with electrical service, restrooms, and showers. The park is located five miles west of Governor Nelson along County Highway M near Middleton. www.countyofdane.com

Lake Kegonsa State Park is located about 20 miles southeast of Governor Nelson off Door Creek Road near Stoughton. The park is situated along the shoreline of Lake Kegonsa and has 96 campsites (29 electric) with restrooms and shower facilities. (See Chapter 40)

WINTER ACTIVITES

Park staff groom six miles of cross-country ski trails for both diagonal and skate skiing. Hiking, snowshoeing, and wildlife watching are also favorite winter activities. Many anglers enjoy ice fishing on Lake Mendota. The park's boat launch provides access to the lake throughout the winter season.

AREA ATTRACTIONS

Madison is one of the state's most popular tourist destinations with many parks, museums, bike trails, restaurants, theaters, a zoo, and other entertainment venues. www.visitmadison.com

39. Governor Nelson State Park

The **University of Wisconsin—Madison Arboretum** is a 1,200-acre nature conservancy located along **Lake Wingra** in Madison. The property showcases hundreds of unique flowers, shrubs, and trees from around the world. The arboretum also has plantings of every ecological plant communities found in Wisconsin, including woodlands, savannas, wetlands, and prairies. The site has a visitor center, 17 miles of hiking trails, natural springs, and ancient effigy burial mounds. The U.W. Arboretum is located about 11 miles south of Governor Nelson off Seminole Highway. www.arboretum.wisc.edu

DIRECTIONS TO THE PARK
Governor Nelson State Park is located north of the city of Madison in Dane County. From either State Highway 113 or Highway 12, exit onto County Highway M to the park entrance.

Governor Nelson State Park
5140 County Highway M
Waunakee, WI 53597
(608) 831-3005

40. LAKE KEGONSA STATE PARK

Shimmering Jewel of Madison's Lakes Region

Lake Kegonsa

PARK SNAPSHOT

Lake Kegonsa State Park is a 342-acre green oasis situated along the shoreline of Lake Kegonsa just southeast of Madison. The property offers camping, hiking, swimming, fishing, boating, nature study, and many other outdoor recreation opportunities. Visitors can also enjoy outstanding sunset views over the shimmering waters of Lake Kegonsa.

GEOLOGY AND HISTORY

The landscape of the park area was shaped by glaciers during Wisconsin's last Ice Age, which began about 30,000 years ago. Geologists have found evidence that Lake Kegonsa was once part of a wide river valley that filled with sand, gravel, and other glacial till by retreating glaciers about 17,000 years ago. Melting glaciers also created an earthen dam just south of Lake Kegonsa, which blocked the flow of the river in this valley. As a result, much of Madison area was submerged beneath a massive glacial lake. Eventually the dam broke apart draining the most of the glacial lake except for four large, deep depressions that would become

40. Lake Kegonsa State Park

Madison's iconic Four Lakes Region: Mendota, Monona, Waubesa, and Lake Kegonsa.

Several Native American cultures have hunted, fished, and gathered wild plants within Madison's chain of lakes, which are connected by the Yahara River, for more than 10,000 years.

The first white settlers arrived in the Madison area in the early 1800s to mine lead and clear land for farmsteads. The Sauk, Fox, and Ho-Chunk people all claimed ownership of much of southern Wisconsin at that time, so ultimately conflicts over mining rights and land ownership arose. This led to the Winnebago War of 1827, after which defeated Native American leaders were forced to cede all their lands south of the Wisconsin River to the U.S. government. In that same year, Judge James Doty, a territorial official and land speculator, purchased a thousand acres of uninhabited land between Lake Mendota and Lake Monona. Doty plotted out streets and building lots for his new town, which he named Madison after President James Madison, who had died earlier that year. In 1836, during the first legislative meeting of the Wisconsin Territory at Belmont in La Fayette County, Doty convinced representatives to move the capitol to Madison in Dane County.

> **Ranger Trivia**
>
> The Yahara River was originally called *Maa'li Yahara*, a Ho-Chunk phase meaning "River of Catfish." The Madison area itself was known by many indigenous people as *Tay-chopera* or "Land of Four Lakes."

Lake Kegonsa was originally built as a roadside rest stop by the State Highway Commission in the early 1960s. The adoption of the Outdoor Recreation Act required that three roadside parks be developed near interstate highways under construction at that time. The Lake Kegonsa site was chosen because of its close proximity to interstate I-90 and I-94 near Madison. Kegonsa Roadside Park was opened as scheduled but almost immediately became a popular recreation area instead. Within a few years, the property was transferred to the Wisconsin Conservation Commission to develop Lake Kegonsa State Park, which opened in 1966.

> **Ranger Trivia**
>
> Why was Madison chosen to be the capitol of the Wisconsin Territory? At the time, it was considered centrally located between Milwaukee, Prairie du Chien, Fort Howard in Green Bay and the important lead mining region of the southwest. Of course Judge Doty's political influence and business connections with other territorial legislators most likely played a role in the final decision.

BEACHES, PICNIC AREAS, AND SHELTERS

The park has a small swimming beach along Lake Kegonsa along with changing stalls, a vault-type restroom, and a playground. The **Boat Launch Picnic Area** and the **Upper Park Picnic Area** have volleyball courts, horseshoe pits, and open-air shelters that can be reserved in advance. The **Williams Knoll Picnic Area** is small secluded picnic spot located on a wooded hilltop across from the **Oak Knoll Trail** parking area.

HIKING, BIKING, AND NATURE TRAILS

There are six miles of hiking trails in the park. The **Oak Knoll Trail** (0.8 miles) leads through a variety of ecosystems including oak forests, restored prairies, and an uncommon wetland called a fen. A side trail crosses over a bridge with scenic views of park's open-water wetland area and leads to the shoreline of Lake Kegonsa.

The **Prairie Trail** (1.3 miles) is surfaced with crushed limestone and loops through a restored grassland area. This trail is especially inviting in early summer when wild sunflowers, compass plants, purple coneflowers, and other prairie plants are in full bloom.

There are no bike routes at Lake Kegonsa but nearby Madison has several excellent biking/hiking trails, including the **Capital City State Trail** (12 miles) and the nearby **Lower Yahara River Trail** (2.5 miles), which begins at **Capital**

Bridge along Oak Knoll Trail

Springs State Recreation Area (See Chapter 34). The **Yahara River Trail** crosses a mile-long wooden bridge and leads to the village of McFarland just north of Lake Kegonsa State Park.

Ranger Trail Pick

The **White Oak Nature Trail** (1.2 miles) is a self-guided nature trail with interpretive stops that highlight the flora, fauna, and human history of the park. The trail leads through an 80-acre old-growth hardwood forest. The massive white oak and cherry trees here give an impression of how the original forests may have appeared to early pioneers. The trail winds through a pine plantation and skirts a grouping of rare burial mounds built by Woodland Effigy Mound builders between 500-1000 A.D.

WILDLIFE VIEWING

The park's forests, prairies, wetlands, and lakeshore attract a wide array of wildlife. White-tailed deer, chipmunks, gray and fox squirrels, raccoons, opossum, muskrats, mink, and even an occasional otter can be seen here.

More than 150 different species of birds either nest or migrate through the park. Upland woodlands are home to many types of songbirds, woodpeckers, blue jays, cardinals, hawks, and owls. Horned larks, bobolinks, bluebirds, tree sparrows, kingbirds, and several types of sparrows can be found in the park's open fields and prairies.

The **Wetland Boardwalk Trail** (0.1 miles) is a short, raised boardwalk and viewing platform that overlooks the park's marshland area. Ducks, geese, cranes, herons, bitterns, and many other wetland birds can be seen here. Several kinds of turtles also live here, including uncommon species such as the Blanding's turtle and the spiny softshell turtle.

INTERPRETIVE PROGRAMS AND FACILITIES

Nature hikes, programs, and special events are offered from time to time throughout the summer season by park volunteer naturalists and guest speakers.

BOATING AND FISHING

Lake Kegonsa (3,200 acres) is large in size but relatively shallow with a maximum depth of 17 feet. Anglers will find good populations of

Ranger Trivia

The lake and park's namesake, *Kegonsa*, is a Ho-Chunk term meaning, "Lake of Many Fishes." The name is certainly an accurate description of this very productive fishing lake.

panfish, largemouth bass, northern pike, walleye, and a few smallmouth bass, catfish, and musky.

The park has three lakeside floating fishing piers and a paved boat launch located in the southeast corner of the park. Outboard motors are allowed on the Lake Kegonsa but boaters need to heed slow-no-wake areas. A canoe/kayak storage rack is located near the boat landing area for overnight campers.

ACCESSIBLE FACILITIES AND TRAILS

The park's visitor entrance station, restrooms, and shower facilities are accessible for mobility impaired visitors. Campsite 56 is a wheelchair-accessible site with electrical service and access to a nearby restroom/shower building.

The **Lakeshore Trail** (0.5 miles) is an asphalt- paved walkway that starts at the beach parking area and follows the shoreline of Lake Kegonsa to the boat launch area. Boardwalk side trails off this trail provide access to the beach and a wheelchair-accessible fishing pier. The **Prairie Trail** (1.3 miles) is a mostly level hiking trail through an attractive restored prairie area. The path was surfaced with crushed limestone but some wheelchair users may still need assistance to navigate this trail.

CAMPING

Lake Kegonsa has 102 campsites (34 electric) located within a forested area of the park. A restroom/shower facility is located near the campground entrance. An RV sanitary and water fill station is located south of the campground along the road to the beach and picnic areas. The park's group campground has six campsites (tent-only) with vault-type restrooms, water fountains, and an open-air shelter building. Each site has a fire ring and picnic tables and can accommodate up to 20 people. The park's campgrounds close at the end of October.

WINTER ACTIVITIES

Park staff groom five miles of cross-country ski trails for both inline and skate skiing. A 1.2- mile trail is reserved for snowshoeing and winter hiking. Ice fishing on Lake Kegonsa is popular. Anglers can expect to catch bluegills, crappie, perch, sunfish, and occasionally northern pike and walleye. There are no vehicle, ATV, or snowmobile access routes onto the lake from the park but anglers can walk or ski to ice fishing spots.

AREA ATTRACTIONS

Madison is one of the top tourist destinations in Wisconsin with many parks, bike trails, museums, cultural centers and unique downtown restaurants and shops to explore. www.visitmadison.com (800) 373-6376

The nearby **Olbrich Botanical Gardens** (16-acres) is a city of Madison nature preserve that features thousands of Midwest-hardy trees, bushes, and flowers. The site has several different themed gardens, walkways, streams, and bridges to explore. A year-round, glass-enclosed conservancy building houses exotic plants, birds, and hundreds of flowering plants. Olbrich Gardens is located on the north end of Lake Monona about ten miles northwest of Lake Kegonsa along Atwood Avenue in Madison. www.olbrich.org

DIRECTIONS TO THE PARK

Lake Kegonsa State Park is located on the northeast shoreline of the lake about 14 miles southeast of Madison. From Interstate I-90, exit onto County Highway N south towards Stoughton. Turn right (west) on Koshkonong Road and then left (south) on Door Creek Road to the park entrance.

Lake Kegonsa State Park
2405 Door County Creek Road
Stoughton, WI 53589
(608) 873-9695

41. LOWER WISCONSIN STATE RIVERWAY

Wisconsin's Longest Free-Flowing River

Lower Wisconsin River

PROPERTY SNAPSHOT

The Lower Wisconsin State Riverway is one of the largest and most diverse properties in the state. This expansive, 79,275-acre property stretches through seven counties for 91 miles, the longest in the state. The free-flowing Lower Wisconsin River flows through the Riverway unhindered by dams or other man-made structures to block its route.

The beauty of the Lower Wisconsin River is enhanced by isolated sandbars, wooded islands, lowland river forests, and towering bluffs that together create a near-wilderness experience for visitors. The Riverway is a recreational paradise for canoeing, kayaking, fishing, camping, hiking, hunting, horseback riding, and wildlife watching.

GEOLOGY AND HISTORY

The Lower Wisconsin Riverway is located within the Driftless Area of southwestern Wisconsin; a region of the state that remained untouched by glaciers during the last Ice Age. Although not directly altered by glaciers, the river valley and bluffs of the Lower Wisconsin River were shaped by their presence nonetheless.

41. Lower Wisconsin State Riverway

Near the end of the Ice Age about 14,000 years ago, a massive rush of meltwater from glacial Lake Wisconsin to the north thundered down the Wisconsin River Valley carrying with it large amounts of gravel, silt, sand, and other glacial debris. This immense flood of water carved out the sandstone formations in the Wisconsin Dells area before continued south through Sauk and Columbia counties, where it collided with a terminal moraine hill created by retreating glaciers. This massive moraine re-routed the river to the southwest, cutting the wide Lower Wisconsin River Valley before emptying into the Mississippi River. The outwash plains, towering hills, and water-sculptured sandstone bluffs of the Lower Wisconsin River Valley are a testament to the power of this catastrophic Ice Age event. Today, the flow of the Lower Wisconsin River is relatively mild and controlled by the hydroelectric dam at Prairie du Sac.

Several Native American cultures have lived within the Lower Wisconsin River Valley for generations including the Menominee, Ojibwe, Potawatomi, and Ho-Chunk.

In 1653, French explorers Father Jacques Marquette and Louis Jolliet became the first Europeans to canoe down the Lower Wisconsin River from Portage on their quest to reach the Mississippi River.

In the early 1820s, lead miners and homesteaders from the East Coast entered the Illinois Territory, which included all of southern Wisconsin. These newcomers were welcomed at first by indigenous people for the trade items they brought with them, but within a few years, clashes over mining rights and land ownership arose. This led to the Winnebago War of 1827, after which the defeated Ho-Chunk and Sauk people were forced to give up all their lands south of the Wisconsin River to the U.S. government. Native Americans were also required to move onto reservation land west of the Mississippi River. One of those forced to cross the river and leave his home village along the Rock River was an influential Sauk leader called Black Hawk. In 1832, Black Hawk, along with 1,600 of his followers broke the treaty agreement by re-crossing the Mississippi River in an attempt reclaim their former homeland in Illinois. This led to the Black Hawk War of 1832, a short but tragic war with the U.S. Army and volunteer militia from both Illinois and Wisconsin.

> **Ranger Trivia**
>
> Father Marquette referred to the Wisconsin River in his 1674 journal as the *Meskousing*, a Miami term meaning "river running through a red place." Historians speculate that this word was used to describe the section of the river that flows through the reddish-colored sandstone bluffs of the Wisconsin Dells. Later, another French explorer renamed it *Quisconsing*, or "Gathering of Waters." The U.S. Congress officially changed the river's name to Wisconsin in 1830.

Black Hawk and his followers retreat through Wisconsin came to a tragic end during the Battle (Massacre) of the Bad Axe River in Vernon County. Although Black Hawk himself escaped, only 150 of his 1,600 followers survived the bloody conflict as they attempted to flee back across the Mississippi River.

During the 1840s, lumber companies cut most of the timber within the Lower Wisconsin River region before moving on to northern forests. European immigrant farmers transformed some of the cut-over forest lands and most of rich prairie soils into cropland. The wetlands, swampy lowland forests, sand bars, and rocky bluff areas along the Wisconsin River remained relatively untouched well into the 20th century, however. In an attempt to preserve this unique wilderness river corridor, the Wisconsin Department of Natural Resources, along with county officials and environmental groups, worked together to establish the Lower Wisconsin State Riverway in 1989.

BEACHES, PICNIC AREAS, AND SHELTERS

Picnic areas can be found at both **Tower Hill** (See Chapter 53) and **Wyalusing State Parks** (See Chapter 55), at local community parks and the **Sauk County Recreation Areas** along the river. There are no designated swimming beaches within the Lower Wisconsin Riverway. **Mazomanie Beach**, a former swimming area along river, is currently closed to both swimming and entry at this time.

HIKING, BIKING, AND NATURE TRAILS

The Lower Wisconsin Riverway has 23 miles of upland hiking trails and 20 miles of marked equestrian trails. Mountain bikes, ATVs, or other off-road vehicles are not allowed on any of the trails in the Lower Wisconsin State Riverway.

Several trails are located within the 815-acre **Black Hawk Unit Recreation Area**. A parking area for this unit is located at the end of Wacher Road off

Cactus Bluff overlook

Highway 78 southeast of Sauk City. An attractive log cabin, an open-air shelter and other buildings associated with equestrian and other public events, are located here. A small parking area along Highway 78 provides access to a hiking trail that leads to the **Wisconsin Heights Battlefield** site of 1832 and a cluster of prehistoric Native American effigy burial mounds.

Ranger Trail Pick

The **Ferry Bluff Trail** (0.8 miles) is a moderately-sloped access trail that leads to the top of **Cactus Bluff**, a sandstone bluff that towers 300 feet above the scenic Wisconsin River Valley below. Interpretive panels here explain the geology, plants, animals, and human history of the Lower Wisconsin River. A side trail from Cactus Bluff leads through a restored oak savanna to the crest of **Ferry Bluff,** which offers spectacular views of the river and surrounding bluffs. Ferry Bluff is located southwest of Sauk City off Highway 12/60. The site is closed from November 1 to April 1 as a refuge for wintering bald eagles.

WILDLIFE VIEWING

The wetlands, bluffs, prairies, and forests along the river are a haven for wildlife. More than 280 species of birds are known to nest or migrate through this area, including most species of waterfowl, songbirds, cranes, herons, egrets, eagles, hawks, and owls. The area is also home to 45 known species of mammals, such as beaver, muskrat, mink, otter, white-tailed deer, raccoon, squirrels, rabbits, coyote, and fox.

There are 20 **State Natural Areas** within the Lower Wisconsin Riverway. These sites provide habitat and protection for at least 60 endangered or threatened species of birds, plants, reptiles, fish, amphibians, insects, and mussels. Some of these species are found only in the Lower Wisconsin River region.

INTERPRETIVE PROGAMS AND FACILITIES

Nature hikes and self-guided interpretive trails are offered at Tower Hill and Wyalusing State parks (See Chapters 53 and 55). Volunteers of the **Friends of the Lower Wisconsin Riverway** (FLOW) occasionally present guided canoe field trips, river-use training sessions, and other special events. Upcoming programs can be found on their website. www.wisconsinriverfriends.org.

BOATING AND FISHING

The Lower Wisconsin is one of the most popular rivers in the state for day-trip canoeing, kayaking, and boating. Every year, thousands of people flock to this area to enjoy an outing on the river. Most paddlers use the upper stretch of the Wisconsin from Prairie du Sac to Spring Green. There are several private canoe rental outlets along the Riverway.

Motorboats can be operated in deeper stretches of the river but canoes and kayaks are the most popular mode of transportation on the river. There are dozens of boat launch areas along both sides of the river. The Lower Wisconsin River is known for its abundant populations of smallmouth bass, northern pike, walleye, crappie, bluegill, and white bass.

CAMPING

Canoe camping is allowed on all state-owned islands and sandbars except for a two-mile section of river from the **Ferry Bluff State Natural Area** downstream to **Grape Island**. Permits are not required but overnight stays are limited to no more than three nights.

Drive-in camping is available at **Tower Hill State Park** (See Chapter 55) near Spring Green in Iowa County. The park has 11 rustic campsites with canoe access to the Wisconsin River.

Victoria Riverside Park has 45 canoe and drive-in campsites with electric service and showers. The park is located along the south shore of the Wisconsin River off Highway 80 near Muscoda. www.muscoda.com/parks

WINTER ACTIVITIES

All hiking trails in the Lower Wisconsin property are open for cross-country skiing and snowshoeing but none are groomed. Fat-tire biking is allowed on the **Black Hawk** and **Millville horse trails** from December 15th to March 1st. Many backwater areas of the river offer great ice fishing in winter for panfish. Each year, thousands of people drive to Prairie Du Sac in winter to observe the dozens of bald eagles who migrate here in winter to catch fish in the open water areas of the Wisconsin River.

AREA ATTRACTIONS

The communities of Prairie du Sac and Sauk Praire are popular tourist centers with many fine riverside parks, bike trails, shops, restaurants, and art centers. www.saukprairie.com

The **Wollersheim Winery & Distillery** is one of the oldest and most visited wineries in the state. The 1850s-era vineyards and original limestone buildings that overlook the Wisconsin River have been designated as National Historic Landmarks. The site offers guided tours and wine tasting year round. Wollersheim is located off Highway 188 near Prairie du Sac. www.wollersheim.com

DIRECTIONS TO THE RIVERWAY

The Lower Wisconsin State Riverway is located in southwestern Wisconsin in parts of Columbia, Crawford, Dane, Grant, Iowa, Richland, and Sauk counties. The property headquarters are at Tower Hill State Park.

Lower Wisconsin State Riverway
5808 County Highway C
Spring Green, WI 53588
(608) 588-7723

42. MacKenzie Environmental Education Center

Outdoor Education Site—A State Treasure

Observation tower

PROPERTY SNAPSHOT

The MacKenzie Environmental Education Center is often mistaken for a state park. This is understandable since the site does have hiking trails, picnic areas, shelters, nature trails, and an indoor group camp and offers interpretive programs similar to most state parks but that's where the similarity ends.

This 500-acre property, located near Poynette in Columbia County, has several facilities not found elsewhere in Wisconsin, including live-animal Wisconsin wildlife exhibits, a logging museum, and an environmental learning center with overnight lodging. The center is also home to the Wisconsin Conservation Museum and state's only game farm for raising ring-necked pheasants.

GEOLOGY AND HISTORY

The landscape of the property was shaped by Green Bay lobe of the glacier that entered Wisconsin about 30,000 years ago. Near the end the last Ice Age, retreating glaciers left behind a maze of small moraine hills, grass-covered prairies, small streams, and wetlands throughout the local Poynette area. A few miles to the

42. MacKenzie Environmental Education Center

west, a new channel of the Lower Wisconsin River was cut by a torrent of melt water released when an ice dam collapsed near Baraboo to the north, draining Glacial Lake Wisconsin.

Many different indigenous people have lived, hunted, and fished throughout this region of the state for thousands of years. When French-Canadian fur traders explored the Poynette area in the 17th century, they found the area occupied by several Sauk and Ho-Chunk villages.

Pierre Pasquette assisted in negotiating the Washington D.C. Treaty of 1829, which compelled the Ho-Chunk people to cede all their lands south of the Fox/Wisconsin River portage to the U.S. government. Soon after, government agents began to sell timber and farm land in the Poynette area to both Yankee settlers and European immigrants primarily from Germany, Norway, and Wales.

> **Ranger Trivia**
>
> Pierre Pasquette (1796-1836) was a French fur trader and early settler of the local area. Pasquette was fluent in several Native American languages and served as Indian interpreter and treaty negotiator for the U.S. government. Pasquette was killed by a Ho-Chunk warrior during treaty proceedings in 1836. The nearby village of Poynette was originally called Pasquette in his honor but the U.S. Postal Service misread the spelling of his name and changed it to Poynette when a post office was established here.

In 1932, the Wisconsin Conservation Department under the direction of Harley MacKenzie purchased 500 acres of property east of Poynette to build the state's first **Fur and Game Farm**. Ruffed grouse, ring-necked pheasants, and fur-bearing animals such as mink, fox, and raccoon were raised here at that time. Today, about 250,000 pheasant chicks are raised at the game farm each spring.

> **Ranger Note**
>
> Harley MacKenzie was Wisconsin's first chief law enforcement warden from 1925 to 1934 and also served as the director of the Wisconsin Conservation Commission until 1942. MacKenzie is credited with the initial development the game farm at Poynette and the formation of the Wisconsin Conservation Congress, a statewide citizen advisory group that assists the Wisconsin Department of Natural Resources in developing hunting, fishing, and other outdoor recreation policies and regulations.

Adult pheasant roosters and hens are released on both public and private lands throughout the state during the fall hunting seasons. In the 1960s, the property was divided between the game farm operation and the Environmental Center. In 1970, it was renamed MacKenzie Environmental Education Center and is now part of the Wisconsin State Parks and Forests system.

BEACHES, PICNIC AREAS, AND SHELTERS

A picnic area with restrooms and an open-air shelter building is located in the southeast section of the property. Picnic tables are also located adjacent to the entrance parking area and at the **Wildlife Exhibit and Logging Museum.**

HIKING, BIKING, AND NATURE TRAILS

There are ten miles of hiking trails to explore at MacKenzie. Bike riding is allowed on park roadways but not on hiking trails. Four trails in the far south end of the property have interpretive guides that can be picked up at the start of each trail. The **Hardwood Trail** (0.5 miles) and the **Ecology Trail** (0.3 miles) both lead through hardwood and conifer forests and descend into a picturesque wooded ravine before looping back to the trailhead. The **Conifer Trail** (0.6 miles) is an asphalt-paved trail that leads over wooden bridges spanning small ravines through a managed pine forest. The **Wildlife Trail** (0.2 miles) is a short side trail off the Conifer Trail with great views of some of the property's largest pine trees. The **Ridge Trail** (0.6 miles) starts near the picnic area in the far western end of the property. This rustic trail leads uphill to a ridgetop forest with great views of the countryside below.

Ranger Trail Pick

The **Nature Trail** (0.4 miles) begins across the road from the entrance parking area. This popular trail leads to an 80-foot-tall observation tower that offers scenic views of the surrounding landscape and the open bison range below. The trail continues into the Wildlife Exhibit, which has several live-animal zoo enclosures. Further along the trail is the **Maple Woods Demonstration Area**, which features a logging museum, sawmill exhibit, maple sugar house, and a small pond.

WILDLIFE VIEWING

The property's diverse landscape of prairies, wetlands, and upland forests of both conifer and hardwood trees attract a variety of wildlife. White-tailed deer, raccoon, squirrels, cottontail rabbits, and chipmunks are commonly seen here, as are several species of woodland songbirds, woodpeckers, upland game birds, waterfowl, and cranes.

The Wildlife Exhibit showcases many of Wisconsin's native animals including gray wolves, white-tailed deer, lynx, fox, ravens, badgers, owls, hawks, and eagles. All animals on exhibit were rescued from life-threatening injuries that prevented them from being released back into the wild. American bison can be viewed within an expansive fenced-in open range area. Interpretive panels near

the observation tower describe the life cycle and historical significance of these iconic American animals.

INTERPRETIVE PROGRAMS AND FACILITIES

The MacKenzie Environmental Center offers some of the finest outdoor education and learning programs in the state. Property staff present courses on several natural resource topics and outdoor skills training primarily to school children and other organized groups. The center has two dormitories with a capacity of 160 people for overnight lodging. A central lodge with a commercial kitchen and a dining room is also available.

Ranger Trivia

Prior to European settlement there were about four million bison in the Eastern United States, including the prairie regions of southern and western Wisconsin. The last two surviving bison anywhere east of the Mississippi were shot by Sioux hunters along the Trempealeau River in southwestern Wisconsin in 1832.

The **Maple Demonstration Area** features a sawmill exhibit and an original 19th-century log cabin that houses the **Nelson Cabin Logging Museum.** Nearby, the **Wallen Sugar House** has seating for interpretive programs and special events, such as the property's annual maple syrup demonstrations in spring and the October Halloween event.

The **Wisconsin Conservation Museum** has dozens of historic displays highlighting Wisconsin's wildlife, fishery, forestry, and other conservation efforts over the years. The museum also has several dioramas featuring mounted specimens of Wisconsin's native mammals, birds, and fish.

BOATING AND FISHING

The nearby **Rowan Creek Fishery Area** (651 acres) has both Class 1 and 2 trout streams that are stocked annually with brown and rainbow trout. The **Hinkson Creek Fishery Area** (233 acres) offers high-quality native brook trout fishing. Both are located about two miles east of the MacKenzie Center.

Lake Wisconsin (7,197 acres) is man-made reservoir of the Wisconsin River that offers great boating and fishing opportunities. Anglers can expect to catch a variety of fish including northern pike, walleye, catfish, trout, bass, and panfish in these waters. Lake Wisconsin is located about 13 miles west of the MacKenzie Center.

ACCESSIBLE FACILITIES AND TRAILS

The **Conservation Museum** and **Wildlife Display** are both wheelchair-accessible and have designated handicap parking areas. The **Conifer Nature Trail** (0.6 miles) has asphalt-paved walkways that loop through a managed forest plantation.

CAMPING

There are no drive-in or backpack campsites at this property. The nearby **Smokey Hollow Campground** has full-hookup campsites and cabins. The campground is located about seven miles west of the MacKenzie Center off County Highway J near Lodi. www.smokeyhollowcampground.com.

WINTER ACTIVITIES

The MacKenzie Center remains open in winter for hiking, wildlife viewing, hiking, snowshoeing, and cross-country skiing. None of the trails are groomed.

AREA ATTRACTIONS

Columbia County has many attractions for visitors, including historical sites, boating, fishing, hunting, biking, and hiking trails. The ever-popular **Wisconsin Dells** is located in the northwest corner of the county.
www.travelcolumbiacounty.net

Old Fort Winnebago Surgeons Quarters and Garrison School are located on the original grounds Fort Winnebago, which was built as a French post in 1816. The fort was purchased by the U.S. Army in 1834 to protect the nearby overland fur trade portage between the Fox and Wisconsin rivers. About 200 soldiers and a commissioned Army surgeon were stationed here until 1845. www.fortwinnebagosurgeonsquarters.org

The **Indian Agency House** was built in 1832 to house the region's U.S. government's Indian agent. ww.agencyhouse.org Both sites are National Register of Historic Places and open for tours. They are located 12 miles north of the MacKenzie Center off U.S. Highway 33 near Portage.

DIRECTIONS TO THE MACKENZIE CENTER

The MacKenzie Center is located near the village of Poynette in Columbia County about 32 miles north of Madison. From Highway 51 or Highway 22, exit onto County Highway CS and follow the signs to the property entrance. The center is open from dawn to dusk year-round.

MacKenzie Center
W7303 County Highway CS
Poynette, WI 53955
(608) 635-8110

43. MERRICK STATE PARK

Mississippi River Park Shines

Fountain City Bay

PARK SNAPSHOT

Merrick State Park is located along the Mississippi River within the coulee region of southwestern Wisconsin in Buffalo County. The coulee area is renowned for its scenic steep-sided valleys, forested ridges, clear-running streams, and some of tallest bluffs in the state. The park entrance is just off the **Great River Road** (Highway 35), the only **National Scenic Byway** in the state. Merrick is known for its great fishing and boating opportunities. The 320-acre park also offers camping, hiking, picnic areas, birdwatching, and biking.

GEOLOGY AND HISTORY

The rocky bluffs and deeply eroded valleys that surround Merrick State Park were shaped by wind, frost, and water erosion over millions of years. Several majestic limestone bluffs are visible from the park including **Eagle Bluff**, a 550-foot-tall monolith that towers over the nearby village of Fountain City.

Merrick State Park is situated within the Driftless Area of the state, a region of southwestern Wisconsin that escaped the crushing effect of glacial ice during the last Ice Age. As a result, the rugged landscape of area looks much as it might have appeared prior to glacial times. Merrick, by contrast, is a relatively flat due to its location along the river. The park is composed of bottomland hardwood forests, wetlands, and upland woods plus numerous Mississippi backwater islands.

Native American people have lived along the banks of the Mississippi for thousands of years. The river area has always provided rich diversity of waterfowl, fish, wild plants, game, and fur-bearing animals. When the first French fur traders canoed down the Mississippi River in the 17th century, they found the land inhabited by primarily by the Ojibwe (Chippewa) and Ho-Chunk (Winnebago) people, but the Santee Sioux often visited this area as well. The Sioux hunted deer, elk, and bison throughout the prairie regions of the Dakotas, Minnesota, and parts of southern and western Wisconsin.

The nearby village of Fountain City is the oldest community in Buffalo County. This historic town dates back to 1839, when Thomas Holmes and his family arrived by boat to set up a fur trading post here. Holmes, along with his wife and the couple's foster child, spent their first winter here living in a nearby cave and cut firewood for steamboats on the Mississippi River. Life in a wilderness did not suit Mrs. Holmes, however. She abandoned her husband the following year and returned downstream to civilization with the couple's foster child. Thomas Holmes remained in the area and eventually married a local Native American woman. Within a few years, his new settlement, now called Holme's Landing, became an important commercial landing for steamboats traveling up and down the Mississippi River.

Beginning in the mid-19th century, European immigrants from Germany, Norway, and Switzerland arrived to set up farmsteads in the Fountain City area. The Swiss were attracted to the rugged hills, valleys, and forests of the area because it reminded them of their homeland. Over time, this part of Wisconsin became known as "The Little Switzerland of America" due to the large number of Swiss living here.

> **Ranger Trivia**
>
> The village of Holme's Landing was renamed Fountain City in 1854. The name was chosen because of the many natural freshwater springs (fountains) found in this area.

In the early 1900s, John Latsch, a wealthy grocer from Winona, Minnesota, purchased more than 18,000 acres of forested acreage on both sides of the Mississippi to provide hunting land for his family and friends. In 1919, Latsch donated 266 acres to the State of Wisconsin to establish a new state park along the Mississippi River. His only condition for the donation was that the park be named in honor George Bryon Merrick, a well-known Mississippi paddlewheel pilot during the mid-1800s. As a result of Latsch's generous donation, Merrick State Park opened to the public in 1932.

BEACHES, PICNIC AREAS, AND SHELTERS

There is no swimming beach at Merrick State Park due to the ever-changing water levels and often dangerous currents of the Mississippi River. The city of Alma, located about 14 miles north of Merrick, has a 300-foot public swimming beach with restrooms, picnic areas, and a bathhouse.

Merrick has several picnic areas and three shelter facilities that can be reserved in advance. The **Nature Center Shelter** is located near the park entrance station and offers great views of the river. The **Round Shelter** is located near the boat launch area and the **Log Shelter** is situated within the **Island Campground**. Both of these historic structures were built by Civilian Conservation Corps (CCC) craftsmen in the 1930s.

HIKING, BIKING, AND NATURE TRAILS

The park has two miles of hiking trails that lead through upland oak and maple woodlands, bottomland forests, and a restored prairie area. A trail in the northeastern side of the park travels along an upland forested peninsula with scenic views of a backwater bay and towering bluffs in the distance. Wooden staircases along this trail descend to observation platforms on the water's edge.

The near-level shady roadways at Merrick are ideal for bike riding. The **Great River State Trail** (24 miles) is a limestone-surfaced hiking/biking route that leads through forests and restored prairies adjacent to scenic backwaters of the Mississippi River. The trail begins 20 miles south of Merrick near **Perrot State Park**. (See Chapter 49)

Ranger Trail Pick

The park has two self-guided canoe trails that loop through the scenic forest bottomlands and backwater islands of the Mississippi River. Both trails begin at the lower boat landing. Trail markers on posts guide paddlers along these water routes. Rental canoes can be reserved at the park office from Memorial Day to Labor Day.

WILDLIFE VIEWING

The park is home to a wide variety of wildlife, including white-tailed deer, raccoon, rabbits, squirrels, coyote, muskrat, mink, and otter, plus several species of turtles, frogs, salamanders, and snakes. Birdwatching at Merrick can be outstanding especially in spring and fall. The Mississippi River is one of the largest bird migration routes in the country with an exceptional variety of ducks, geese, cranes, herons, songbirds, swans, hawks, eagles, osprey, owls, and other raptors.

The **Whitman Dam State Wildlife Area** (2,253 acres), located adjacent to the park, preserves hundreds of remote backwater islands, several restored prairies, and upland hardwood forests. The area is a haven for white-tailed deer, rabbits, squirrels, wild turkeys, ruffed grouse, and many other wildlife species. Whitman Dam also has great fishing, hunting and birdwatching opportunities.

INTERPRETIVE PROGAMS AND FACILITIES

Interpretive programs and hikes are offered throughout the summer season. Evening programs are held at the **Merrick Nature Center Shelter**.

Guided canoe night hikes through the backwaters of the Mississippi are occasionally offered.

Ranger Wildlife Trivia

Merrick State Park is located in *Buffalo* County, just south of the village of *Buffalo*. To the north of the park is the Buffalo River, which empties into the Mississippi River near the city of Alma, the county seat of Buffalo County.

So where are all the buffalo? Free-roaming herds of Buffalo (American Bison) were once numerous in this part of Wisconsin. French voyager Father Louis Hennepin named the nearby Buffalo River, "Rivere der Boeufs" in 1680 due to the large number of bison he encountered in this area. Unfortunately, the last two bison east of the Mississippi River were shot by Sioux hunters just south of Merrick along the Trempealeau River in 1832.

BOATING AND FISHING

Merrick has two boat launches along **Fountain City Bay**, a backwater of the Mississippi River. Anglers can expect to catch walleye, sauger, channel catfish, bass, northern pike, bluegill, and crappie on the river. **Eagle Creek**, located just north of the park, is a popular fly-fishing area. The upper eight miles of this stream is a Class III brown trout stream. Brook trout are occasionally caught here as well.

ACCESSIBLE FACILITIES AND TRAILS

Most of the park's picnic areas, shelters and restrooms are accessible for mobility-impaired visitors. The Merrick Nature Center has paved walkways to the building.

Campsite 6 is wheelchair-accessible and has electric service. Some of the park's hiking trails have steep inclines with wooden staircases or erosion timbers, which may be challenging for visitors with walking difficulties.

CAMPING

Merrick has 67 campsites (22 electric). The **North Campground** has all the electric sites and a restroom/shower facility. Campsites here are well-spaced and

Mississippi River boating

most have shade trees. The **South Campground** has several sites located along the waterfront with docking areas for watercraft. There are five walk-in campsites here as well.

The **Island Campground** (tent only) has five walk-in shoreline campsites and a group campsite that can accommodate up to 50 campers. A road bridge provides access to the island parking area.

WINTER ACTIVITIES

Hiking, snowshoeing, wildlife viewing, and cross-country skiing are popular off-season activities. Trails are not groomed. Many anglers enjoy ice fishing the backwaters of the Mississippi in winter when ice conditions allow.

AREA ATTRACTIONS

Buffalo County is known for its outstanding natural beauty and quaint historic river towns. www.buffalocounty.com

The **Great River Road** is Wisconsin's only **National Scenic Byway**. This 250-mile stretch of Highway 35 is often described as the most scenic road in the Midwest. Much of this roadway travels between the towering rock bluffs and the shoreline of the Mississippi River. The route passes through 33 historic river towns and villages from Prairie du Chien north to Prescott. The **Great River Road Visitor and Learning Center** is located at **Freedom Park** along Monroe Street in Prescott. www.freedompark.org

DIRECTIONS TO THE PARK

Merrick State Park is located along Highway 35 in Buffalo County about 40 miles north of La Crosse and three miles north of the village of Fountain City.

Merrick State Park
S. 2965 State Road 35
Fountain City, WI 54620
(608) 687-493

44. MILL BLUFF STATE PARK

Ancient Island of Glacial Lake Wisconsin

Bee Bluff

PARK SNAPSHOT

Mill Bluff State Park was named for the 120-foot-tall sandstone bluff that towers above the forest canopy of the park. The property offers camping, picnicking, a swimming beach, and several hiking trails, including an uphill climb to the very top of Mill Bluff itself. The 1,603-acre park is part of the federal **Ice Age National Scientific Reserve,** which preserves several sandstone bluffs and other unique rock formations in south-central Wisconsin.

GEOLOGY AND HISTORY

The castle-like bluffs and towering pinnacles found at Mill Bluff are some of the largest in Wisconsin. Geologists refer to the wide flat-topped bluffs as *mesas* (Spanish for "table") and the narrower bluffs as *buttes*. The towering bluffs here are composed of sandstone rock, which formed from sand deposits beneath an ancient sea nearly **600 million years ago.** Wind, ice, and rain erosion has reduced most of this sandstone into small hills and level plains except for a few towering buttes like those at Mill Bluff State Park. These rocky megaliths are still standing today thanks to their chemically hardened sandstone caps that have resisted erosion over time.

About 18,000 years ago, a stalled Ice Age glacier 40 miles to the south near Baraboo created an ice dam that blocked the route of the Wisconsin River. As a result, a massive glacial lake was formed and flooded most of central Wisconsin. During this period, most of Mill Bluff State Park would have been submerged under 60 feet of water except for the sandstone bluffs, which would have appeared as islands or sea stacks rising above the surface of the lake. For the next 3,000 years, the sides of these seas stacks were reshaped by the forces of wind, waves, and collisions with icebergs. Near the end of the Ice Age about 14,000 years ago, the ice dam suddenly collapsed, draining the entire glacial lake. Left behind were sand deposits, small streams, wetlands, and the battered sea stacks, now towering bluffs, above the extinct lake bottom.

Mill Bluff was considered a sacred site by many Native American cultures over time. Early inhabitants of the area created rock carvings (petroglyphs) on the sides of sandstone bluffs throughout the park and local area. The Menominee and Ho-Chunk people lived in this part of Wisconsin for generations but their hold on the land came to an end after the Black Hawk War of 1832. Tribal leaders at that time were forced to cede their lands to the U.S. government and move to reservations west of the Mississippi. Many did leave Wisconsin but not all. A village of about 500 mostly Ho-Chunk people were recorded as still living near Mill Bluff as late at 1872.

In the early 19th century, the towering bluffs of the area served as colossal trail markers for immigrant settlers traveling west across the Wisconsin frontier. Pioneers on these early wagon trains often wrote about the immense bluffs that could be seen for miles away in their personal journals. The nearby community of Camp Douglas was founded in 1864 when two railroad lines crossed through that area. Steam locomotives required lots of firewood to feed their engines so the area's forests were an important fuel source.

The local area surrounding Mill Bluff was settled primarily by Norwegian and Irish immigrants, who set up farmsteads and cut timber from the surrounding forests. The park's iconic sandstone bluff was named after a 19th-century saw "mill" that once operated near the base of the bluff. The park property was initially purchased to build a highway rest area in the early 1930s by the State Highway

Ranger Trivia

The village of Camp Douglas was named after James Douglas, one of the first lumbermen to sell firewood to the railroad in 1864. As other timber companies began to compete with his firewood sales, Douglas erected a sign along the railroad track that read, "Camp Douglas" to distinguish his business from his competitors. The name has stuck ever since.

Commission. It was later transferred to the Wisconsin Conservation Commission to develop Mill Bluff State Park in 1936.

BEACHES, PICNIC AREAS, AND SHELTERS

Mill Bluff has an attractive white-sand beach located along a 2.5-acre spring-fed pond. Adjacent to the beach is a picnic area with restrooms, changing stalls, and an open-air shelter. A second picnic area is located east of the park office at the base of Mill Bluff. This area has a unique A-frame shelter made of local stone and timber was built by Civilian Conservation Crews in the 1930s

Swimming beach

HIKING, BIKING, AND NATURE TRAILS

The **Mill Bluff Summit Trail** (0.2 miles) is a somewhat strenuous 120-foot climb up **223** stone steps to the flat crest of the bluff. An observation deck on top provides panoramic views of the forests, farms, and nearby bluffs of the area. The **Mill Bluff Nature Trail** (0.4 miles) encircles the lower perimeter of Mill Bluff. Interpretive signs along this trail describe the flora, fauna, and geology of the area.

There are twelve named sandstone buttes within the park including Bee, Round, Long, Sugar Bowl, and Wildcat bluffs. Most can't be accessed by vehicle or marked trails but adventuresome hikers can reach them by foot and compass if desired.

44. Mill Bluff State Park

The **Omaha Bike Trail** (12.5 miles) is an asphalt-paved bike trail that connects the nearby village of Camp Douglas to the community of Elroy. The bike trail travels past rock bluffs, scenic farmsteads and forests, and through a 300-foot-long former railroad tunnel built in 1876. The trail can be accessed three miles southeast of the park in the village of Camp Douglas. www.co.juneau.wi.gov/trails

---

Ranger Trail Pick

The **Camels Bluff Trail** (1.5 miles) loops between the two massive sandstone buttes or "humps" of Camels Bluff. Also along this trail are **Cleopatra's Needle**, a tall, thin rock pinnacle and a sandstone formation called **Devils Monument**. The access paths to these rock formations are not signed so hikers need to watch for well-worn side trails to reach them. Parking for Camels Bluff is along Funnel Road (County Road W) on the north side of Interstate Highway 90/94.

WILDLIFE VIEWING

The varied landscape of Mill Bluff provides habitat for many different wildlife species to thrive here. The park's pond and wetland areas attract a variety of waterfowl, cranes, and herons. The area's oak, maple, and pine forests are home to white-tailed deer, gray squirrels, chipmunks, rabbits, raccoon, coyote, and fox.

The **Mill Bluff State Natural Area** (485 acres) protects vital habitat for dozens of species of woodland songbirds, such as the eastern wood pewee, ovenbird, and yellow-throated vireo, plus several types of woodpeckers, hawks, and owls. Turkey vultures can often be seen soaring over the tall bluffs of the park. Prairie plants such as wild lupine, rock cress, and Indian grass can be found in drier sand barrens. Sandstone bluffs harbor several species of rare ferns and uncommon grasses.

INTERPRETIVE PROGRAMS AND FACILITIES

Informational signs and panels describing the flora, fauna, and geology of the park are located along the Mill Bluff Nature Trail and at the observation platform on top of Mill Bluff. A Wisconsin Historical Marker located in the picnic area highlights the history and geology of Mill Bluff.

BOATING AND FISHING

Mill Bluff's swimming pond is not stocked with fish. **Lake Tomah** (245 acres), located ten miles northwest of Mill Bluff, offers great fishing and slow-no-wake boating. Anglers can expect to catch panfish, bass, and northern pike. Shore fishing on Lake Tomah is available at **Winnebago** and **Lake** parks.

ACCESSIBLE FACILITIES AND TRAILS

Most picnic areas, restrooms, and shelters at Mill Bluff are accessible for mobility impaired visitors. Campsite **18** is wheelchair-accessible campsite. A section of the Mill Bluff Nature Trail near the park office has a crushed gravel base and is wheelchair accessible. Other trails in the park may be challenging for visitors with walking difficulties due the steep topography and sandy soil of this area.

CAMPING

Mill Bluff has **21** campsites (6 electric) located just north of the swimming beach and picnic area. The area has vault-type restrooms and a hand pump for water. The campground is open from late May to the end of September. Shower facilities are available for a small fee at the nearby KOA campground and the Road Ranger and Love's Travel Stop gas stations located three miles north of the park. The KOA campground also has full-service campsites for campers who prefer these amenities.

> **Ranger Note**
>
> Mill Bluff State Park is situated between Interstate I-90/94 and Highway 12/16 and the Soo Line Railroad. Some campers find the highway and railroad noise disturbing but the park's forests trees do help muffle the sound somewhat.

WINTER ACTIVITIES

The interior roads at Mill Bluff are closed in the off-season but visitors can still access hiking trails from a parking area along Funnel Road. Snowshoeing, hiking, wildlife viewing, and cross-country skiing are favorite activities in winter. None of the trails are groomed for skiing.

> **Ranger Tip**
>
> Some of the best views from the top of Mill Bluff can be seen in late autumn after the leaves have fallen and again in early spring before the trees leaf out. The picturesque snow-covered bluffs in winter are enchanting to see but visitors are advised not to climb the stone staircase during icy conditions.

AREA ATTRACTIONS

The Juneau and Monroe County area has some of the most dramatic scenery in the state, along with small, quaint towns, local museums, parks, and other attractions to explore.

www.juneaucounty.com

www.gomonroecounty.com

The nearby **Wisconsin National Guard Museum** has several interesting exhibits featuring Wisconsin's National Guard from its pre-Civil War beginnings to the present day. The museum also has dozens of vintage airplanes, fighter jets, tanks, artillery, and other military equipment on display. The museum is located within

the Camp Williams/Volk Field Air Base near the community of Camp Douglas.
www.wvmfoundation.com/wi-national

DIRECTIONS TO THE PARK
Mill Bluff State Park is located north of the village of Camp Douglas within Monroe and Juneau counties. From Interstate I-90/94 take either the Oakdale exit #48 or the Camp Douglas exit #55 to U.S. Highway 12/16 and follow the signs to the park entrance.

Mill Bluff State Park
15819 Funnel Road
Camp Douglas, WI 54618
(608) 427-6692

45. MIRROR LAKE STATE PARK

Wisconsin Dells Oasis

Mirror Lake's Bluewater Bay

PARK SNAPSHOT

Mirror Lake State Park is located along its namesake, a serene flowage that often reflects a "mirror" image of the surrounding shoreline. This popular 2,192-acre property offers camping, swimming, fishing, boating, picnicking, nature study, and miles of hiking and off-road bike trails. For many visitors, Mirror Lake serves as a peaceful green oasis from the hustle and bustle of nearby Wisconsin Dells, one of the busiest tourist destinations of the Midwest.

GEOLOGY AND HISTORY

Mirror Lake is at the eastern edge of the Driftless Area, a region of southwestern Wisconsin that escaped the crushing effects of continental glaciers that pushed through most of the state 30,000 years ago. About 18,000 years ago, a glacier stalled near the Baraboo Hills south of the park, blocking the flow of the Wisconsin River. As a result, Mirror Lake, along with 1,800 square miles of central Wisconsin became submerged beneath Glacial Lake Wisconsin. When the glacial ice dam finally collapsed, a torrent of melt water filled with sand, gravel, and rocks emptied the lake and cut new river channels as it thundered south toward the Mississippi River. When the flood subsided and the rivers and streams fell to their current levels, an astonishing assortment of water-carved sandstone bluffs, pillars, and gorges was revealed. Many of these towering rock sculptures can still be seen today within the **Dells of the Wisconsin River** and the **Dell Creek Gorge** at the Mirror Lake State Park.

The abundance of fish and game animals in the area attracted several Native American cultures to live here over the course of thousands of years. Early indigenous people left their imprint on the landscape through their effigy burial mounds and petroglyph carvings inscribed on sandstone bluffs. When early European explorers, fur traders and settlers arrived here, the Ho-Chunk (Winnebago) people were the dominant tribe of region. Conflicts over land ownership ultimately led to the Winnebago War of 1827 and finally the Black Hawk War of 1832.

> **Ranger Trivia**
>
> The term *Dells* was derived from a French word *"dalles,"* which means "a narrow gorge where river rapids run through rocky walls."

After the wars, Ho-Chunk leaders were forced to sell all their remaining land in the Wisconsin Territory to the U.S. government and agree to move to reservation land west of the Mississippi River. This decision was reversed in 1873 and allowed the Ho-Chunk people to return to Wisconsin and file claims on certain parcels of the land in the Dells area. In addition to rebuilding their villages, Ho-Chunk elders began to revive their traditional festivals and pow-wows. These ethnic dances, along with the natural beauty of the towering sandstone formations of the Wisconsin River, were the cornerstones of the early tourism industry in the Wisconsin Dells.

Mirror Lake is one of the oldest man-made lakes in Wisconsin. It was created in 1866 by Horace LaBar when he installed a log dam across Dell Creek to power his flour mill. A much larger lake was created in 1925 by W. J. Newman, a Chicago millionaire. Newman built an 18-foot-high dam to raise the water level of Mirror Lake for the benefit of resort guests at his exclusive Dell-View Hotel. In the early 1960s, the Wisconsin Conservation Commission began to purchase several properties in the Dells area to develop Mirror Lake State Park, which opened in 1966.

BEACHES, PICNIC AREAS, AND SHELTERS

Mirror Lake has a swimming beach along the Bluewater Bay section of the lake. Adjacent to the beach is a picnic area with restrooms and an open-air shelter. Other picnic areas and shelters are located at the boat launch and within the **Bluewater Campground**. All shelters have electrical service and can be reserved in advance.

HIKING, BIKING, AND NATURE TRAILS

Mirror Lake has more than 20 miles of hiking trails to explore. The **Pulpit Rock Nature Trail** (0.6 miles) is located within the **Dell Creek State Natural Area.** The trail leads to a sandstone ridge overlook area with scenic views of Dell Creek below. Parking is available along Fern Dell Road about a half mile west of the park entrance.

The **Northwest Trail** (2.3 miles), **Wildwood Trail** (0.6 miles), and **Ringling Pass Trail** (0.3 miles) all loop through mature forest areas across the lake from the swimming beach. The **Newport Trail** (1.0 miles), **Lake View Trail** (0.2 miles), and the **Kilbourn Trail** (0.2 miles) are located between the beach/picnic areas and the campgrounds.

Ranger Trail Pick

The **Echo Rock Trail** (0.6 miles) and **Ishnala Trail** (2.2 miles) are perched high above Mirror Lake in the northeast area of the park. Both trails lead through oak/pine forests and past sandstone formations with great views of the lake. A 150-foot-long bridge crosses a deep sandstone gorge between these trails. A side trail north of the bridge leads to one of Wisconsin's most iconic restaurants, the **Ishnala Supper Club**. The restaurant is located on a beautiful sandstone bluff overlooking Mirror Lake and is known for its indoor pine trees that grow right through the roof of the building.

The term **Ishnala** is a Ho-Chunk term for "by itself alone," an apt description of the scenic sandstone point that juts out into Mirror Lake here.

Echo Rock hiking trail

Mirror Lake has nine miles of off-road bike trails located south of Fern Dell Road. The nearby **400 State Trail** (29.5 miles) is a popular hiking/biking trail that features scenic rock outcrops along the Baraboo River between the communities of Elroy and Reedsburg. Trail parking is available along South Walnut Street in Reedsburg about 11 miles west of the park.

WILDLIFE VIEWING

The park's forests, wetlands, prairies, and lakeshore provide habitat for many wildlife species. White-tailed deer, squirrels, muskrat, beaver, mink, wild turkey, chipmunks, rabbits, hawks, blue jays, crows, and chickadees are commonly seen in the park. More nocturnal animals such as fox, raccoon, coyote, and several kinds of owls are occasionally spotted here as well.

The **Pine/ Oak Forest State Natural Area**, located in the northwest corner of the park, is a great place to observe wildlife, especially uncommon birds and rare plants. This wilderness area has sandstone cliffs, wetlands, and upland forests of maple, oak, and pine. The natural area can be accessed off Lakeview Road near the Town of Delton's boat launch south of Highway 23.

The nearby **International Crane Foundation** is a nationally recognized refuge for cranes from around the world. The site is the only place on earth to see all 15 species of cranes. The Crane Foundation is located along Shady Lane Road off Highway 12 about nine miles south of the park. www.savingcranes.org

INTERPRETIVE PROGRAMS AND FACILITIES

Interpretive panels highlighting the flora, fauna, and geology of the park are located along the shoreline of Mirror Lake and along nature trails. Nature hikes are offered throughout the season and evening programs held at the park's amphitheater.

BOATING AND FISHING

Mirror Lake (137 acres) is a man-made flowage with good populations of panfish, bullhead, bass, walleye, and northern pike. Much of the lake is fairly shallow except a few sections of Dell Creek, which can reach 14 feet in depth. The park has a paved boat launch area. Motor boats are limited to slow-no-wake operation on Mirror Lake. A private concessionaire rents boats, canoes, and kayaks. Dell Creek is a 10-mile cold-water stream that flows into Mirror Lake. The creek is stocked with fingerling trout annually.

ACCESSIBLE FACILITIES AND TRAILS

Most of the park's restrooms, picnic areas, amphitheater, and shelters are accessible for mobility-impaired visitors. A wheelchair-accessible fishing pier and boardwalk are located adjacent to the boat launch area. A boarding pier for visitors who require assistance to access to their watercraft is also available. A wheelchair-accessible pontoon motorboat is available for rent at the boat landing.

Campsites **32** and **63** are wheelchair accessible and have electrical service. Both have paved walks to a restroom/shower facilities. The park's **Cabin in the Woods** is an indoor camping option reserved for visitors with physical disabilities. The cabin features a barrier-free kitchen, living room, bedroom, and a wheel-in shower/restroom.

The **Echo Rock Trail** (0.6 miles) is an asphalt-paved trail that begins near the boat launch area and gradually ascends to a scenic overlook of Mirror Lake. A wheelchair-accessible connector trail leads across a 150-foot bridge over a scenic sandstone gorge to the **Ishnala Trail.**

> **Ranger Note**
>
> A one-of-a-kind indoor camping experience is available at Mirror Lake. The Seth Peterson Cottage, located on a bluff overlooking the lake, was designed by famed architect Frank Lloyd Wright in 1958. It is the only Wright structure of its kind available for overnight rental to the public. The cottage is open for tours from time to time as well.
> www.sethpeterson.org

CAMPING

Mirror Lake has **159** campsites (**50** electric) located within three campgrounds. Each campground has its own restroom/shower facility. An RV water fill/sanitary station is located along the park's entrance road. A group campground (tent only), located north of the Bluewater Bay Campground, has seven campsites that can accommodate up to **20** campers each.

WINTER ACTIVITIES

Park staff groom **19** miles of cross-country ski trails for both skate and classic skiing. Marked snowshoe routes are located within the **Sandstone** and **Wild Rice** trails system.

Ice fishing on Mirror Lake is popular in winter especially for panfish. Winter camping is available within the **Sandstone Ridge Campground.**

AREA ATTRACTIONS

The famous Wisconsin Dells area has several large water parks, fine restaurants, museums, gift shops, and many other recreational offerings to choose from. www.wisdells.com

The **Dells Boat Tours** offer great views of the majestic water-sculptured sandstone rock formations of the Wisconsin River. The 15-mile **Upper River Tour** is the most popular route. This tour includes an onshore walk through a fern-lined sandstone canyon and the ever-popular dog jump to **Stand Rock**, a towering sandstone rock pinnacle. Boat tour passes can be purchased online or at a downtown office along Broadway Street in Wisconsin Dells. www.dellsboats.com

DIRECTIONS TO THE PARK
Mirror Lake State Park is located in northeast Sauk County about nine miles south of Wisconsin Dells. From Interstate I-90-94, exit onto Highway 12 (south) and turn onto Fern Dell Road and follow the signs to the park entrance.

Mirror Lake State Park
E 10320 Fern Dell Road
Baraboo, WI 53913
(608) 254-2333

46. NATURAL BRIDGE STATE PARK
Wisconsin's Largest Sandstone Arch

Sandstone arch and rock shelter

PARK SNAPSHOT

Natural Bridge State Park abounds in natural beauty and human history. The property features the largest natural rock arch in Wisconsin and preserves one of the oldest known human habitation sites in the Midwest. This quiet out-of-the-way state park is rarely crowded, making it a great place to relax, have a picnic or enjoy a hike on a trail.

GEOLOGY AND HISTORY

Few places in Wisconsin have such an intact landscape as Natural Bridge State Park. The sandstone rock of the park was formed from deposits of sand, silt, and minerals beneath a shallow inland sea that once covered this area about 1.6 million years ago. The park is situated within the Driftless Area of the state, an area that escaped the crushing effect of glaciers that pushed through most of Wisconsin 30,000 years ago. Near the end of the Ice Age, a massive glacier located only 18 miles northeast of the park came to a standstill at **Devil's Lake State Park.** As a result, the rock formations and landscape of Natural Bridge remained intact and only shaped by wind, frost and water erosion over time.

46. Natural Bridge State Park

The most prominent feature of Natural Bridge is the park's impressive 35-foot-wide and 25-foot-high sandstone arch. The top layer of the arch is composed of very hard sandstone imbedded with even harder pebbles of quartzite. Much of the softer sandstone below the crest of the arch has eroded away over time, leaving behind this unique natural bridge.

Beneath the sandstone arch is a 60-foot long cave referred to as the **Raddatz Rock Shelter**. The shelter was first excavated by archeologists from the Wisconsin Historical Society in 1957. Later excavations and radiocarbon dating of artifacts revealed that prehistoric people made use of the rock shelter as far back as 9,000 B.C. This discovery proved that prehistoric people were hunting, gathering plants, and living in Natural Bridge area at the same time glaciers still covered most of the state.

Artifacts excavated from the rock shelter included many types of plants, clam shells, hand tools, and the skeletal remains of over 50 animal species, including turkey, elk, wolf, bobcat, fisher, mountain lion, and the now-extinct passenger pigeon. Archeologists consider the Natural Bridge rock shelter to be one of the oldest human

> **Ranger Note**
>
> Natural stone arches are very rare in Wisconsin. Geologists theorize that the arch at Natural Bridge State Park was formed primarily by the natural weathering over millions of years. More recent studies, however, hint that the rock bridge may actually have been the rooftop of a long-exposed and eroded cave system.

Raddatz Rock Shelter

occupation sites in the Midwest. The site was added to the National Register of Historic Places in 1978.

Several different Native American cultures have in the lived in the park area and throughout Sauk County for thousands of years, including the Fox, Ho-Chunk and the Sac (or Sauk) people.

The first European settlers arrived in the Natural Bridge area 1838; only one year after the Sauk and Ho-Chunk leaders were forced to sell their lands to the U. S. government and leave the state following the end of the Black Hawk War. Many more immigrants arrived during the late 19th century, including Carl Raddatz, who arrived from Germany in 1880. Raddatz, along with his wife Hanna, raised crops and grazed cattle on land that would eventually become Natural Bridge State Park.

The sandstone arch on the Raddatz farm was a popular gathering area for people as far back as the 1870s. A local newspaper, *The Sauk County News*, described Natural Bridge in 1888 as "a wonderful and beautiful production (park) that is being visited nearly every day." The site hosted annual festivals for many years, including May Day gatherings, fall harvest parties, and July 4th celebrations. These social gatherings were attended primarily by local people and featured food, drinks, games, and dancing.

Eventually, this unique site was discovered by outsiders as well, which led to damage to the sandstone arch by graffiti carving and painting and unlawful digging for Native American artifacts. In 1957, the Wisconsin Historical Society completed an archeological study of the rock shelter and led the effort to save the site from abuse and neglect. As a result, the State of Wisconsin purchased 530 acres of the property to create Natural Bridge State Park in 1973.

> **Ranger Trivia**
>
> Sauk County was named after the Sauk tribe who established villages in this region of the state for generations. In 1766, John Carver, an early English explorer, encountered a large Sauk settlement located along the Wisconsin River about 13 miles southeast of Natural Bridge. He described the village as an "Indian City" and recorded it "having more than 60 multi-family houses." Today, the nearby city of Sauk Prairie is located near the original site of this Sauk settlement.

BEACHES, PICNIC AEAS, AND SHELTERS

A picnic area and a vault-type restroom are located adjacent to the parking area along County Highway C. There are no water fountains in the park so visitors must bring their own drinking water. Natural Bridge has no lakes, ponds, or streams but **Devil's Lake State Park** (See Chapter 37) located northeast of the park has two swimming beaches and picnic areas.

HIKING, BIKING, AND NATURE TRAILS

There are five miles of hiking trails at Natural Bridge. The most popular route is the short quarter-mile walking trail that leads to the base of the park's iconic sandstone arch formation and rock shelter. Many visitors leave the park after seeing the sandstone arch but the property's other trails are well worth the hike.

The **Whitetail Hiking Trail** (2.0 miles) leads through open prairie areas and woodlands with the chance to observe undisturbed wildlife and view the park's native flowers and grasslands. This trail is a favorite for birdwatchers throughout the year but especially during the spring and fall migration periods. There are no bike trails at Natural Bridge, but **Great Sauk State Trail** (10.5 miles) can be accessed at nearby Prairie du Sac. This popular asphalt-paved hiking/biking trail leads through the **Sauk Prairie State Recreation Area.** (See Chapter 52)

Ranger Trail Pick

The **Indian Moccasin Nature Trail** (1.0 miles) begins and ends at the park's picnic area. This route leads through mature hardwood forests to the natural bridge and then ascends to a perched sandstone ridge high above the arch formation. A side trail about half way along this trail leads to an overlook area with a panoramic view of the park and rural landscape below. Interpretive panels along this trail describe how Native American people used local plants for food, medicine, and other needs.

WILDLIFE VIEWING

The park's oak, hickory, and maple forests are home to many species of wildlife, including white-tailed deer, raccoon, rabbits, chipmunks, squirrels, skunks, fox, and coyotes. Birdwatchers enjoy viewing dozens of species of warblers, grassland sparrows, and woodland songbirds. Chickadees, nuthatches, blue jays, cardinals, and pileated woodpeckers are often spotted here, along with ruffed grouse and wild turkey. Bald eagles, turkey vultures, and several different types of hawks and owls live here as well.

Sandstone rock outcrops harbor uncommon plants such as prickly pear cactus, walking ferns, purple cliff brake ferns, and rare species of goldenrod. The park's prairie areas have flowering plants such as blazing star and native grasses like little blue stem and Indian grass.

INTERPRETIVE PROGRAMS AND FACILITIES

Interpretive panels at the natural bridge site describe the unique geology of the arch and archeological discoveries found at the rock shelter. The **Indian Moccasin Nature Trail** has several interpretive signs along its route as well.

BOATING AND FISHING

Honey Creek, located about two miles northwest of Natural Bridge, is a Class II trout stream with populations of brook and brown trout. **White Mound Lake** (93 acres) located 12 miles west of the park, has a public boat launch. Anglers can expect to catch panfish, largemouth bass, and northern pike in this lake.

CAMPING

Natural Bridge is a day-use park so it doesn't have camping facilities. **Devils Lake State Park**, located northeast of the park, has 423 campsites (154 electric) with restroom and shower facilities. (See Chapter 37)

WINTER ACTIVITIES

Wildlife watching, hiking, snowshoeing, and cross-country skiing are popular winter activities at Natural Bridge. None of the trails are groomed.

AREA ATTRACTIONS

Sauk County is known for its outstanding natural beauty and nationally renowned attractions, such as the **Wisconsin Dells, Circus World Museum**, and the **International Crane Foundation**. www.co.sauk.wi.us/general/visit

The **Mid-Continent Railroad Museum** is a living museum that depicts the Golden Age of railroading from period between the Civil War and World War II. The site has more than 100 restored locomotives and railroad cars on display. Visitors can "ride the rails" along a seven-mile-long train ride through the scenic valleys of the Baraboo Hills aboard an authentic passenger car of that era. The museum is located about 10 miles north of Natural Bridge off Museum Road near North Freedom. www.midcontinent.org

DIRECTIONS TO THE PARK

Natural Bridge State Park is located in Sauk County about 18 miles southwest of Baraboo. From U.S. Highway 12, take County Road C west about 10 miles to the park entrance.

Natural Bridge State Park
E 7792 County Road C
North Freedom, WI 53951
(608) 356-8301 (Devil's Lake State Park)

47. NELSON DEWEY STATE PARK

Mississippi River Park—A Scenic Gem

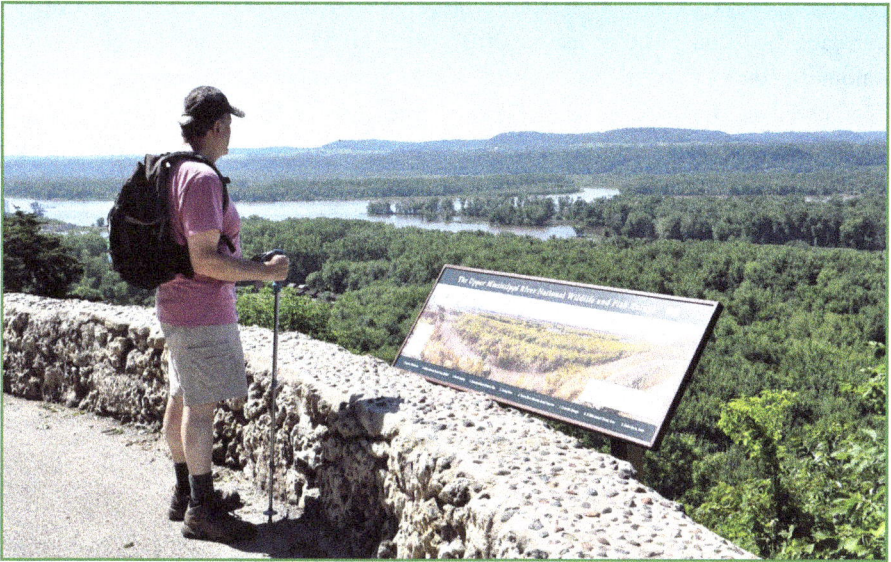

Overlook of the Mississippi River Valley

PARK SNAPSHOT

Nelson Dewey State Park is perched on a hilltop more than 800 feet above the Mississippi River in southwestern Grant County. The park is known for its panoramic views of the Mississippi River Valley and scenic bluffs. Nelson Dewey offers camping, picnicking, hiking trails, wildlife viewing, nature study, and many more outdoor activities.

The park is also home to the **Stonefield Historic Site**, an outdoor museum featuring early 19th-century historic buildings and the **State Agriculture Museum.**

GEOLOGY AND HISTORY

Nelson Dewey lies within what geologists refer to as the Driftless Area, a region of southwestern Wisconsin that escaped the crushing effect of continental glaciers during the last Ice Age. The bluffs and valleys here were primarily shaped by water, wind, and frost over millions of years. Near the end of Wisconsin's Ice Age about 19,000 years ago, massive amounts of melt water from glaciers far to the north thundered down the Mississippi River Valley eroding the riverside bluffs of the park. The exposed sandstone and limestone rock seen in the park's bluffs today were formed from deep layers of sand, silt, clay, and sea life deposited beneath an ancient sea that once covered this area 600 million years ago.

271

Several Native American cultures have lived along the banks of the Mississippi River in this region for untold generations including the Kickapoo, Meskwaki (Fox), Menominee, Potawatomi, and the Ho-Chunk people. Archeological surveys of two prehistoric village sites within Nelson Dewey revealed artifacts believed to be at least 7,000 years old. Three groups of ancient burial mounds, some of which are believed to have been built 2,000 years ago, are also preserved at the park. Nelson Dewey State Park is located in Grant County, which was established in 1836, a few years after the Black Hawk War. At that time, the Ho-Chunk and other native people were forced to cede all their remaining lands east of the Mississippi River to the U.S. government. Some of the first white settlers to move to this area came from the East Coast. One of these was Nelson Dewey, a lawyer for a land speculation company who moved to the nearby village of Cassville from Connecticut in 1836. Dewey also served as a territorial assemblyman at that time and later became the District Attorney for Grant County. Nelson Dewey was also elected as a senator and served as Wisconsin's first governor from 1848 to 1852.

Dewey, along with his wife Catherine, purchased a 2,000-acre tract of land north of Cassville in the mid-1850s, where they built an elaborate Gothic revival mansion and developed a plantation-style farm called Stonefield. The couple raised milk cows, beef cattle, horses, pigs, and chickens and planted several acres of fruit orchards and vineyards on their farm. Unfortunately, Nelson Dewey fell into hard times beginning in 1873 when a house fire nearly destroyed his mansion and he lost most of his fortune during the nationwide financial panic that same year. His estate went into foreclosure in 1878, but he continued to live in Cassville and practice law until his death in 1889.

Nelson Dewey's Stonefield property changed ownership several times over the next several decades but eventually his mansion and farm fell into disrepair. In 1935, the Wisconsin Conservation Commission, along with Grant County and the village of Cassville, purchased the house along with 720 acres of the estate to develop Nelson Dewey State Park in 1936 and Stonefield Historic Site in 1953.

BEACHES, PICNIC AREAS, AND SHELTERS
The park has three picnic areas that are all located along scenic bluff areas. The **Dewey Heights Picnic Area** has a beautifully-preserved shelter built by WPA (Works Progress Administration) craftsmen in the 1935 using local limestone and timber. The shelter has two fireplaces and electrical service. Nearby is the open-air **Prairie Shelter**, which offers an outstanding view of the Mississippi River Valley. Both shelters can be reserved in advance. The park does not have a swimming beach but an outdoor public pool is available in nearby village of Cassville. A popular swimming beach located along a sandy stretch of the

Mississippi River is available at the **Wyalusing Recreation Area**, a county park about 20 miles north of the park.

HIKING, BIKING, AND NATURE TRAILS

Nelson Dewey has two miles of hiking trails to explore. The **Mound Point Trail** (0.6 miles) follows the edge of the bluff with benches that offer great views of the Mississippi River Valley below. A more inland part of the trail leads past several ancient Native American burial mounds. Some of the mounds are dome-shaped while others are linear in design. Archeologists believe these mounds were built by Late Woodland Indians around 500-900 A.D. The **Oakwood Trail** (0.4 miles) is relatively easy walk through a mature oak, hickory, and maple hardwood forest. A paved section of this trail passes connects the park's campground to the picnic area.

Ranger Trail Pick

The **Prairie Trail** (0.2 miles) and the **Cedar Trail** (0.2 miles) are both short paths that can be hiked in combination with each other. The trails follow the outer edge of bluff areas over 800 feet above the river valley below. A section of the Prairie Trail leads through the 27-acre **Dewey Heights Prairie State Natural Area**. This site preserves remnant prairie plants such as little blue stem and Indian grass, pasque flowers, compass plants, and orange butterfly weed. The Cedar Trail section offers a bird's eye view of the **Stonefield State Historic Site** and the Wisconsin State Agricultural Museum below.

WILDLIFE VIEWING

Nelson Dewey is a birdwatcher's paradise with over 80 species of birds known to nest in the park, including pileated woodpeckers, scarlet tanagers, indigo buntings, rose-breasted grosbeaks, turkey vultures, and several types of hawks. The nearby Mississippi River Valley is known as one of the best bald eagle viewing sites in the Midwest. A **Bald Eagle Days** celebration is held in the nearby village Cassville each winter. More than 100 eagles can often be viewed roosting and catching fish in the open waters beneath **Lock and Dam Number 12** along the river throughout the winter season.

Wildlife abounds in the park's forests and bluff areas, including white-tailed deer, raccoon, chipmunks, rabbits, and squirrels, plus reptiles such as prairie ring-necked snakes, garter snakes, and black rat snakes. Gray tree frogs and wood frogs are found in forested areas, while green frogs, leopard frogs, and several types of turtles inhabit wetland areas adjacent to the Mississippi River.

INTERPRETIVE PROGRAMS AND FACILITIES

Interpretive programs and nature hikes are presented throughout the season by park volunteers. Interpretive panels are located at several overlook areas. The **Woodbine Nature Trail** (0.3 miles) follows an easy-walking asphalt-paved path through both woodlands and prairie areas with great views of the Mississippi River Valley.

BOATING AND FISHING

A public boat launch with access to the Mississippi River is available in **Riverside Park** in the nearby village of Cassville. Shoreline fishing and a riverside walking path are also available at this park. Anglers can expect to catch walleye, sunfish, bluegill, northern pike, catfish, and other species of fish that inhabit the Mississippi River.

ACCESSIBLE FACILITIES AND TRAILS

The park's entrance station and the **Mound Point Picnic** overlook and shelter areas are accessible for mobility-impaired visitors. Campsite 12 is wheelchair-accessible and located near the park's shower and flush-toilet building. The steep topography and numerous steps along bluff-side hiking trails may be challenging for visitors with walking difficulties. The paved sections of the Oakwood Trail and Woodbine Nature Trails are somewhat wheelchair accessible with assistance.

CAMPING

Nelson Dewey has 45 campsites (21 electric), which are located within well-shaded hardwood forest areas. Four walk-in campsites are situated along the bluff area with exceptional views of the Mississippi River Valley. The campground has a central restroom/shower facility and an RV water fill/sanitary station. An outdoor group camp with three campsites is located in lower region of the park adjacent to Dewey Creek. Each campsite has 50 amp electric service and can be reserved for both tent and trailer/RV camping units. The campground has vault-type restrooms, water fountains, a shelter building, and can accommodate up to 120 people in total.

WINTER ACTIVITIES

Nelson Dewey's entrance road is closed in winter but the entrance station parking area is generally plowed open for vehicles. Visitors can hike, snowshoe, or cross-country ski into the park. None of the trails are groomed.

AREA ATTRACTIONS

Western Grant County is known for its outstanding scenic beauty, historical sites, outdoor recreation, and welcoming small towns such as Cassville. www.cassville.org/tourism www.grantcounty.org/tourism

The **Stonefield Historical Site** is located in the lower section of Nelson Dewey State Park adjacent to the Mississippi River. This outdoor museum features a collection of 30 historic buildings that represent a small farming community as it would have appeared in the 1890s. Tours are available inside the Gothic revival-style mansion built by Wisconsin's first governor, Nelson Dewey, in 1868. Adjacent to Stonefield is the **Wisconsin State Agriculture Museum**, which features the largest collection of antique farm equipment in the state. Both museums are operated by the Wisconsin Historical Society. www.stonefield.wisconsinhistory.org

DIRECTIONS TO THE PARK

Nelson Dewey is located about one mile north of the village of Cassville in western Grant County. From Highway 133, exit onto County Highway VV and travel north to the park entrance.

Nelson Dewey State Park
12190 County Road VV
Cassville, WI 53806
(608) 725-5374

Stonefield Historical Site and the State Agriculture Museum

48. NEW GLARUS WOODS STATE PARK
Doorway to America's "Little Switzerland"

Picnic area playground

PARK SNAPSHOT

New Glarus Woods State Park is located in Green County within the scenic rolling hills and valleys of southwestern Wisconsin. This attractive 435-acre park offers rustic camping, hiking trails, and a picnic area and playground. The park's old-age hardwood forests and restored prairies is a favorite destination for wildlife viewing, especially birdwatching. A short hookup trail from the state park leads to the Sugar River State Trail in the nearby Swiss-themed village of New Glarus.

GEOLOGY AND HISTORY

The landscape of the park marks the divergence of two separate geologic regions found in Wisconsin. The massive continental glaciers that pushed south through most of northern and eastern Wisconsin during the last Ice Age stopped short of the park area by only a few miles. As a result, this area and much of Green County became a division line between the glaciated region of the state and non-glaciated Driftless Area of southwestern Wisconsin. As the last remnants of glacial ice continued to retreat out of Wisconsin about 12,000 years ago, several different Native American cultures began to repopulate this area. By the time white settlers and miners arrived in the New Glarus area in 1828 to set up trading posts and dig for lead deposits in the area, the Sauk, Fox, and Ho-Chunk people had been living here for generations.

In 1845, a group of men representing the Swiss Emigration Society of the Canton of Glarus, Switzerland crossed the Atlantic to find a suitable location to build a new Swiss settlement in America. After scouting the New Glarus area, the delegates purchased 1,200 acres of land on behalf of the organization. That same year, the first 108 Swiss immigrants arrived to set up farmsteads in the area. Each homesteader family was given a section of land to farm and a forest plot within the nearby New Glarus Woods for firewood and lumber. These early Swiss immigrants set up dairy farms and built some of the first cheese factories in the state. Their dream of establishing a new American Swiss colony became a reality in 1901 when the community of New Glarus became an incorporated village.

> **Ranger Note**
>
> A section of an historic trading route used by Native American people for hundreds of years (now County Highway NN) is located within New Glarus Woods State Park. Early lead miners widened this trail into a dirt road to transport lead ore by ox cart to the village of Mineral Point for smelting. During the early years of Wisconsin, this rustic road became an important link in a network of frontier and military road connections between the Mississippi River and Lake Michigan. In 1931, a monument to mark this historic **Old Lead Trail** was erected at New Glarus Woods.

The Swiss culture of the area has been well preserved over the years. Today, the New Glarus area has become one of Wisconsin's most popular tourist destinations and is known throughout the country as "America's Little Switzerland." In 1934, a section of the forest acquired by the Swiss Emigration Society was purchased by the State of Wisconsin to develop New Glarus Woods State Park.

BEACHES, PICNIC AREAS, AND SHELTERS
The park has a picnic area that features an award-winning children's playground and an open-air shelter with electrical service. The shelter can be reserved in advance for groups up to 64 people. A public outdoor swimming pool is available in the nearby village of New Glarus at the community's Village Park.

HIKING, BIKING, AND NATURE TRAILS
The park has 13 miles of hiking trails to explore. The **Bison Nature Trail** (0.8 miles) loops through a restored prairie area in the north end of the park with

interpretive signs that identify many of the prairie plants found here. One interpretive panel tells the story of the American bison and elk that once roamed the prairie regions of Wisconsin. A life-sized statue of a bison located along this trail serves as a monument to the many extirpated mammals and birds no longer found here.

The **Sugar River Spur Trail** (1.5 miles) leads from New Glarus Woods north to the village of New Glarus and the headquarters trailhead of the **Sugar River State Trail.**

The **Havenridge Nature Trail** (4.2 miles) leads through a mature oak and hickory forest, a restored prairie, and a lowland meadow. A 36-stop trail guidebook explains the flora, fauna, and natural features of the park. The **Basswood Nature Trail** (0.4 miles) located adjacent to the picnic area has interpretive signs that identify native trees and plants and explains the early history of New Glarus Woods.

> **Ranger Trivia**
>
> The Sugar River was named after a Native American term, *Toon-a-Sook-Ra* or "Sugar River." Historians speculate that the river may have gotten its name from the glistening, tan-colored sand found on its river bottom that resembles the maple sugar once made here by indigenous people. Maple trees were tapped for sap and collected in birch-bark baskets in early spring. The sweet sap was boiled down into dry maple sugar, which was used throughout the year to season food.

> **Ranger Trail Pick**
>
> The **Sugar River State Trail** (24 miles) is one of Wisconsin's premier bicycle and hiking trail. The trail begins at the historic train depot located in the village of New Glarus and heads south and west along a former railroad grade adjacent to the Little Sugar River. The trail passes through four small communities and crosses 14 trestle bridges, including a beautiful wooden covered bridge before reaching the village of Broadhead. Bicycle rentals are available at the train depot visitor center along Railroad Street in New Glarus.

WILDLIFE VIEWING

The park's upland forest areas, prairies, wetlands, and the nearby Little Sugar River provide ideal habitat for the hundreds of bird species that either nest or migrate through this area. Birdwatchers can expect to see dozens of species of woodland and grassland songbirds, plus pileated woodpeckers, bald eagles, sandhill cranes, and several types of owls, hawks, and waterfowl. Mammals such as raccoon, deer, fox, coyote, squirrels, chipmunks, opossum, and rabbits are commonly seen in the park.

The park's forests include oak, maple, walnut, hickory, cherry, and basswood trees. Some of the larger trees in the park are believed to be at least 250 years old, remnants of a much larger forest that once covered this hill in pre-settlement days.

INTERPRETIVE PROGRAMS AND FACILITIES

Nature hikes are presented by park volunteers throughout the season. Evening programs are occasionally offered at the **Chattermark Amphitheater** located between the family and group campgrounds. Interpretive panels are spaced along the Bison and Havenridge nature trails.

The nearby **Swiss Historical Village and Museum** has more than a dozen historic ethnic buildings and interpretive displays that highlight the Swiss immigrant settlement of the region during the 1850s. The museum is located on 7th Avenue in New Glarus. www.swisshistoricalvillage.org

BOATING AND FISHING

Belle View Lake (44 acres) is located about nine miles northeast of the park near the village of Belleville and has good populations of panfish, crappie, largemouth bass, walleye, and northern pike. Green County has dozens of trout streams, including the nearby **Hefty Creek** and **Bushnell Creek** which offer brown and brook trout fishing. The upper section of the Little Sugar River is classified as a Class II trout stream. The Sugar River is also popular for its canoeing, kayaking, tubing, and fishing opportunities. Several watercraft rental services and fishing guide services are available locally.

ACCESSIBLE FACILITIES AND TRAILS

Due to the hilly topography of the park, some of the trails may be challenging for visitors with walking difficulties. The **Walnut Trail** (0.5 miles), located just south of the family campground, is relatively flat and leads through an attractive forested area of the park. A side trail leads to a unique accessible wildlife observation blind. Campsites 4 and 53 are wheelchair accessible. Both have electrical service and paths to accessible vault-type restrooms.

CAMPING

The park has **38** rustic campsites. Most sites are well shaded but many are small in size, making them more suitable for tents or small trailers. None of the campsites at New Glarus Woods State Park have electrical service except those reserved for mobility-impaired campers.

A group campground with six (tent-only) campsites is located north of the family camping area. Each site can accommodate up to 20 campers. Campsites are situated in both wooded and open grassy areas. There are 14 additional walk-in campsites located along the Sugar River Trail Spur in the northeast section of the park.

WINTER ACTIVITIES
Cross-country skiing, snowshoeing, wildlife watching, and hiking are favorite winter activities in the park. None of the trails are groomed. All campgrounds are closed from November 30th through March 31st.

> **Ranger Tip**
>
> There are no flush restrooms or showers at **Nelson Dewey** but these facilities are available for a small fee at the Village Park public swimming pool located along 2nd Street in New Glarus.

AREA ATTRACTIONS
The New Glarus area is known for its Swiss culture, architecture, museums, and European-inspired restaurants, cheese, and chocolate shops.
www.swisstown.com www.greencounty.org

The **New Glarus Brewing Company**, located just across the street (Highway 69) from the park, has become a tourist destination in itself. This facility is home to the popular "Spotted Cow" beer and the first brewery to be founded and operated by a woman in the United States. Visitors can take a self-guided tour of the brewery and visit the tasting room and a beer pub any day of the week. Guided tours are offered on Saturdays.
www.newglarusbrewing.com

DIRECTIONS TO THE PARK
New Glarus Woods is located about 28 miles southwest of Madison in Green County. From Madison, take Highway 18 west and exit onto Highway 69/39 south to New Glarus. The park entrance is off of County Highway NN just south of the village of New Glarus.

New Glarus Woods State Park
W5508 County Highway NN
New Glarus, WI 53574
(608) 523-4427

49. Perrot State Park

Scenic Park Towers above the Mighty Mississippi

Trempealeau Mountain

PARK SNAPSHOT

The lofty bluffs and panoramic vistas found at Perrot State Park have drawn people to this scenic spot for untold generations. The park's bluffs tower 520 feet above the junction of the Mississippi and Trempealeau rivers. Today, visitors still flock to this site to enjoy the awe-inspiring view of the Mississippi River Valley. This popular 1,270-acre property offers camping, picnicking, nature study, biking, boating, fishing, wildlife viewing, and some of the most scenic hiking trails in the state.

GEOLOGY AND HISTORY

The exposed rock of the park's bluffs was formed more than 600 million years ago when most of Wisconsin was submerged beneath a shallow inland sea. Over time, underwater deposits of sand, silt, and sea life were transformed into deep layers of sandstone and limestone. When the sea receded, most of the sandstone deposits eroded over time into small hills and flat plains due to the action of wind, frost, and rain. The towering bluffs of the park survived thanks to a hard limestone cap which protected them from erosion.

Perrot lies within the Driftless Area of the state, a region of southwestern Wisconsin that escaped the crushing effect of glaciers during the last Ice Age. About 12,000 years ago, melting glaciers to the north sent torrents of water carrying sand, rocks, and other glacial debris down the Mississippi River. These massive flood events eroded and reshaped all the bluffs along the river, including the 425-foot-high Trempealeau Mountain at Perrot State Park.

Native American people canoed up and down the Mississippi River to hunt, fish, and conduct trade with other tribes for untold generations prior to its "discovery" by European explorers. Trempealeau Mountain was an important landmark along this water route. The Dakota (Santee Sioux), Ho-Chunk, Menominee, and other native people all camped along its shoreline while traveling on the river. Evidence of even earlier inhabitants of this region has been discovered in the many burial and effigy mounds found in this area. Some of these prehistoric structures were built more than 4,000 years ago.

> **Ranger Note**
>
> Trempealeau Mountain is the only bluff along the Mississippi River entirely surrounded by water. Early French explorers named it *La montagne qui tempe a l'eau* , meaning "a mountain whose foot is bathed in water."

> **Ranger Trivia**
>
> How do you pronounce Perrot? Online phonetic pronunciations range from "Pay-Roat" to "Pay-Ho," but the most common use today is "Purr-Row."

Perrot State Park was named as a tribute to Nicholas Perrot, the first known European to explore this area. In the fall of 1685, Perrot and his men set up a winter camp at a site along the Mississippi River near what is today the park entrance. Perrot chose this encampment to take advantage of towering bluffs, which provided shelter from the cold winter wind. This area also had abundant game, fish, and firewood, vital to the survival of Perrot's expedition.

The site of Perrot's encampment later became an important fur trading post and rendezvous site for Native American trappers and hunters. In 1735, French soldiers built a fort at this site. European immigrants from Norway, Scotland, Poland, and Germany began to arrive in the mid-19th century to work for lumber companies and set up farmsteads in the area. The nearby village of Trempealeau, established in 1852, became an important Mississippi River port for shipping wheat and lumber from the local area.

49. Perrot State Park

Perrot State Park was established in 1918 after 1,010 acres of land were donated to the State of Wisconsin by John Latch, a businessman from Winona, Minnesota. Much of the early development of the park was completed by the Civilian Conservation Corps (CCC), a federal Depression-era work program for unemployed young men in the 1930s.

BEACHES, PICNIC AREAS, AND SHELTERS

A large picnic area and attractive indoor shelter building are located above the shoreline of Trempealeau Bay. There is no swimming beach in the park but an outdoor municipal swimming pool is located in the nearby village of Trempealeau along Fourth Street.

HIKING, BIKING, AND NATURE TRAILS

There are 15 miles of hiking trails to explore at Perrot. The **Black Walnut Nature Trail** (0.5 miles) interprets how early Native American people lived off the land here. A sandstone rock shelter once used by early nomadic hunters is located along this trail. The **Riverview Trail** (2.5 miles) is a fairly level route that begins near the campground and follows the upper shoreline of Trempealeau Bay. Along the trail is an overlook area with a spotting scope for great views of the river valley.

> **Ranger Tip**
>
> A short walk north of the Riverview Trail Overlook leads to a picturesque rock ledge called **Horseshoe Falls**. A small waterfall tumbles over this ledge during rainy periods.

Perrot Ridge Trail (1.5 miles) starts from the **Mounds parking area** near the park office and leads uphill toward the Perrot Ridge summit 520 feet above the river valley. The trail continues west along the crest of the ridge to **Reed's Peak** and then meanders downhill to the **Nicholas Perrot Encampment Historical Marker**. From here a short stretch of the Riverview Trail leads back to the Mounds parking area.

The **Voyageurs Canoe Trail** (3.4 miles) starts at the park's canoe landing near the campground area. This water route glides through the waters of beautiful Trempealeau Bay. Canoe and kayak rentals are available at the park office.

Bike riding is available along the nearby **Great River State Trail** (24 miles). The trail follows a former railroad grade past 500-foot bluffs, restored prairie areas, two national wildlife refuges, and Mississippi River backwater areas. Great River Trail links the city of La Crosse to the town of Marshland, located about seven miles northwest of Perrot.

Ranger Trail Pick

The most popular trails in the park all lead to the top of **Brady's Bluff,** which towers 520 feet above the river valley. The scenic **East Trail** (0.7 miles) begins at the Mounds Parking Area and leads through a section of **Brady's Bluff Dry Prairie State Natural Area**. This trail offers spectacular views of Trempealeau Mountain and the Minnesota bluffs across the Mississippi River. Near the summit of Brady's Bluff is a small stone and timber shelter built by Civilian Conservation Corps (CCC) in the 1930s. From Brady's Bluff, the **West Trail** (0.5 miles) leads down along stone and wooden staircases past sandstone outcrops to a more gradual descent to the park's boat launch area. A short hike to the east along the Riverview Trail leads back to the Mounds parking area. The **North Trail** (1.0 miles) offers a more gradual uphill route through a forested area to the summit of Brady's Bluff. This trail begins near the park shop area.

Brady's Bluff - East Trail

WILDLIFE VIEWING

White-tailed deer, squirrels, raccoon, chipmunks, coyote, and fox are often seen in the park's upland forests. Wetland animals such as beaver, muskrat, otter, and mink and painted, snapping, soft-shelled, and wood turtles can often be spotted in backwater areas of the Mississippi. Bluffs areas harbor fox snakes, blue racers, bull snakes, and (rarely) timber rattlesnakes. Remnant short-grass openings called

Goat Prairies are present along upper bluff areas. These rare ecosystems support prairie plants such as puccoon, lead plant, pasque flowers, and coneflowers. Several types of prairie grass and forbs normally found in more western states also grow here.

Adjacent to Perrot State Park is **Trempealeau National Wildlife Refuge**, a 6,200-acre wildlife area and important part of the **Great Mississippi River Flyway**. More than 200 species of birds are known to nest or migrate through this area including woodland songbirds, shorebirds, cranes, herons, turkey vultures, osprey, bald eagles and several species of hawks, owls, and waterfowl.

INTERPRETIVE PROGRAMS AND FACILITIES

The **Perrot Nature Center** is located inside the **Trempealeau Bay Picnic Area** shelter. The center has several interesting displays highlighting the flora, fauna, and human history of the area. Interpretive programs, hikes, guest speakers, and special events are offered throughout the summer season.

BOATING AND FISHING

A boat launch along Trempealeau Bay provides access to the Mississippi River. A separate canoe/kayak launch is located south of the camping area. Anglers can expect to catch panfish, perch, bass, crappie, northern pike, walleye, catfish, and other Mississippi River fish. Shoreline fishing is available along **Third Lake (35 acres)**, a backwater flowage of the Mississippi River, located south of the nearby village of Trempealeau.

ACCESSIBLE FACILITIES AND TRAILS

Most of the park's buildings, restrooms, and shower facilities are accessible for mobility-impaired visitors. Campsite 4 is wheelchair-accessible and has electrical service. A paved walkway leads to a nearby flush restroom facility. A specially-adapted kayak for use by people with physical disabilities can be reserved free of charge for boating on Trempealeau Bay. Due to the steep and rocky topography of the park, most hiking trails here are challenging for visitors with walking impairments.

CAMPING

Perrot has 107 campsites (38 electric) located east of Trempealeau Bay in the northern section of the park. The campground has a restroom/shower facility and an RV water-fill and sanitary station. Four walk-in group campsites (tents only) are located adjacent to the family campground. Group sites can accommodate up to 20 people. Wheeled carts are available to transport camping gear to campsites.

WINTER ACTIVITIES
Park staff groom nine miles of cross-country ski trails for both traditional and skate skiing. Winter camping, hiking, wildlife watching, and snowshoeing are popular off-season activities. Snowshoes are available for rent at the park office.

AREA ATTRACTIONS
The Trempealeau area is known for its outstanding scenery and many outdoor recreational opportunities. www.co.trempealeau.wi.us

  Lock and Dam #6 is an interesting local attraction to visit. A public observation platform here allows visitors a chance to see how a Mississippi River lock and dam operate. Tugboats, pleasure watercraft, and massive river barges regularly travel through this lock. Built in 1936, the dam has 2,600 feet of earthen embankment, a 600-foot long concrete lock, and 10 gates. The site is located along the riverfront in the nearby village of Trempealeau. www.mvp.usace.army.mil/lock-dam 6

DIRECTIONS TO THE PARK
Perrot State Park is located 24 miles north of La Crosse. From Interstate I-90 in La Crosse, exit onto Highway 53 North through the community of Holmen and then turn west onto Highway 35 to the village of Trempealeau. Drive through town on Main Street to the Mississippi River and then turn north onto First Street/ Sullivan Road to the entrance to the park.

Perrot State Park
W26247 Sullivan Road
Trempealeau, WI 54661
(608) 534-6409

50. ROCHE-A-CRI STATE PARK

Majestic Bluffs and Ancient Rock Art

Roche-A-Cri Mound

PARK SNAPSHOT

Roche-A-Cri State Park has been open to the public for more than 70 years but many have yet to discover this scenic jewel. The 604-acre park was named after Roche-A-Cri Mound, one of the tallest sandstone bluffs in Wisconsin. The park offers camping, hiking, picnicking, wildlife viewing, and nature study and is one of only a few sites in the Midwest where ancient Native American rock art can be seen. Visitors can reach the summit of the mound via a staircase to enjoy outstanding views of the surrounding area.

GEOLOGY AND HISTORY

The park's 300-foot-tall sandstone bluff was formed millions of years ago from deposits of sand, silt, and sea life that settled to the bottom of a great inland sea that once covered most of Wisconsin. Geologists have discovered fossils of extinct sea life imbedded within the park's rock formations, including the very top of the mound. Roche-A-Cri Mound has survived nearly intact for millions of

287

years thanks to a very hard and resilient sandstone cap that has shielded it from wind, rain, and frost erosion over time.

The earth-crushing glaciers that pushed through much of the rest of the state during the last Ice Age never reached Roche-A-Cri but the mound was engulfed by Glacial Lake Wisconsin, which submerged 1,800 square miles of central Wisconsin about 15,000 years ago. At that time, Roche-A-Cri Mound would have appeared as an island rising more than 150 feet above the surface of the glacial lake. Submerged sections of the bluff were shaped by wave action and collisions from floating icebergs during that period. Near the end of the Ice Age about 14,000 years ago, the glacial dam that created Lake Wisconsin near the Wisconsin Dells suddenly collapsed. This caused the entire lake to drain through Wisconsin River Valley, located about ten miles west of the park. Today, the sandy plains, hills, and wetlands surrounding Roche-A-Cri are all that remains of this extinct glacial lake.

Native American cultures are known to have lived in the Roche-A-Cri region for centuries prior to European settlement. In 1851, a government surveyor discovered several Native American carvings and paintings on an exposed rock wall near the south end of Roche-A-Cri Mound. Many of these ancient works of art depicted birds, foot tracks, and other avian subjects. Archeologists speculate this early form of rock art may have represented a connection between the mythical Thunderbird and man.

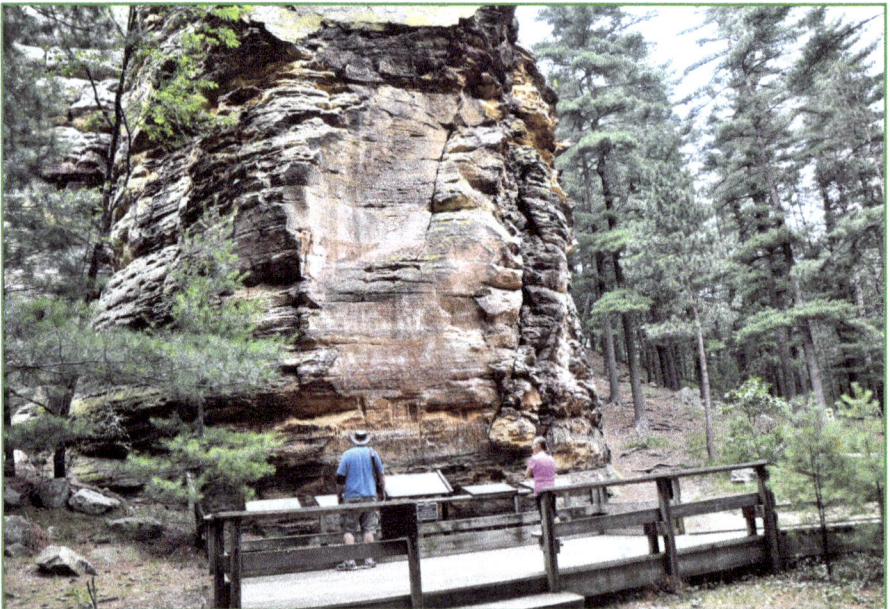

Native American rock art site

Ranger Note

"Rock Art" refers to both petroglyphs (rock carvings) and pictographs (rock paintings) created by early Native Americans artists. Archeologists believe the rock carvings at Roche-A-Cri were etched into the face of the bluff sometime before 900 A.D. while the paintings may have been completed as early as 400 years ago. Some of the artwork could have been done by ancestors of the Ho Chunk people who still consider Roche-A-Cri a sacred site.

Early 17th-century French explorers gave the towering sandstone bluff the name, *La Roche-A-Cri* or "the crevice in the rock," a reference to the notch-like opening in the top of the bluff, which could be seen for miles around. Another French interpretation reads, "The shouting or screaming rock," perhaps due to the cries of eagles that once nested on this bluff. Immigrant European settlers attempted to raise crops in the local area in the late 19th century but found the sandy soils of the region unsuitable for agriculture. A few enterprising farmers discovered that the wetlands of the region were ideal for growing cranberries, still the leading agricultural crop of the area today.

In 1937, the State Highway Commission purchased the Roche-A-Cri property to create a rest stop for motorists along the newly developed State Highway 13. The initial development of the park was conducted by work crews from the Depression-era Civilian Conservation Corps (CCC) in the 1930s. The property was later transferred to the State Park System to establish Roche-A-Cri State Park in 1948.

BEACHES, PICNIC AREAS, AND SHELTERS

The **Park Entrance Picnic Area** has tables, grills, a playground, and an open-air shelter with electrical service. A second picnic area is located near the start of the **Mound Trail**, which leads to the top of the Roche-A-Cri bluff. The park does not have a swimming beach but one is available at **Friendship Lake Park** located 1.5 miles south of the park along North Main Street in the village of Friendship.

HIKING, BIKING, AND NATURE TRAILS

There are 6.5 miles of hiking trails at Roche-A-Cri. The **Mound Trail** (0.3 miles) leads to a somewhat challenging 300-step staircase to the summit of Roche-A-Cri Mound. A boardwalk on top of the bluff provides access to an observation deck with a panoramic view of the fields and forests below. Interpretive panels here highlight local landmarks and explain the origin of the mound.

Acorn Trail (3.5 miles) leads through upland forests of oak, pine, and maple trees along the outer perimeter of the park. The **Turkey Vulture Trail** (0.9 miles) loops off the Acorn Trail to a restored prairie area on the park's southern border. The **Spring Peeper Trail** (0.25 miles) is a short path that follows the edge of **Carter Creek** and links up with the Acorn Trail. The trail is named after a common woodland frog called a spring peeper, known for the loud "peeping" call they make in early spring.

Ranger Trail Pick

The **Mound Trail** ascent to the top of Roche-A-Cri Bluff is a must for those willing and able to climb the steep staircase, but the nearby **Chickadee Rock Nature Trail** (0.3 miles) is a fun hike as well. The trail leads through a relatively flat oak and jack pine forest with several interpretive signs along the way. Near the trail's junction with the Acorn Trail is an impressive 25-foot-high sandstone formation called **Chickadee Rock**. Nearby are smaller rock formations and ledges, which are ideal for low-level rock climbing. A wooden bench located at the base of Chickadee Rock is a great place to sit, relax, and enjoy the sights and sounds of nature.

WILDLIFE VIEWING

The park's diverse landscape of forests, prairies, wetlands, and bluffs provide ideal habitat for wildlife. Woodland birds such as blue jays, cardinals, owls, thrushes, chickadees, woodpeckers, and warblers can be seen throughout the park. Hawks and turkey vultures are frequently seen soaring above the bluff area during the warm season. White-tailed deer, squirrels, chipmunks, rabbits, fox, coyote, and many other mammals make their home here as well. Several rare species of plants, ferns, and fungi find ideal growing conditions on the rocky crevices of the sandstone formations.

Ranger Tip

Most of Roche-A-Cri Mound is concealed from view within the main park area due to heavy forest cover. A better view of the bluff can be seen from a parking lot located along Czech Avenue south of the park entrance.

Roche-A-Cri Mound State Natural Area (25 acres) encompasses the entire mound and surrounding old-growth pine and oak forest. The **Roche-A-Cri Woods State Natural Area** (442 acres) is a rarely visited area of the park that features relatively undisturbed woodlands and prairie areas. The site is located northeast of the park entrance across Highway 13.

INTERPRETIVE PROGRAMS AND ACTIVITIES

An interpretive kiosk located at the beginning of Mound Trail describes the construction the mound staircase and the flora and fauna of the park. Interpretive panels are also located at the Native American rock art site, the bluff observation deck and along the Chickadee Rock Nature Trail. Interpretive programs and hikes are presented by park volunteers throughout the use season

BOATING AND FISHING

Shore fishing for sunfish and brook and brown trout is available along Carter Creek in the south section of the park. **Friendship Lake** (125 acres) located just south of the park, offers shoreline and watercraft (slow-no-wake) fishing for sunfish, bass, northern pike, and walleye. A public boat landing is located along North Main Street in the village of Friendship. A Class 1 trout stream is located along the eastern edge of this lake.

ACCESSIBLE FACILTIES AND TRAILS

The **Park Entrance Picnic Area** has an accessible picnic table, grill, and water fountain. The Native American rock art display has a wheelchair-accessible walkway and ramp that leads to a wooden platform for viewing the ancient pictographs and interpretive panels. The **Chickadee Nature Trail** (0.68 miles) is a relatively level and graveled walking path that may be suitable for wheelchair use with assistance.

CAMPING

Roche-A-Cri has 42 rustic campsites (6 electric) located within a well-shaded pine and oak forest area. Most sites can accommodate tents and small trailers but larger camping units may have difficulty setting up in some areas. An RV water fill and sanitary station is located near the park office. The Roche-A-Cri campground does not have flush toilets or shower facilities. A list of local vendors that offer shower facilities for a small fee is available at the entrance station. Campers who prefer more modern amenities will find them at **Pentenwell Lake County Park** located about 16 miles northwest of the park along Big Horn Drive near the community of Arkdale. Pentenwell Lake has 500 sites, most with electric service and shower facilities. www.adamscountywi.com.

WINTER ACTIVITIES

Snowshoeing, hiking, wildlife watching, and cross-country skiing are popular winter activities at Roche-A-Cri. None of the trails are groomed. The main entrance road is closed during the off-season but visitors can enter the park from the parking area off of Czech Avenue along the south boundary of the park.

AREA ATTRACTIONS

Adams County is a popular vacation destination with many parks, lakes, wildlife areas, and outdoor activities to enjoy. www.visitadamscountywi.com

Quincy Bluff State Natural Area is one of the best known natural areas in Wisconsin. This 4,870-acre nature preserve features undisturbed wetlands, upland forests, and sandstone formations. The towering 200-foot Quincy Bluff can be reached by hiking the Lone Rock Trail, a 6.6-mile loop that ascends 120 feet uphill to the crest of the bluff with outstanding views of the unbroken wilderness below. Quincy Bluff is located about 19 miles south of Roche-A-Cri, west of Highway 13 along 14th Court near the community of Adams. www.dnr.wi.gov/topic/lands/naturalareas

DIRECTIONS TO THE PARK
Roche-A-Cri State Park is located in Adams County along Highway 13 just south of Highway 21 between Interstates I-90/94 and I-39/51, about 1.5 miles north of the village of Friendship.

Roche-A-Cri State Park
1767 Highway 13
Friendship, WI 53934
(608) 339-6882

51. ROCKY ARBOR STATE PARK

A Natural Rock Garden Wonder

Rocky Arbor Nature Trail

PARK SNAPSHOT

Rocky Arbor State Park is located less than two miles from the Wisconsin Dells, one of the busiest and most popular vacation destinations in Wisconsin. For many, this small 245-acre state park serves as a quiet green oasis from the hustle and bustle of waterparks, resorts, amusement rides, and other crowded Dells attractions. The park's towering oak and white pine woodlands gives an ambience of a remote northern forest despite being in central Wisconsin. Rocky Arbor has camping, picnicking, hiking, and exceptional birdwatching, plus attractive fern-covered bluffs and rare sandstone formations.

GEOLOGY AND HISTORY

Most all Wisconsin state parks and forests have scenic landscapes but very few were established primarily due to their unique geology alone. According to written historical accounts, Rocky Arbor was set aside as a state park specifically "to protect the property's sandstone outcrops, walls and ledges." The park's exposed rock was formed more than 500 million years ago beneath an inland sea that once covered most of Wisconsin. Sand and other sediment deposits beneath this ocean were compressed and chemically transformed into sandstone over time. As the sea receded, geologists believe an ancient version of the Wisconsin River cut through sandstone layers at Rocky Arbor, creating the 90-foot-high bluffs and other water-sculptured rock formations seen in the park today.

The park's landscape was transformed again during the last Ice Age when a mile-high glacier stopped short of Rocky Arbor by only a few miles to the northeast. The stalled glacier created a massive ice dam, causing more than 15,000 acres of central Wisconsin to become submerged under what became known as Glacial Lake Wisconsin. When the ice dam finally collapsed about 14,000 years ago, a massive surge of water filled with sand, rocks, and chunks of glacial ice was released downstream. This thundering surge of melt water excavated a "new" Wisconsin River channel more than a mile east of Rocky Arbor, leaving its former river channel in the park high and dry.

Several different Native American cultures inhabited the local area over time, including the Chippewa and Menominee but when the first European explorers reached the Dells area in the 17th century, the Ho-Chunk people were the dominant tribe of this region. After the Black Hawk War of 1832, the Ho-Chunk people were forced to cede all their remaining land in the Wisconsin Territory to the U.S. government. Soon after, lumber companies began to cut and float untold millions of pine logs down the Wisconsin River from northern forests. The logs were cut into lumber at dozens of sawmills that lined the riverbank, including several near the Wisconsin Dells.

> **Ranger Note**
>
> The Wisconsin Dells area was known as The Dalles long before it was settled. Early French maps indicated "The Dalles of the Wisconsin River" as far back as the 17th century. The term "Dalles" was a French word used to describe fast-moving river rapids within a rock-lined gorge.

In 1852, railroad executive Byron Kilbourn built one of the first railroad lines across the state to transport lumber and agricultural products to urban markets and deliver much needed supplies back to rural areas. Kilbourn built a new town south of Rocky Arbor along the Wisconsin River that became known as Kilbourn City. The town was later renamed Wisconsin Dells in 1931.

The State Highway Commission purchased the Rocky Arbor property in the early 1930s as part of the construction route for Highway 12. When news of the plan to build a highway through this rare, undeveloped natural area came to light, scientists and citizens from throughout the state voiced their objections. Thanks to their efforts, the state was convinced to reroute the new highway and Rocky Arbor State Park was established in 1932.

BEACHES, PICNIC AREAS, AND SHELTERS

Rocky Arbor's picnic area is located near the visitor entrance station within a towering old-growth stand of white pine trees. The picnic area has a playground and an open-air shelter that can be reserved in advance. There is no swimming beach at Rocky Arbor but **Mirror Lake State Park,** (See Chapter 45) located about ten miles south of the park, has a wide sandy beach.

HIKING, BIKING, AND NATURE TRAILS

There are 2.5 miles of hiking trails at Rocky Arbor, including a connector trail that serves to link the picnic area and campground.

The Dells of the Wisconsin River State Natural Area (1,300 acres) preserves a 5-mile-long section of sandstone canyons along the river. The **Chapel Gorge Trail** (2.0 miles) travels through this site and leads to one of the only natural sand beaches along the Wisconsin River. The Dells Natural Area is located about two miles south of Rocky Arbor off of Highway 12 and River Road near Wisconsin Dells. www.dnr.wi.gov/topic/lands/naturalareas

Ranger Trail Pick

The **Rocky Arbor Nature Trail** (1.0 miles) showcases many of the park's most attractive water-sculptured sandstone bluffs. The lower section of the trail parallels a small spring-fed creek that runs through a sandstone-lined gorge. Many types of ferns, some quite rare, grow adjacent to the trail and along the rocky bluffs. Two wooden staircases lead to the top of the bluff with panoramic views of the lush, green forested gorge below. The upper bluff section of the trail leads through an oak and white pine area of the park. Interpretive signs along the trail explain the geology of the area and identify some of the plants that grow here.

WILDLIFE VIEWING

Rocky Arbor's diverse landscape of wetlands, creeks, sandstone bluffs, and upland forests of oak and pine offers ideal habitat for white-tailed deer, chipmunks, rabbits, raccoons, squirrels, fox, and coyotes. The park is a favorite site for birdwatchers, especially during the spring and fall migration period when dozens of species of woodland songbirds, hawks, owls, and many other birds can be

Water-sculpted sandstone rock

spotted here. Hundreds of native tree and shrub species and other plant life thrive here, including several rare and threatened plants and ferns that only grow on sandstone bluffs.

Nearly a third of Rocky Arbor State Park is located across Highway 12 to the east. This little-visited region of the park is great place to observe undisturbed wildlife, sandstone rock ledges, and bluff formations.

INTERPRETIVE PROGRAMS AND FACILITIES
Rocky Arbor Nature Trail has interpretive panels that describe the flora, fauna, and geology of the park. Interpretive programs and hikes at Rocky Arbor are coordinated with staff at Mirror Lake State Park.

BOATING AND FISHING
Mirror Lake (137 acres) and **Lake Delton** (249 acres) are located south of the park along Highway 12. Both of these man-made lakes have boat launch areas and watercraft rentals. Anglers can expect to catch northern pike, bass, walleye, perch, crappie, catfish, crappie, bluegills, and other panfish in these waters. The **Hulbert Creek State Fishery Area** located south of Rocky Arbor is a Class 1 stream for brook trout and Class 2 for brown trout.

ACCESSIBLE FACILITIES AND TRAILS

The park's picnic area and shelter facility are both accessible for mobility-impaired visitors. Campsite 35 can accommodate wheelchairs and has electrical service. A paved access route leads to an accessible restroom/shower facility. Due to the steep topography of the park, travel on hiking trails may be difficult for visitors with walking conditions. The lower section of the Rocky Arbor Nature Trail is somewhat accessible for visitors with moderate walking disabilities.

CAMPING

Rocky Arbor has 89 campsites (19 electric) located within a well-shaded oak and pine forest area in the northwest section of the park. The campground has a centrally located flush toilet and shower facility. An RV water fill and sanitary station is located near the campground entrance road.

Rocky Arbor's campground is only open from Memorial Day weekend through Labor Day. During the off-season, campers are directed to camp at **Mirror Lake State Park,** which is open all year. (See Chapter 45)

> **Ranger Tip**
>
> The Rocky Arbor campground is located between Highway 12 to the east and Interstate I-90/94 near its western edge. Campers who find traffic noise disturbing may want to occupy campsites near the middle or eastern section of the campground. The trees, brush, and bluffs in this area help muffle highway sounds somewhat.

WINTER ACTIVITIES

The entrance road to Rocky Arbor is closed during the off-season, but parking is allowed at the gate area along Highway 12. Visitors can snowshoe, ski, or hike into the park from this area throughout the winter season.

AREA ATTRACTIONS

The Wisconsin Dells area is well-known for its world-class resorts, waterparks, museums, and scenic boat rides on the Wisconsin River that lure thousands of visitors to this small town daily. www.wisdells.com

The **H. H. Bennett Studio and Photography** is a fascinating museum dedicated to displaying the photography equipment and iconic images taken by H. H. Bennett, one of the nation's premier photographers from 1865 to 1908. Bennett's stereoscopic (near 3-D) images of the area's sandstone bluffs, steamboats, and lumber rafts along Wisconsin River have been viewed by millions of people around the world and were instrumental in the Dells becoming a tourist destination. A reproduction of Bennett's studio is housed in the original red brick

building he had built for that purpose in 1875. The museum is operated by the State Historical Society and is located about two miles south of Rocky Arbor along Broadway Street in downtown Wisconsin Dells.
www.hhbennettstudio.wi.history.org

DIRECTIONS TO THE PARK
Rocky Arbor State Park is located in Juneau and Sauk counties about 1.5 miles north of Wisconsin Dells. From Interstate I- 90/94 take the U.S. Highway 12/16 exit and travel east about 0.8 miles to the park entrance.

Rocky Arbor State Park
N101 Highway 12/16
Wisconsin Dells, WI 53965
(608) 254-8001

52. SAUK PRAIRIE RECREATION AREA

A Diamond in the Rough

Wildlife pond

RECREATION AREA SNAPSHOT

Sauk Prairie Recreation Area is located along the southern foot of the Baraboo Bluff Range just south of **Devil's Lake State Park** in Sauk County. The 3,391-acre property was once the site of a U.S. Army ammunitions plant but visitors find little evidence of this today except for a few roads and remnant building foundations. This relatively new recreation area is undergoing several restoration projects, including reestablishing the long-grass prairie that once flourished here. The property offers hiking, horseback riding, birdwatching, nature study, dog training, and hunting opportunities. A paved section of the **Great Sauk State Trail** travels through the property as well.

GEOLOGY AND HISTORY

The landscape of Sauk Prairie has changed dramatically over time both by natural forces and development by man. Near the end of the last Ice Age about **14,000** years ago, a lobe of a massive glacier stalled here just south of the nearby **Baraboo Hills.** For the next few centuries, this massive sheet of ice continued to push forward but only at the same rate as it was melting. As a result, an earthen moraine ridge made up glacial sand and gravel was deposited along the leading edge of this glacier. Geologists refer to this glacial formation as the **Johnson Terminal Moraine.** Today, it appears as an elevated grassy hill that leads through the center of the Sauk Prairie Recreation Area.

At about the same time another glacier located northeast of the Wisconsin Dells formed a massive ice dam, which blocked the flow of the Wisconsin River, submerging most of central Wisconsin under 160 feet of water. When this dam eventually collapsed, a torrent of melt water and glacial debris was sent downstream carving out a new path for the Wisconsin River. The overflow of this ancient flood filled what geologists called **Glacial Lake Merrimac** located east of the recreation area. A nearly flat washout plain west of the Wisconsin River was also deposited at that time and became the 14,000-acre **Great Sauk Prairie**, part of a vast tall-grass prairie that once covered much of southern Wisconsin.

> **Ranger Trivia**
>
> The Sauk Prairie Recreation Area, Sauk County, and the nearby communities of Sauk City and Prairie du Sac were all named after the Sauk nation. In 1766, Johnathan Carver, an early English explorer, recorded what he called, "a large village of 90 well-built houses and more than 300 Sauk and Fox people living along the (Wisconsin) river," at what is now the nearby community of Sauk City.

Several Native American cultures hunted, fished, and lived along the Wisconsin River within the Sauk Prairie region over time, including the Ho-Chunk, the Fox (Meskwaki) and the Sauk (Sac) people.

Many European immigrants settled in the Sauk Prairie region during the 19th century. Most were farmers who found the dark, rich soils of the prairie ideal to grow crops and graze cattle. When America entered World War II, the U.S. Army decided to build an ammunition plant south of Baraboo within the Sauk Prairie area. Almost immediately, federal government agents began to condemn private land to start construction of the ammunition plant in 1942. Within a few months, 80 farm families were evicted from their homes and 7,000 acres of prime agricultural land was acquired by the Army.

When completed, the Badger Army Ammunitions Plant consisted of 1,400 buildings and several miles of roads and railway spurs, making it the largest facility of its kind in the world. At its peak more than 6,000 people worked seven days a week around the clock to produce nitrocellulose-based rocket propellant and smokeless gun powder for the war effort. The ammunitions plant continued in operation throughout the Korean and Vietnam wars before being decommissioned in 1997.

52. Sauk Prairie Recreation Area

Ranger Note

Local farmers had handed down family farmsteads from one generation to the next well into the 20th century throughout the Sauk Prairie area. Ironically, just as the Native American people before them, they too were forced to sell their land and leave their homes by order of the U.S. government.

In 2011, the State of Wisconsin acquired 3,391 acres of the now-abandoned ammunitions plant to establish the Sauk Prairie Recreation Area. About 2,200 acres of cropland was assigned to the **Federal USDA Dairy Forage Research Center.** The previously evicted Ho-Chunk Nation acquired 1,550-acres of the property. They are currently restoring a long-grass prairie on their property in hopes of possibly re-introducing bison here.

Today, the combined 7,000-acre property is managed jointly between Wisconsin State Parks and Forests, the Ho-Chunk nation, and the United States Department of Agriculture. Nearly all of the ammunition plant structures have been removed and clean-up of contaminated areas has been completed.

Ranger Note

The Ho-Chunk and the Dairy Forage Research Center properties are closed to public access. A color-coded property map available at the property entrance is helpful in identifying closed areas and serves as an excellent guide to explore the state recreation area.

BEACHES, PICNIC AREAS, AND SHELTERS

Day-use facilities such as restrooms, picnic areas, and shelter facilities will be developed at Sauk Prairie over time. **Devil's Lake State Park**, (See Chapter 36) located just north of the recreation area, offers all these amenities and has two swimming beaches.

HIKING, BIKING, AND NATURE TRAILS

The **Southern Trails** of the recreation area are designated for hiking and off-road bike riding. The **Central Trails** and **Northern Trails** are open for horseback riding, off-road biking, and hiking. When completed, the recreation area will offer 20 miles of hiking trails, 12 miles of equestrian trails, and a 10-mile mountain bike trail.

Ranger Note

Flush toilets and an outside drinking water tap are available for public use at the Badger Ammunition Plant Museum located near the vehicle entrance to the recreation area along Highway 12.

Great Sauk State Trail

Ranger Trail Pick

The **Great Sauk State Trail** (10.5 miles) is an asphalt-paved biking/hiking trail that begins in the nearby community of Sauk City and leads north adjacent to the Wisconsin River through the village of Prairie du Sac. The trail travels through the center of Sauk Prairie Recreation Area with outstanding views the Baraboo Hills to the north.

WILDLIFE VIEWING

White-tailed deer, raccoon, fox, coyote, cottontail rabbits, and wild turkey are commonly seen at the recreation area. Turkey vultures, bald eagles, and red-tailed hawks are often spotted soaring over prairie regions and the nearby Baraboo Hills. The property's ponds and wetlands area attract many species of waterfowl, shoreline birds, cranes, and herons and are home to mink, muskrats, and several types of frogs and turtles.

In an unexpected twist of fate the construction of the **Badger Ammunition Plant** in the 1940s actually improved the habitat for some wildlife species. Although large sections of the property were developed into roads, railroad lines, factory buildings, and parking lots, many of the farm fields were left untouched. As a result, thousands of acres the idle crop land reverted back into grasslands; similar to the prairies that once dominated this area. Meadowlarks, bobolinks, bobwhite quail, dickcissels, and many other uncommon birds were able to flourish here at a time when other grassland birds were disappearing in rural areas due to intense agricultural practices.

INTERPRETIVE PROGRAMS AND FACILITIES
More than 20 interpretive sites have been identified throughout the property, including scenic overlooks, ponds, historic cemeteries, restored prairies, old farmsteads, and remnants of the former Badger Army Ammunition Plant. The locations of these sites are indicated on property maps available at the entrance to the recreation area.

Volunteers from the **Sauk Prairie Conservation Alliance**, a not-for-profit support group for the site, have installed interpretive signs and built rest stops along the Great Sauk State Trail. Members of the group conduct informative hikes for the general public from time to time. A listing of their programs can be found on their website. www.saukprairievision.org.

BOATING AND FISHING
Future development plans include a carry-in boat launch along **Weigand's Bay** of the Wisconsin River located on the east side of the recreation area. An accessible fishing pier, picnic area, shelter, and restrooms are also planned for this site. **Devil's Lake** (374 acres) located north of the recreation area offers both shoreline and watercraft fishing (electric motors only). The lake has good populations of northern pike, bass, walleye, brown trout, and panfish.

ACCESSBILE FACILITIES AND TRAILS
The **Great Sauk State Trail** is asphalt-paved and wheelchair-accessible. The trail travels through the center of the recreation area with scenic views of the expansive grasslands and majestic hills of the **Baraboo Range** to the north.

CAMPING
Sauk Prairie Recreation Area is a day-use-only property. Camping is available at **Devil's Lake State Park.** (See Chapter 37) The park's campgrounds are some of the busiest in the state, so advanced campsite reservations are recommended.

WINTER ACTIVITIES
The recreation area's roads are closed in winter but visitors can hike, snowshoe, or ski into the recreation area from the parking lot along Highway 12.

AREA ATTRACTIONS

Baraboo and Sauk County are popular tourist destinations with outstanding scenery and recreational opportunities plus world-class attractions such as the **Circus World Museum** and the **Wisconsin Dells**. www.saukprairie.com www.co.sauk.wi/general/visit

Pewit's Nest State Natural Area (36 acres) is a small but enchanting nature preserve. A short trail leads to an overlook above a scenic 40-foot-deep sandstone gorge that contains small waterfalls and rapid areas of **Skillet Creek**. The natural area is located ten miles north of the recreation area along County Highway W, west of Baraboo. www.dnr.wi.gov/topic/lands/naturalarea/pewitsnest

> **Ranger Trivia**
>
> **Pewit's Nest** was named after an eccentric 19th-century waterwheel mill operator who lived inside a rock opening in the gorge above the creek. Local citizens thought his unusual home looked like the nest of a "peewit," a common woodland bird now called a phoebe.

DIRECTIONS TO THE RECREATION AREA

Sauk Prairie Recreation Area is located about eight miles south of Baraboo and nine miles north of Sauk City. The property's entrance is on the east side of Highway 12 at the former entrance gate of the Badger Army Ammunitions Plant.

Sauk Prairie Recreation Area
Devil's Lake State Park (Mailing Address)
S5975 Park Road
Baraboo, WI 53913
(608) 356-8301 (ext. 111 for Sauk Prairie Information)

53. TOWER HILL STATE PARK
Wisconsin River Historical Gem

Lead shot smelter house

PARK SNAPSHOT
Tower Hill State Park preserves a rare early 19th century lead shot tower and smelting house. This small 77-acre park is located atop a towering sandstone bluff adjacent to the Wisconsin River in Iowa County. In addition to interpreting the historic shot tower, the park has a rustic campground, hiking trails, a picnic area, and offers nature study, fishing, canoeing, and kayaking.

GEOLOGY AND HISTORY
The rock-crushing force of glaciers that pushed through much of Wisconsin during the last Ice Age stopped short of Tower Hill State Park by only by a few miles to the east. As a result, the park's landscape remained untouched by glacial ice. The sandstone bluffs of the park were eroded somewhat, however, by glacial meltwater, which thundered through the Wisconsin River Valley from the north near the end of the Ice Age about 12,000 years ago.

Archeologists have found evidence that several Native American cultures have lived within the Southern Wisconsin River Valley for thousands of years, including the Sauk, Meskwaki (Fox), Potawatomi, and Ho-Chunk people. Father

Jacques Marquette and Louis Joliet may have been the first Europeans whom local indigenous people had ever encountered along the Wisconsin River during the explorers' epic journey by canoe from Green Bay to the Mississippi River in 1673.

A large influx of white settlers to this region occurred in the early 19th century when Scottish and English immigrants arrived to clear land for farmsteads, cut timber, built dams, and work in local lead mines. Conflicts over land use and ownership with Native American inhabitants eventually led to military confrontation. In 1829, the U.S. government forced Ho-Chunk leaders to relinquish ownership of all lands south of the Wisconsin Rivers, including the Tower Hill region. One year later, Daniel Whitney, a businessman from Green Bay, moved here to build a lead smelter and shot tower.

Daniel Whitney hired Thomas Shaunce, a lead miner from Galena, Illinois, to build his shot tower. It took Shaunce two years of hard labor to bore a 120-foot-deep vertical shaft through the sandstone bluff using only hand tools and black powder. A 60-foot wooden enclosure was built alongside the face of the bluff, making the entire shot tower 180 feet in height. Shaunce also excavated a 90-foot-long horizontal tunnel at the base of the bluff to collect the shot.

Whitney's lead shot business flourished for over 30 years and led to the establishment of the small community of Helena. The shot industry came to an end in the early 1860s, mostly due to the poor national economy at that time. The village of Helena struggled when the shot tower closed and was abandoned altogether when the railroad bypassed the town. The vacated town was purchased in 1889 by Reverend Jenkin Jones, a Unitarian minister from Chicago. Jones developed the site into the **Tower Hill Pleasure Company**, a religious retreat for ministers and their families.

> **Ranger Note**
>
> Shot are round lead pellets used primarily in shotguns. They were made by melting lead and pouring it through an iron ladle with different sized holes from the top of the shot tower.
>
> As drops of melted lead tumble through the air, they form into round balls and harden when they fall into a collection tank filled with water below. Smaller shot was used primarily for hunting small game and waterfowl. Larger shot, known as buckshot, was used to hunt deer or for personal defensive in scatterguns.

> **Ranger Trivia**
>
> Reverend Jones' nephew was the famed architect Frank Lloyd Wright. Wright purchased land on a nearby hill in 1911, where he built his iconic Taliesin home and studio, now a world-class tourist attraction.

53. Tower Hill State Park

After Reverend Jones' death in 1918, his wife Edith donated the property to the State of Wisconsin to establish Tower Hill State Park, which opened in 1922.

BEACHES, PICNIC AREAS, AND SHELTERS

Tower Hill's picnic area features an historic log gazebo and an indoor shelter building. Both structures were built more than 100 years ago for Reverend Jones' religious retreat. The park does not have a swimming area but a public beach is available along the sandy shoreline of the Wisconsin River at **Peck's Landing**, located northwest of the park.

HIKING, BIKING, AND NATURE TRAILS

Tower Hill has two miles of hiking trails. The **Old Ox Nature Trail** (0.5 miles) begins near the picnic area and travels uphill through a forested area to the smelter house on top of the bluff. This route was once used by teams of oxen to haul wagonloads of 75-pound lead bars called "pigs" to the smelter house. The lead bars were melted and formed into shot when dropped down the 120-foot shaft. Shot was sorted by size, bagged, and hauled overland by ox teams or shipped by boat down the Wisconsin and Mississippi rivers to southern and eastern markets.

The **Smelter House Trail** (0.25 miles) is a steep, asphalt-paved walkway that leads directly uphill to the smelter house. The nearby **Campground Trail** (0.3 miles) provides more gradual ascent to the smelter house via a series of steps.

Ranger Trail Pick

From the picnic area, take **Old Ox Nature Trail** up to the smelter house on top of the bluff. After viewing the interpretive displays inside, take the **Shot Tower Tunnel Trail** (0.2 miles) down a series of staircases to the shot collection tunnel along the Mill Creek at the bottom of the bluff. Visitors can enter the tunnel to view the lead shot collection water tank. Hike back up to the smelter house, then take the **Smelter House Trail** back down to the picnic area.

WILDLIFE VIEWING

The park's woodlands, wetlands, bluffs, and the nearby Wisconsin River provide prime habitat for a variety of wildlife. Tower Hill is known as an outstanding birdwatching area. During the fall and spring migration periods, the forests and backwater areas of the Mississippi come alive with dozens of species of ducks, geese, shorebirds and songbirds, especially warblers. Bald eagles, hawks and turkey vultures are often seen soaring above the park area. White-tailed deer,

Lead shot collection tunnel

squirrels, cottontail rabbits, and chipmunks are commonly seen in upland areas. Wetland areas are home to muskrat, mink, raccoon, otter and beaver.

The **Tower Hill Bottoms State Natural Area** (481 acres) is a backwater floodplain hardwood forest located adjacent to the park. Woodland birds such as pileated woodpeckers, tufted titmice, warblers, and many other rarely seen birds can often be viewed here, especially when exploring by canoe or kayak.

INTEPRETIVE PROGRAMS AND FACILITIES

The shot tower was designated as a **National Historic Place** in 1973. Interpretive displays inside the smelter house describe how the tower was built and how lead shot was made here. Other displays explain the history of the once thriving community of Helena. An interpretive panel is also located at the entrance of the shot collection tunnel at the bottom of the shot tower

BOATING AND FISHING

A carry-in canoe and kayak launch located along Mill Creek provides access to the backwaters of the Wisconsin River. Boaters with larger watercraft can access the river at **Peck's Landing** near the Highway 23 Bridge northwest of the park.

The Wisconsin River offers excellent fishing opportunities. Anglers can expect to catch walleye, northern pike, white bass, catfish, smallmouth bass, crappie, bluegill, and other panfish. Many visitors enjoy taking day trips paddling the Wisconsin River. There are several rental canoe and kayak outfitters along the river, including the **Wisconsin Canoe Company** located along Highway 14 in the nearby village of Spring Green.

ACCESSIBLE FACILITIES AND TRAILS
The park's entrance station, picnic area, restrooms, and indoor shelter facility are accessible for mobility-impaired visitors. Due to the hilly topography of the park, most trails can be challenging for visitors with walking disabilities. The **Smelter House Trail** is an asphalt-paved walkway but is quite steep and does not meet ADA wheelchair standards.

CAMPING
Tower Hill has 11 rustic campsites (no electric) located near the Mill Creek channel of the Wisconsin River. The campground has a vault-type restroom, a water fill station, and is open from May until about mid-October. Campers who prefer more modern amenities will find them at **Governor Dodge State Park** located about 12 miles south of Tower Hill along Highway 23. (See Chapter 38)

WINTER ACTIVITIES
Tower Hill is open in winter for hiking, wildlife watching, snowshoeing, and cross-country skiing. None of the trails are groomed.

AREA ATTRACTIONS
Iowa County and the village of Spring Green are well known for their natural beauty, outdoor recreation, and historical sites, such as the **Frank Lloyd Wright Visitor Center and Taliesin.** www.iowacounty.org www.springgreen.com

The **Spring Green Preserve State Natural Area** (1,104 acres) is often referred to as "Wisconsin's Only Desert." This unique site has dry sandy prairies, open sand dunes, prickly pear cactus, pocket gophers, blue racer lizards, and other plants and animals normally found in western states. A 1.6-mile hiking trail ascends through a short-grass prairie and an oak savanna to the crest of a bluff with great views of the grassy hillside and rare flowering prairie plants of the area. The preserve is located about five miles north of the Tower Hill along Jones Road north of Spring Green.
www.dnr.wi.gov/topic/lands/naturalareas/springgreenpreserve

DIRECTIONS TO THE PARK

Tower Hill State Park is located along the Wisconsin River and Mill Creek in northern Iowa County about five miles south of the village of Spring Green. From State Highways 23 or 14, exit on County Highway C and follow the signs to the park entrance.

Tower Hill State Park
5808 County Highway C
Spring Green, WI 53588
(608) 588-2116

54. WILDCAT MOUNTAIN STATE PARK

Kickapoo River Scenic Gem

Kickapoo Valley overlook

PARK SNAPSHOT

Wildcat Mountain State Park is located in west central Wisconsin within the heart of the **Kickapoo River Valley**, one of most scenic regions in the state. The park's towering bluffs, forested ridges, deep ravines, and clear-running streams have attracted visitors to this scenic gem for generations. Wildcat Mountain offers camping, hiking, nature study, and picnic areas overlooking the Kickapoo River Valley below. The park is also a popular home base for paddling the **Kickapoo River** or biking the nearby **Elroy-Sparta State Trail**.

GEOLOGY AND HISTORY

Wildcat Mountain is located within the Driftless Area, a 20,000 square-mile region of southwestern Wisconsin that escaped the crushing effect glaciers during the last Ice Age. The Kickapoo River flows through the property on its way south to join the Wisconsin River. Geologists believe this ancient stream may be one of the oldest river systems in the world. The park's towering 242-foot-high bluffs are composed of sandstone and dolomite limestone formed beneath a shallow ocean more than 500 million years ago. Erosion from wind, frost, and water have cut deep ravines and sculpted these steep-sided bluffs over the course of millions of years. The most prominent of these bluffs are **Mount Pisgah** and the park's namesake, **Wildcat Mountain.**

Native American people have hunted, fished, and gathered wild plants in the Kickapoo River Valley for thousands of years. Archeologists have discovered dozens of rock shelters, burial mounds, rock art, and artifacts dating as far back as 2000 B.C. Early French fur traders to region referred to the Kickapoo River as the "River of Canoes" because of its importance as a water highway to the Wisconsin River and Mississippi trade routes.

When the first permanent white settlers arrived here in the early 1800s, they found the area inhabited by the Sauk (Sac) and Meskwaki (Fox) people. These tribes were later displaced by the Ho-Chunk people, who were themselves forced to relocate when their lands were seized through treaty agreements with U.S. government in 1837. Near the end of the Civil War, families from the war-torn southern states, along with immigrants from Norway, Germany, Scotland, and Ireland, relocated to this area to cut timber and develop farmsteads in the Kickapoo Valley area.

In 1938, Amos Saunders, a long-time local citizen, donated 20 acres of his land to the State of Wisconsin. Saunders' only stipulation was that the property remain open to the public so that everyone could enjoy the natural beauty of the Kickapoo River area as he had done for so long. In 1948, a 60-acre parcel owned by Vernon County was added to Saunders' donation to create Wildcat Mountain State Park. Since then, the park has expanded to 3,821 acres in size.

> **Ranger Trivia**
>
> Wildcat Mountain was named by a group of local hunters who shot a large wildcat (bobcat) on this bluff in the early 1800s. The cat had been stalking small livestock in the area and had killed several sheep on a nearby farm. Mount Pisgah was named after a Hebrew word meaning "peak" or "cliff." The reference was taken from the biblical account of Mount Nebo, the mountain top from which Moses saw the "Promised Land" for the first time.

BEACHES, PICNIC AREAS, AND SHELTERS

The **Upper (Bluff) Picnic Area** has a playground and a shelter facility that can be reserved in advance. A nearby bluff overlook offers a beautiful view of the Kickapoo River Valley below. The **Lower Picnic Area** located along the Kickapoo River, has an open-air shelter and a canoe/kayak launch site. There is no swimming beach at Wildcat Mountain but an outdoor swimming pool is available in the nearby city of Viroqua.

HIKING, BIKING, AND NATURE TRAILS

There are 6.5 miles of hiking trails at Wildcat Mountain. The **Old Settler's Trail** (2.5 miles) begins in the Upper Picnic Area and travels through a hardwood forest and a pine plantation with an uphill climb of 390 feet to the crest of Wildcat Mountain. A short 0.4-mile side trail leads to the **Taylor Hollow Overlook,** which offers a bird's eye view of the river valley and the village of Ontario below.

The **Prairie Trail** (0.2 miles) follows the edge of a bluff through a restored prairie area high above the Kickapoo River. The **Ice Cave Trail** (0.75 miles) located in the south end of the park, follows a stretch of Billings Creek and leads to a picturesque sandstone cave. A small waterfall spills over the top of the cave during rainy periods.

> ### Ranger Trivia
>
> The 32-mile **Elroy-Sparta State Trail** was developed in 1967 along an abandoned railroad grade. It was considered the first ever "rails-to-trails" hiking/biking trail in the country.

Wildcat Mountain has 17 miles of equestrian trails that are rated as some of the best in the state. Horseback riders have seven different trails to choose from, all of which can be accessed from the horse campground. The nearby **Elroy-Sparta State Trail** (32.5 miles) is one of the most popular bike trails in Wisconsin. This scenic trail leads through three 150-year-old railroad tunnels, including one 3,800 feet in length. Trail passes and bike rentals are available at the trail headquarters located inside a restored railroad depot about 14 miles northeast of the park in the village of Kendall.

> ### Ranger Trail Pick
>
> The **Hemlock Nature Trail** (1.3 miles) is a beautiful, if somewhat challenging trail that begins along the shoreline of the Kickapoo River at the lower picnic area. The trail ascends along several switchbacks 365 feet to the top of **Mount Pisgah**. Interpretive signs along this route highlight the unique wildlife and geology of the river valley and bluffs.

WILDLIFE VIEWING

The park's rocky bluffs, forests, cool ravines and streams are a wildlife haven for hundreds of species of birds, mammals, reptiles, amphibians, and plants. The cool north-facing bluff of Mount Pisgah have allowed white pine, hemlock, and other more northern forest trees to grow there. This area of the park has never been logged so many impressive old age trees are

> ### Ranger Note
>
> The park's namesake, the secretive bobcat (wildcat), has been seen in the park area in recent years after being absent for decades.

found here. White-tailed deer, squirrels, raccoon, red-tailed hawks, turkey vultures and many kinds of woodland songbirds are commonly seen in the park. Lesser-seen animals such as mink, otter, coyote and fox make their home here as well.

INTERPRETIVE PROGRAMS AND FACILITIES
Wildcat Mountain has a small nature center that is staffed on occasion by park volunteers. Interpretive programs, hikes, and special events are offered through-out the use season. Evening programs are held at the park's amphitheater located near the main campground.

Ranger Trivia

The Kickapoo River was named after the Kickapoo people who moved to Great Lakes region from the Atlantic seaboard to escape harassment from the Iroquois in the 17th century. The Kickapoo fought with the English against the United States during the Revolutionary War and again during the War of 1812. After the English defeat the Kickapoo were forcefully removed from the Wisconsin Territory. Some relocated to reservations in Kansas and Oklahoma while others fled to Texas.

BOATING AND FISHING
The Kickapoo River is renowned for the scenic riverside forests and sandstone bluffs. The river is often called the "most crooked" stream in North America due

Canoeing the Kickapoo River

to the many twists and turns along its 130-mile route to the Wisconsin River near the village of Wauzeka in Crawford County. The Kickapoo offers some of the best canoeing, tubing, and kayaking opportunities in the state. A small craft launch is located at the park's lower picnic area. Watercraft rentals are available in the nearby town of Ontario for both day trips and multi-day camping outings on the river. Anglers can find excellent fishing on the river as well, especially for brown, brook, and rainbow trout.

ACCESSIBLE FACILITIES AND TRAILS

The Upper Picnic Area's shelter, picnic tables and overlook area along the bluff are accessible for mobility-impaired visitors. Campsite 30 is designed for wheelchair use with paved access to restroom/shower facilities. The steep topography of the park renders most parks trails challenging for visitors with walking disabilities.

CAMPING

Wildcat Mountain has 72 campsites (22 electric) with 25 vehicle access sites and 20 walk-in campsites. Wheeled carts are provided to transport camping gear to walk-in sites. A restroom/shower facility is located adjacent to the RV water fill and sanitary station in the north end of the main campground. The park's group campground has three sites which can accommodate from 25 to 50 campers each. The group site has vault-type restrooms and a shared shelter facility. A primitive canoe campsite that can only be reached by watercraft is located near the canoe landing along the Kickapoo River. Use of this site is free and available on a first-come, first-serve basis.

Wildcat Mountain has one of the finest equestrian campgrounds in the state. The campground has 24 individual campsites, each with its own electrical service, hitching post, fire ring, and picnic table. The campground has a corral, loading ramp, vault-type restrooms, and water fountains.

WINTER ACTIVITIES

Hiking, skiing, snowshoeing, wildlife watching, and winter camping are favorite off-season activities at Wildcat Mountain. Park staff groom six miles of cross-country ski trails. A 2.5-mile snowshoe trail is located along a loop of the Old Settlers Trail. The park also has 16 miles of snowmobile trails.

AREA ATTRACTIONS

Vernon County is located adjacent to the Mississippi River in the heart of the Wisconsin's scenic coulee country, known for its many small historic towns, picturesque farms, and scenic forests and streams. www.vernoncounty.org

The **Kickapoo Valley Reserve** is an 8,600-acre natural area that borders the Kickapoo River for more than 25 miles south of Wildcat Mountain. The reserve is managed by the State of Wisconsin and the Ho-Chunk Nation and offers canoeing, hiking, wildlife viewing, biking, and primitive camping. The **Kickapoo Valley Reserve Visitor Center** is a modern nature and learning center with many interactive displays highlighting the flora, fauna, geology, and history of this unique part of Wisconsin. The center is located along Highway 131 about a mile north of the village of La Farge. www.kvr.state.wi.us

DIRECTIONS TO THE PARK
Wildcat Mountain is located adjacent to Highway 33 south of the village of Ontario in Vernon County, about 40 miles southeast of La Crosse.

Wildcat Mountain State Park
E 13660 State Highway 33
Ontario, WI 54651
(608) 337-4775

55. Wyalusing State Park

Pearl of the Mississippi and Wisconsin Rivers

Point Lookout

PARK SNAPSHOT

Wyalusing State Park is located along the crest of 600-foot bluffs perched high above the confluence of the Wisconsin and Mississippi rivers in northwestern Grant County. This scenic 2,674-acre park offers camping, picnicking, hiking, nature study, and exceptional birdwatching opportunities. The backwater areas of the Mississippi River in the park offer great canoeing, kayaking, and fishing opportunities as well.

GEOLOGY AND HISTORY

The landscape of Wyalusing is mix of steep cliffs, deep ravines, upland woods, prairies, and river-bottom hardwood forests. The parks rocky bluffs were formed beneath a shallow ocean that once submerged most of the state about 600 million years ago. Sand, silt and sea life deposited beneath this ancient sea were transformed over time into the sandstone and limestone outcrops seen in the park today. Evidence of this ancient sea life can be seen in the many fossil-imbedded rocks found throughout the park.

Wyalusing is located in the Driftless Area of southwestern Wisconsin, a region that escaped the crushing effect of continental glaciers that pushed through the state during the last Ice Age about 30,000 years ago. Glacial drift (sand, silt, gravel, and rocks) deposited elsewhere in Wisconsin is not found at Wyalusing. The rocky bluffs of the park have been eroded by frost, wind, and water over millions of years. Although not touched by glacial ice, the bluff areas were shaped by floodwaters of melting glaciers to the north that thundered down the Wisconsin and Mississippi rivers about 14,000 years ago.

Archeologists have found evidence of 14 different Native American cultures that lived or hunted within the park area. The most visible relics left by these ancient people are burial mounds created by the Effigy Mound Builders who lived here more than 1,600 years ago. Dozens of well-preserved conical, linear, and effigy (animal-shaped) mounds can be viewed along the **Sentinel Ridge Trail** overlooking the Mississippi River.

The first Europeans to explore this area were Father Jacques Marquette and Louis Joliet, who came ashore here during their epic canoe journey from Green Bay to the confluence of the Wisconsin River and the Mississippi River in 1673. Their expedition journal mentions that both men hiked to the top of a tall bluff, possibly at Wyalusing, to get a better view of both rivers. Over the next century, the French, British and American military fought over the control the Mississippi and Wisconsin rivers. These river highways were important transportation routes for the fur trade, settlers, and to supply military outposts. Conflicts between white settlers and local indigenous people over land use erupted as more and more immigrants arrived in the early 19th century to work in timber camps or farm the rich soils of the region. In 1829, Ho-Chunk and other tribal leaders were induced to sign a peace treaty in the nearby community of Prairie du Chien ceding their land in southwest Wisconsin to the U. S. government and agreeing to move onto reservation land west of the Mississippi River.

One of the earliest settlers of the area was Robert Glenn, who arrived from Pennsylvania in the 1840s. Glenn purchased a large section of bluff-top property, which he farmed for many years before selling it to the State to develop a

> **Ranger Note**
>
> A small band of Ho-Chunk people led by Chief Green Cloud refused to accept the conditions of the Prairie du Chien Treaty and continued to live in the park area, now the **Green Cloud Picnic Area**, for more than 50 years before finally crossing the Mississippi River to reservation lands in 1882.

state park. When opened in 1917, the new park was called Nelson Dewey State Park after Wisconsin's first governor. It was later changed to Wyalusing State Park when Governor Dewey's former estate near Cassville became a state park in 1936. (See Chapter 47)

BEACHES, PICNIC AREAS, AND SHELTERS

Wyalusing has several picnic areas located along bluff overlooks. Many have attractive stone and timber shelters built by Civilian Conservation Corps (CCC) and Works Progress Administration (WPA) craftsmen in the 1930s. Both the CCC and WPA were Depression-era federal work programs for young unemployed men. The **Peterson Picnic Shelter**, located near the **Wisconsin Ridge Campground**, has three stone fireplaces, electrical outlets, and houses the parks concession stand. The **Green Cloud Picnic Area** offers a panoramic view of the Mississippi and Wisconsin rivers. Nearby is the **Passenger Pigeon Monument** that commemorates the extinction of this once abundant bird species.

The park does not have a swimming beach but a popular beach along the Mississippi River is located about a mile south of the park at the **Grant County Wyalusing Recreation Area**.

HIKING, BIKING, AND NATURE TRAILS

There are 18 miles of hiking trails at Wyalusing. **Sand Cave Trail** (2.4 miles) starts near the **Paul Lawrence Nature Center** and descends through a scenic ravine with access to both **Little** and **Big Sand Cave**. The **Sentinel Ridge Trail** (1.6 miles) follows the crest of a ridge overlooking the Mississippi River and descends to park's boat launch area. Interpretive signs along this trail highlight a procession of rare effigy mounds located here. The park's **Whitetail Meadows Trail** (3.1 miles) and **Mississippi Ridge** Trail (1.8 miles) are open for off-road biking.

The **Bluff Trail** (0.9 miles) leads past several scenic overlooks in the upper **Wisconsin Ridge Picnic Area** and then descends below the bluff with access to **Treasure Cave**. From here the trail heads east below the bluff area with a hookup trail that ascends to the **Knob Shelter** adjacent to the campground.

The **Old Wagon Road Trail** (0.8 miles) and **Immigrant Trail** (1.3 miles) follow an historic roadway used by westward-bound pioneers in covered wagons to access the small settlement of **Walnut Eddy**. A ferry boat here provided commercial passage over the Wisconsin River for early settlers.

Pictured Rock Cave

WILDLIFE VIEWING

Wyalusing's diverse landscape of upland woodlands, bluffs, prairies, and bottomland hardwood forests provide ideal habitat for wildlife. At least 47 mammals, 27 reptiles, 15 amphibians, and 284 bird species have been recorded here. Birdwatching, especially during the spring and fall migration period, can be outstanding in the park.

The **Walnut Forest and Hardwood Forest State Natural Areas** harbor some of the rarest plants in the state, including several species of ferns that may be remnants of glacial times. In spring, wildflowers such as shooting rockets, Dutchman's breeches, and white trilliums light up forested bluff areas.

INTERPRETIVE PROGRAMS AND FACILITIES

Nature hikes and programs are offered throughout the use season. The **Paul Lawrence Interpretive Center** is located along the park entrance road north of the visitor entrance station. The **Huser Astronomy Observatory**, built by a local group of called the Starsplitters, regularly offer star-gazing programs using high-powered telescopes.

BOATING AND FISHING

A boat launch is located at Glen Lake, a backwater of the Mississippi River. Fishing here can be excellent at times. Anglers can expect to catch bluegill, crappie, bass, walleye, northern pike, catfish, and many others species. A six-mile signed canoe trail begins at the boat launch area. The trail meanders through wooded backwater areas to the main channel of the Mississippi River, where it turns south before returning to the boat launch area. Canoe and kayak rentals are available through the park's concession stand.

ACCESSIBLE FACILITIES AND TRAILS

Most of the park's buildings, restrooms, and larger shelters are accessible for mobility-impaired visitors. Campsites 136 and 214 are wheelchair accessible. Both have electrical service with access to restrooms and showers. A wheelchair-accessible fishing pier is located adjacent to the boat launch on Glen Lake. The half-mile **Sentinel Ridge Nature Trail** is surfaced with crushed gravel and mostly level.

CAMPING

The park has 114 campsites (33 electric) within the **Wisconsin Ridge Campground** situated on a bluff overlooking the river valley and the **Homestead Campground** located west of the park visitor station. Both have restroom and shower facilities. An RV water fill and sanitary station is located along the entrance road to the Homestead Campground.

The park's **Outdoor Group Camp** has five (tent-only) campsites that can accommodate up to 26 campers each. The group camp has water fountains and vault-type restrooms. The **Hugh Harper Indoor Group Camp** has four sleeping dormitories with bunk beds, restrooms, and shower facilities. A central lodge facility has large meeting/dining room with a fireplace and commercial-type kitchen. The camp can be reserved for groups up to 108 people.

WINTER ACTIVITIES
Park staff groom eight miles of cross-country ski trails for both traditional and skate skiing. Snowshoeing, winter camping, and ice fishing are popular off-season activities. A hike to one of the park's sandstone cave "frozen" waterfalls can be especially rewarding in winter.

AREA ATTRACTIONS
The Prairie du Chien area is a popular vacation destination known for its outstanding scenery, historic sites, and many outdoor recreational opportunities. www.prairieduchien.org   www.grantcounty.org/tourism

Villa Louis State Historic Site preserves a Victorian-style mansion built in 1871 by Villa Louis, the son of Hercules Dousman, a wealthy fur trader. Also located here is the site of **Fort Shelby**, built by the U.S. Army during the War of 1812. Tours of Villa Louis are conducted by the State Historical Society. The property is located along Villa Louis Road in the city of Prairie du Chien. www.villalouis.wisconsinhistory.org

DIRECTIONS TO THE PARK
Wyalusing is located in northwestern Grant County. From U.S. Highway 18/35 turn west onto County Highway C to its junction with County Highway X. Turn right on Highway X about one mile to the park entrance.

Wyalusing State Park
13081 State Park Lane
Bagley, WI 53801
(608) 996-2261

56. YELLOWSTONE LAKE STATE PARK

If You Build a Park, They Will Come

Yellowstone Lake

PARK SNAPSHOT

Yellowstone Lake State Park is located along a 455-acre lake of the same name in Lafayette County. This 1,000-acre property has a popular swimming beach, shady campsites, hiking trails, and offers excellent boating and fishing opportunities.

Most state parks were built around natural features such as towering bluffs, scenic waterfalls, or picturesque rivers and lakes. Yellowstone was an exception to that rule. The entire park, including the lake itself, was intentionally built in hopes of drawing more visitors this rural region of southwestern Wisconsin. It worked. Today, after more than 70 years of development, Yellowstone has become one of the most popular state parks in the state.

GEOLOGY AND HISTORY

The original landscape of the park was a mix of forested hills, steep valleys, prairies, and the Yellowstone River. This rugged topography was shaped by frost, wind, and water erosion over millions of years. The park is within a region of Wisconsin known as the Driftless Area. The massive continental glacier that pushed through most of Wisconsin during the last Ice Age never reached this part of the state. As a result, there are no glacial drift (sand, gravel, and rock) deposits, moraines, or natural lakes in this area.

Several Native American cultures have lived, hunted, and fished in the **Yellowstone River Valley** for generations. The first Europeans to enter the area were French fur traders looking to trade for lead from the Ho-Chunk miners in the early 18th century. Reports of exposed lead veins in this region induced lead miners from Galena, Illinois, to enter the area and establish a small community southwest of the park called **New Diggings** in 1824. Early miners worked their own claim and smelted raw ore into lead bars, which were then transported back to Galena for sale. The massive influx of thousands of miners into Ho-Chunk tribal lands inevitably led to conflict and the Winnebago War of 1827. Postwar treaties forced Ho-Chunk leaders to sell their lead-rich lands in southwestern Wisconsin to the U.S. government in 1829.

Lead and zinc mining in southern Wisconsin attracted both Yankee and European immigrants to the area. The nearby community of **Blanchardville** was settled in 1842 by members of the Church of Jesus Christ of Latter Day Saints. Church leaders named their new settlement *Zarahemlas* (City of God) after the name of an ancient American city first described in the Book of Mormon. The Mormons mined lead as well, but were also the first to plant crops in the rich prairie soils of the area. In 1856, Alvin Blanchard, a merchant and businessman from Dodgeville, purchased a gristmill along the nearby Pecatonica River from the Mormons. Blanchard built a sawmill, store, and the first cheese factory in the area. He also helped plat the city streets of Zarahemlas for homes and businesses. The village would later be renamed Blanchardville in his honor.

> **Ranger Trivia**
>
> The Mormons left the local area within twenty years of their arrival, most likely due to the intolerance of their religious beliefs from the large influx of Christian immigrants from Ireland, Switzerland, Germany, and Norway.

In the 1940s, local businessmen from the Blanchardville area lobbied state lawmakers to build a new state park near their community in hopes of attracting visitors and improving the local economy. With no major natural lake in the area, state planners opted to purchase several farms within the Yellowstone River Valley to build an earthen dike and concrete dam. Work on the new man-made lake, campground, and other park facilities was completed in 1954 and Yellowstone Lake State Park opened to the public.

56. Yellowstone Lake State Park

BEACHES, PICNIC AREAS, AND SHELTERS
A wide, sandy swimming beach and large picnic area are located along the east shoreline of Yellowstone Lake. A picnic area near the park entrance station has a playground and an open-air shelter with electrical service, which can be reserved in advance.

HIKING, BIKING, AND NATURE TRAILS
Yellowstone Lake has 12 miles of hiking trails to explore. The **Wildlife Loop Trail** (1.0 mile) located near the visitor entrance station, circles the perimeter of a waterfowl refuge area along the northwestern shore of Yellowstone Lake. The **Oak Ridge Trail** (1.3 miles) and the **Prairie Loop Trail** (0.8 miles) both begin at the picnic area north of the entrance station. Many visitors enjoy walking both trails to experience the contrast between the oak/ hickory forest and the open grasslands of the restored prairie area.

Off-road bike riding is allowed on **Oak Grove Trail** (2.1 miles) and **Windy Ridge Trail** (1.7 miles). Both trails have steep climbs through oak forests and open grasslands. The **Military Ridge State Trail** (40 miles), located about 20 miles north of the park, is a hiking/biking trail that travels east from Dodgeville to the city of Fitchburg near Madison.

There are no horse trails in the park but the nearby **Yellowstone Lake State Wildlife Area** has a 30-mile network of hiking and equestrian trails. The wildlife area also has picnic tables, hitching rails, and horse watering areas.

Ranger Trail Pick

The **Blue Ridge Trail** (3.5 miles) offers the best overall view of the varying landscape found at Yellowstone. The trail ascends through an old-age oak forest and prairie area to a ridgetop overlook area with scenic views of the Yellowstone Lake Valley below.

WILDLIFE VIEWING
Yellowstone Lake is known for its outstanding birdwatching opportunities. Over 170 species of birds have been identified that either nest or migrate through the park, including osprey, cormorants, common loons, wild turkey, and dozens of species of woodland songbirds, hawks and owls. Waterfowl and marshland birds

Ranger Note

Yellowstone is home to thousands of little brown bats that return to the park each summer. Park volunteers maintain over 30 bat houses in the park that provide safe haven for the bats and nurseries for mother bats to raise their pups. Each evening, right after sunset, park visitors enjoy watching the spectacle of hundreds of bats leaving these shelters to take flight into the night sky in search of mosquitoes, moths, and other night-flying insects.

such as wood ducks, mallards, geese, herons, and sandhill cranes are commonly seen here as well. Bald eagles nest in the nearby Yellowstone Lake State Wildlife Area.

White-tailed deer, raccoon, woodchucks, rabbits, chipmunks and squirrels are commonly seen throughout the park. More secretive animals such as coyote, mink and occasionally otter are spotted here as well.

INTERPRETIVE PROGRAMS AND FACILITIES

Interpretive panels highlighting the parks flora, fauna, and history are located throughout the park. Nature hikes featuring prairie ecosystems, wildlife, star gazing, and evening bat watching are offered during the summer season. Evening programs are held at the park's amphitheater located near the campground.

BOATING AND FISHING

For many visitors, Yellowstone Lake itself is the central attraction of the park. Boating, tubing, fishing, and swimming are popular activities on this 2.5-mile-long, 455-acre man-made lake. Two boat launch facilities are located on the eastern shore the lake. Visitors with canoes, kayaks, or paddleboards can opt to use a small landing near the visitor entrance station. Fishing on can be excellent at times

Tubing on Yellowstone Lake

on Yellowstone Lake. Anglers can expect to catch largemouth and smallmouth bass, bluegill, crappie, perch, bullhead, northern pike, muskie, and walleye.

ACCESSIBLE FACILITIES AND TRAILS
Most of the park's buildings, restrooms, and picnic areas are accessible for mobility-impaired visitors. Campsites 42 and 49 can accommodate wheelchairs. Both sites have electrical service and access to restroom and shower facilities. A wheelchair-accessible fishing pier is located adjacent to the southeast boat launch. The **Wildlife Loop Trail** located near the visitor entrance station is accessible for visitors with moderate walking disabilities. Most other hiking trails in the park have steep grades and/or stairways, which make them challenging for visitors with physical limitations.

> **Ranger Tip**
>
> Motorboats, row boats, canoes, and kayaks can be rented at the park's boat concession located along the southeast corner of the lake near the dam.

CAMPING
Yellowstone Lake has 134 campsites (41 electric and 14 walk-in sites), all within a well-shaded forest area of the park. A shower/restroom facility is located in the main campground. The park's group campground has five campsites (three electric) and can accommodate between 25 and 50 campers each. An RV water fill and sanitary station is located north of group camp area.

> **Ranger Tip**
>
> Campsite cooking is part of the camping experience but for those looking for a culinary break, the **Yellowstone Lake Dairy Restaurant and Ice Cream Shop Concession** serves up an excellent breakfast each morning during the summer season. The concession also sells camping supplies, refreshments, fast food, snacks, and "real" Wisconsin scoop ice cream in several flavors.

WINTER ACTIVITIES
Hiking, snowshoeing, wildlife watching, and winter camping are popular off-season activities at Yellowstone. Park staff groom four miles of cross-country ski trails for inline skiing. The park has seven miles of snowmobile trails that connect to the Lafayette County snowmobile trail system. Many anglers enjoy ice fishing on Yellowstone Lake for walleye, northern pike, crappie, and bluegills.

AREA ATTRACTIONS

Lafayette County is well known for its scenic farmsteads, historic sites, and small welcoming communities. www.lafayettecounty.org/explore

The **Pendarvis State Historical Site** offers tours of two homes built from local timber and limestone by Cornish immigrants in the 1830s. Most immigrants to the community of Mineral Point arrived from Cornwall in southwestern England to work the lead and zinc mines of the area. Mining equipment is displayed at the nearby **Merry Christmas Mine**, an abandoned 19th-century zinc mine. Pendarvis is operated by the State Historical Society along Shake Rag Street in Mineral Point, about 16 miles northwest of the park. www.pendarvis.wisconsinhistory.org

DIRECTIONS TO THE PARK

Yellowstone Lake is located in the northeast Lafayette County. From Highway 18/151 west of Madison, take Highway 78 south to Blanchardville. Turn right (west) onto County Highway F and travel about eight miles to the park entrance on Lake Road.

Yellowstone State Park
8495 Lake Road
Blanchardville, WI 53516
(608) 523-4427

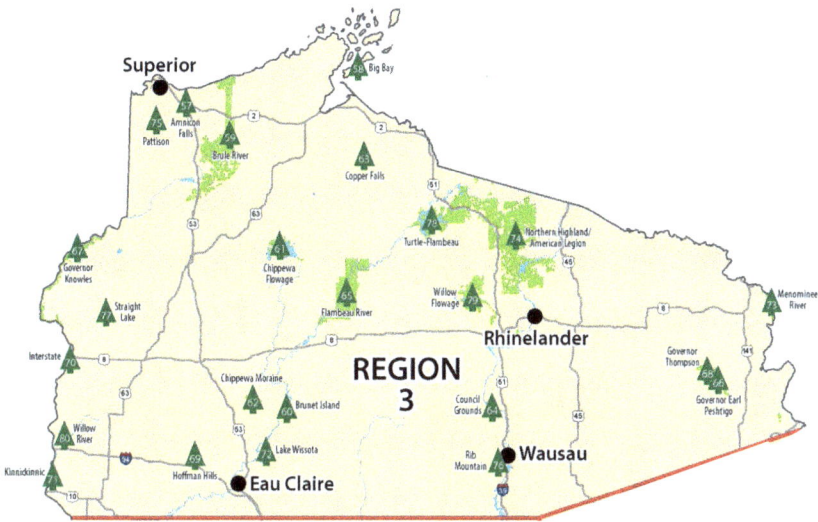

Superior

Big Bay
Amnicon Falls
Pattison
Brule River
Copper Falls
Governor Knowles
Chippewa Flowage
Turtle-Flambeau
Northern Highland/American Legion
Straight Lake
Flambeau River
Willow Flowage
Menominee River
Interstate
Chippewa Moraine
Brunet Island
REGION 3
Council Grounds
Governor Thompson
Willow River
Lake Wissota
Governor Earl Peshtigo
Kinnickinnic
Hoffman Hills
Rib Mountain
Rhinelander
Wausau
Eau Claire

Region Three
Northern Forests, Parks and Scenic Water Flowage Areas:
Northern Highlands/American Legion Forest and Lake Superior area, Wausau, Marinette, Chippewa Moraine, Apostle Islands, and Brule, Flambeau, Brule, Peshtigo, Willow, St. Croix, Menomonee, and the Upper Wisconsin rivers.

57. Amnicon Falls State Park
Wisconsin's Natural Water Park

Horton Covered Bridge

PARK SNAPSHOT

Amnicon Falls is one of Wisconsin's most enchanting state parks. The 825-acre property is known for its picturesque waterfalls and tumbling rapids along the Amnicon River. The river flows over volcanic basalt rock and colorful sandstone outcrops through the center of the park on its way to Lake Superior. Amnicon offers rustic camping, picnic areas, nature study, and hiking trails. It's also a one-of-a-kind natural water park for those willing to wade or swim in the cool, refreshing waters of the Amnicon River.

GEOLOGY AND HISTORY

Amnicon Falls is one of the few sites in the state where exposed rock formations can tell the fantastic story of the earth's geology. The 30-foot-high brownish stone seen in the upper falls of the Amnicon River is basalt rock formed from molten lava when volcanic eruptions were active throughout the Lake Superior Basin 1.1 billion years ago. Farther downstream at the lower falls, the flow of the Amnicon River cut through a different kind of rock called Lake Superior sandstone. This reddish-colored sedimentary rock was formed from compressed sand and iron deposits that settled beneath a vast inland sea that once covered most of Wisconsin about 600 million years ago.

About 500 million years ago, a rock-crushing fracture in the earth's crust occurred in the basin due to earthquakes caused by a massive collision of tectonic plates along the eastern seaboard of the United States. This event created a 62-mile crack in the earth in northern Wisconsin known to geologists as the Douglas Fault. Uplifted sections of bedrock from this fault can be seen at the park's upper waterfall. This same geologic uplift caused the Amnicon River to drop 640 feet in elevation on its 30-mile route to Lake Superior. During Wisconsin's last Ice Age period about 30,000 years ago, Amnicon Fall's landscape was reshaped once again, this time by massive glaciers that pushed through this area. Granite boulders and other glacial debris picked up by moving ice sheets in Canada were deposited at Amnicon Falls by receding glaciers.

> **Ranger Trivia**
>
> Amnicon Falls State Park was named after the Amnicon River, which flows through the property before emptying into Lake Superior. The term *Amnicon* was derived from the Ojibwe word *Aminikan*, meaning "where fish spawn." The Amnicon River was, and still is, an important fish spawning area for many Great Lakes fish.

Several Native American cultures are known to have lived in the park area over thousands of years. One of the earliest were the Copper Culture people, who mined exposed veins of copper in this area to fashion tools, weapons, fish hooks, and trade items. Later cultures, including the Woodland and Ojibwe (Chippewa) people, built permanent villages, raised crops, hunted game animals and caught fish in the Amnicon River.

French fur traders reached the Amnicon River area in the 17th century to trade with the Ojibwa people for mink, beaver, and otter pelts. The fur trade industry lasted into the early 19th century when commercial interest in mining copper ore increased. Evidence of open-pit copper mining can still be seen within Amnicon Falls today. Timber companies began to cut the pine forests of the local area in the mid-1800s. Logs were floated down the Amnicon River to sawmills during high water times in spring. The shallow rapids and waterfalls in the park area led to massive log jams, which caused considerable risk to loggers who had to untangle them.

The beauty of Amnicon Falls attracted visitors long before it became a state park. In 1932, James Bardon, a long-term resident of the area, donated some of his land and sold the rest to Douglas County to create Bardon Park. Early development of the park was completed by crews from the Works Program

Administration (WPA), a 1930s-era federal jobs program for unemployed young men. The Town of Amnicon maintained and expanded the park until 1961, when the property was transferred to the State to become Amnicon Falls State Park.

BEACHES, PICNIC AREAS, AND SHELTERS

Many people visit Amnicon Falls just to enjoy wading or swimming in the cool, refreshing river on warm summer days. During dry periods, water levels may only be knee-deep or less in many stretches of the river except for deeper pools beneath waterfall areas.

> **Ranger Note**
>
> River levels can fluctuate rapidly depending on precipitation so caution must be taken at all times when entering the river. Diving or jumping from rocks or waterfall ledges is not permitted in the park.

Picnic areas are situated along both sides of the Amnicon River. An open-air shelter with electric service is located across the river from the park entrance station. The shelter can be reserved in advance for groups up to 50 people.

HIKING, BIKING, AND NATURE TRAILS

Amnicon Falls has 2.75 miles of marked hiking trails with numerous side trails to waterfall overlooks and other scenic spots. One of the most photographed structures in the park is the **Horton Covered Bridge.** This picturesque bridge spans the Amnicon River between the Upper and Lower Falls. The 55-foot "bowstring" bridge was designed by well-known architect Charles M. Horton of La Crosse in the late 1800s and served as a local highway bridge for many years. The bridge was dismantled and moved to its current site in 1930 to provide access to the park's river island. A crew from the Brule Civilian Conservation Corps (CCC) added the iconic roof to the bridge in 1939.

The **Thimbleberry Nature Trail** (0.8 miles) travels through a forested area along a section of the Amnicon River with a short side trail that leads to a scenic water-filled brownstone quarry. Brownstone is dark-colored sandstone rock that was quarried for use in building some of the largest city buildings and elegant mansions throughout the 19th and early 20th centuries.

The **Osaugie Trail** (10.3 miles) is a popular hiking/biking route located in the nearby city of Superior. The trail follows the shoreline of Superior Bay, one of the busiest harbors on the Great Lakes. Along the trail are historical sites such as the S.S. *Meteor* Whaleback Ship Museum and the Burlington Ore Dock, where the *Edmund Fitzgerald* set out on its on its fatal voyage on Lake Superior in 1975. The Osaugie Trail starts at Bear Creek Park about nine miles northwest of Amnicon Falls off Highways 2 and 53 near Superior. www.ci.superior.wi.us.trails

Upper Falls of the Amnicon River

Ranger Trail Pick

The **Waterfalls Trail** (0.25 miles) is located on an island within a divided stretch of the Amnicon River. This easy-walking trail is accessible by two bridges and leads past scenic rock formations, river cascades with views of both the Upper and Lower waterfalls. A short side trail leads to Snake Pit Falls, a deep, narrow cauldron within the Douglas fault. A section of the Amnicon River thunders through this split-rock gorge.

WILDLIFE VIEWING

Amnicon Falls is within the Lake Superior's Coastal Plain Region, which contains some of the most diverse wildlife habitat in Wisconsin. The area's mix of wetlands, prairies, lakeshore, upland hardwoods, and rare boreal forests is home to many wildlife species not regularly found in other regions of the state. Gray wolves, pine martins, and occasionally moose can be found here. In winter, snowy owls and great gray owls from northern tundra areas often migrate through this region as well. White-tailed deer, red squirrels, chipmunks, snowshoe rabbits, beaver, otter, bobcat, and black bear can all be found in the park area, along with several species of woodland songbirds, woodpeckers, ruffed grouse, ravens, owls, and hawks.

57. Amnicon Falls State Park

INTERPRETIVE PROGRAMS AND FACILITIES

Interpretive panels highlighting the park's unique geology can be found along river and waterfall overlook areas. An interpretive guidebook describing the flora and fauna of the park is available for the Thimbleberry Nature Trail.

BOATING AND FISHING

The park's waterfalls and shallow rapids are not suitable for canoeing or kayaking but deeper sections of the Amnicon River located outside the park area are popular with paddlers. Anglers will find excellent fishing opportunities downstream from the park, where the river widens and deepens before reaching Lake Superior. Several Great Lakes fish migrate up the Amnicon River to spawn including walleye, steelhead trout, and salmon. Lake Superior charter boat services are located within the harbor area in the city of Superior. Barker's Island Marina located along Marina Drive has a public boat launch, a fish cleaning station, and other amenities for boaters.

ACCESSIBLE FACILITIES AND TRAILS

Wheelchair-accessible vault-type restrooms are available in picnic areas and near campsites 8 and 18 in the campground. Due to the rocky and uneven topography of the park, some trails may be challenging for visitors with walking impairments.

CAMPING

Amnicon Falls has 36 rustic campsites (2 walk-in) located within a forested area. There are no flush restrooms, showers, or electric service in the campground. Pattison State Park, located about 17 miles southwest of the park, has 62 campsites (18 electric) with restroom and shower facilities. (See Chapter 75)

WINTER ACTIVITIES

Hiking and wildlife watching are popular off-season activities in the park. The waterfalls and rapids of the Amnicon River take on a special beauty when cloaked in snow and ice in winter. A 1.5-mile marked snowshoe route and cross-country skiing are available but trails are not groomed.

AREA ATTRACTIONS

Douglas County and the Superior area have many historical sites and tourist amenities in addition to having some of the most pristine wilderness areas, streams and hiking trails in the state.
www.superiorchamber.org
www.douglascountywi.org

Wisconsin Point Park and Lighthouse are located on the largest freshwater sandbar in the world, a natural feature that serves as harbor break for both the cities of Superior and Duluth in Minnesota. The 229-acre park has hiking trails and three miles of beachfront to explore and preserves an historic 1913 lighthouse. A sacred 17th-century Chippewa burial ground is also located here. The park is located about 15 miles northwest of Amnicon Falls on Wisconsin Point Road. www.ci.superior.wi.us/parks-and-forestry

DIRECTIONS TO THE PARK
Amnicon Falls State Park is located in Douglas County about 15 miles southeast of the city of Superior. From the junction of Highways 2 and 53, follow Highway 2 east about one mile and exit onto County Highway U north to the park entrance.

Amnicon Falls State Park
4279 South County Highway U
South Range, WI 54874
(715) 398-3000

58. BIG BAY STATE PARK

Island Park—A Lake Superior Jewel

Lake Superior shoreline

PARK SNAPSHOT

Big Bay is one of the most unique state parks in Wisconsin. This scenic 2,474-acre park is located on Madeline Island, one of 22 islands that make up the legendary Apostle Islands of Lake Superior. Madeline Island can only be reached by boat or ferry service from the port of Bayfield, Wisconsin's northernmost city. Big Bay has a modern campground, swimming beach, picnic areas, hiking trails, and other developed facilities, yet retains an aura of wilderness situated on an island within the largest freshwater lake in the world.

GEOLOGY AND HISTORY

Madeline Island is composed of sedimentary sandstone rock that was formed beneath the Lake Superior Basin more than 600 million years ago. Since then, at least four glaciers from the Canadian Arctic have advanced and retreated through this region over the past two million years, each time reshaping the Apostle Islands and deepening Lake Superior. Near the close of the last Ice Age about 17,000 years ago, a massive glacial lake covered all but the highest points of the Apostle Islands, depositing sand and deep layers of red clay.

Today, after thousands of years of ice and wave erosion, the shoreline of Big Bay State Park is a mix of sand beaches, reddish sandstone cliffs, and water-sculptured caves. The park also features a rare Great Lakes bog lagoon. Once an open Lake Superior bay, powerful lake currents have deposited large amounts of sand onshore here creating a barrier ridge isolating the lagoon from the lake.

Madeline Island is considered the spiritual home of the Ojibwa people, also known as the Lake Superior Band of the Chippewa. The Ojibwa lived here for hundreds of years prior to the arrival of European fur traders.

French fur traders, along with Jesuit priests including Father Jacques Marquette, arrived at the Apostle Islands in the mid-1600s. The Jesuits established missionaries on Madeline Island and many other locations throughout the region. They are believed to have named the Apostle Islands after the biblical Twelve Apostles of Jesus Christ. A French fur trading post was established on Madeline Island at La Pointe in 1693. This site was used as a fur trade and military post for more than 150 years by the French, English and eventually the Americans.

> **Ranger Trivia**
>
> The Ojibwa referred to Madeline Island as *Mooning-wane-kaaning* or "Place of abundant golden-breasted woodpeckers." This familiar Wisconsin bird is now known as the yellow-shafted flicker.

> **Ranger Note**
>
> In 1808, Michael Cadotte, a fur trader stationed at the La Pointe trading post, married a Native American woman by the name of Ikwesew, the daughter of Chief White Crane, a revered Ojibwa leader of the island. Ikwesew was also known by her Christian name, Madeline, and was a beloved citizen of La Pointe her entire life. The island was renamed Madeline as a tribute to her in 1828.

The Ojibwa people were compelled to leave Madeline Island by order of the U.S. government as a condition of the Treaty of La Pointe in 1854. Most resettled on the Bad River Reservation near Ashland or the Red Cliff Reservation on the northern tip of the Bayfield Peninsula. Soon after, hundreds of American and European immigrants moved to the island to set up farmsteads, work in logging camps, or the commercial fishing industry. Others found jobs quarrying brownstone rock, the dark-colored sandstone used to construct commercial buildings and expensive mansions throughout the 19th century.

The spark to create a state park on Madeline Island began in 1939 as a recommendation from the National Park Service. Efforts to build the park continued for many years until 1963, when a joint resolution between the State of Wisconsin and Ashland County was passed to establish Big Bay State Park.

Barrier beach along Lake Superior

BEACHES, PICNIC AREAS, AND SHELTERS

A 1.5-mile-long white sand beach is located along the northeast shoreline of the park. A nearby picnic area has restrooms, changing stalls, and a boardwalk with easy access to the beach. Lake Superior is the coldest of all the Great Lakes. Water depth and sand bars shift daily and dangerous rip currents can occur at times. Swimmers need to be aware of water conditions prior to entering the lake.

The **Point Picnic Area** is located on a peninsula that juts out into Lake Superior in the southeast corner of the park. This day-use area offers some of the best views of the lake.

HIKING, BIKING AND NATURE TRAILS

Big Bay has 8.5 miles of hiking trails to explore. The **Barrier Beach Boardwalk and Nature Trail** (1.3 miles) is a wooden boardwalk that follows a sand ridge between the Lake Superior and Big Bay Lagoon. The trail has wildlife observation platforms with interpretive panels. At the north end of the trail, a bridge provides access to **Big Bay Town Park**, a local recreational area that has hiking trails and a campground. www.bigbaytownpark.com.

The **Lagoon Ridges Trail** (2.6 miles) follows the western perimeter of Big Bay Lagoon and terminates at Big Bay Town Park. The trail leads through an upland hardwood forest and a lowland area of cedar and spruce. Water-resistant hiking boots are recommended for this trail. The **Woods Trail** (.50 miles) provides access to the beach area from the group camp and links the Bay View and Point Trails to the **Barrier Beach Parking Area.** Just outside the park, a 6-mile paved **Bike Lane** adjacent to Highway H (Middle Road) to the village of La Pointe is a fun ride for bicycle riders.

Ranger Trail Pick

The **Bay View Trail** (1.3 miles) and **Point Trail** (1.7 miles) are located in the southeast area of the park. Both trails offer dramatic views of the Lake Superior and its colorful sandstone rock ledges and shoreline caves.

WILDLIFE VIEWING

Many wildlife species found on Madeline Island are linked to other islands and the Bayfield mainland. Animals such as white-tailed deer, otter, mink, red squirrels, snowshoe hares, and muskrats travel between islands either by swimming or crossing over the ice in winter. More secretive animals including coyote, black bear, and even gray wolves and moose, can move freely between the islands and the mainland as well.

Ranger Trivia

Interestingly, some of Wisconsin's most common animals, such as the eastern chipmunk, striped skunk, and raccoon are not found on any of the Apostle Islands.

The Apostle Islands are important flyway resting spots for migrating birds that cross Lake Superior. Over 200 kinds of songbirds, gulls, terns, shorebirds, sandpipers, and waterfowl congregate here during the spring and fall migration period. The **Big Bay Sand Spit and Bog State Natural Area** (963 acres) is one of the richest sphagnum-sedge bogs in the Lake Superior region. Blueberry, bearberry, huckleberry, false heather, and many unique bog plants can be viewed here.

INTERPRETIVE PROGRAMS AND FACILITIES

Nature programs and hikes are offered throughout the summer season. Interpretive panels highlighting the park's unique flora, fauna, and geology are located adjacent to the **Barrier Beach Nature Trail.** The **Madeline Island State Historical Museum** has displays and artifacts pertaining to the life of early Native Americans, fur traders, and the first settlers of the island. The museum is operated by the State Historical Society and is located along Colonel Woods Avenue in the village of La Pointe. www.madelineislandmuseum.wisconsin

BOATING AND FISHING

Big Bay is part of the 400-mile-long **Wisconsin Lake Superior Water Trail**, which begins near the city of Superior and ends at the Montreal River along the Michigan border.

Park visitors can launch canoes or kayaks along the park's swimming beach to paddle the quiet waters of Big Bay Lagoon or follow the rocky shoreline of Lake Superior on calm days. Canoes and rowboats can be rented from a private concessionaire at the nearby Big Bay Town Park

Fishing for panfish and northern pike is available on the **Big Bay Lagoon** (130 acres). Lake Superior sport fishing for lake trout, king salmon, brown trout, and Coho salmon is available through charter boat services on Madeline Island or the Bayfield harbor.

ACCESSIBLE FACILITIES AND TRAILS

The **Barrier Beach Boardwalk** (1.3 miles) is wheelchair-accessible. A specially designed beach wheelchair with balloon-type tires can be reserved at the park office free of charge.

Campsite 29 is wheelchair-accessible with electrical service and access to a restroom/shower facility.

CAMPING

Big Bay has 63 campsites (15 electric, 7 walk-in sites) with a centrally located restroom/shower facility. A 20-person outdoor group campground is located near the park's south shoreline area. The park also has a rustic indoor camp off County Road H (Big Bay Road) that can accommodate up to 20 people. Big Bay does not have an RV sanitary station but one is available about five miles west of the park along Big Bay Road.

WINTER ACTIVITIES

Hiking, snowshoeing, and winter camping are popular off-season activities. Park staff groom five miles of cross-country ski trails for inline skiing.

AREA ATTRACTIONS

Madeline Island is the largest and the most visited of the Apostle Islands. The picturesque village of La Pointe and the port city of Bayfield are both top Wisconsin tourist destination communities.

www.madelineisland.com   www.bayfield.org

The **Apostle Islands National Lakeshore** encompasses 21 islands managed by the National Park Service. Commercial island boat tours are available to view these beautiful islands as well as the wave-sculptured caves along the Lake Superior shoreline. Lighthouse tours and backpack camping are offered on certain islands. The property also has a mainland hiking trail perched above the Lake Superior shoreline. The **Apostle Islands Visitor Center** is located along Washington Avenue in Bayfield. www.nps.gov.apostleislands

> **Ranger Note**
>
> The ferry service to Madeline Island shuts down in winter, but adventurous park visitors can "drive" to the island across a plowed ice road between Bayfield and La Pointe. A propeller-driven wind sled bus is also available for passenger service to the island in winter.

DIRECTIONS TO THE PARK

From Highway 2 west of Ashland exit onto Highway 13 north to the city of Bayfield. Board the Madeline Island Ferry to cross the channel to the village La Pointe on Madeline Island. After exiting the ferry, turn right (south) onto Main Street and take Highway H (Middle Road) east about four miles to Hagen Road and the park's entrance.

Big Bay State Park
2402 Hagen Road
La Pointe, WI 54850
(715) 747-6425

> **Ranger Note**
>
> The Madeline Island Ferry service operates from about April to the end of December depending on ice conditions. The ferry transports vehicles, bicycles, passengers, RV motorhomes, and trailers. The ferry is first come-first serve and does not accept advance reservations. For rates and departure schedules visit www.madferry.com or call (715) 747-2051.

59. BRULE RIVER STATE FOREST

River of Presidents Runs Through It

Bois Brule River

FOREST SNAPSHOT

The Brule River State Forest is Wisconsin's northernmost forest and encompasses the entire 44-mile route of the Brule River from its headwaters near the Upper St. Croix Lake to its confluence with Lake Superior. The state forest offers rustic campsites, hiking trails, backpacking, wildlife watching, and outstanding North-woods scenery. The Brule River is renowned as a world-class trout stream and one of the most popular canoe and kayak routes in Wisconsin.

GEOLOGY AND HISTORY

The landscape of the Brule River region was shaped by several glaciers that pushed through northern Wisconsin over the past two million years. During the last Ice Age, a colossal ice dam caused Lake Superior to rise 500 feet higher than its present-day level submerged the entire Brule River region under its icy waters. When the earth's climate began to warm again about 20,000 years ago, the ice dam collapsed releasing a torrent of glacial water raging south carving out both the Brule and the St. Croix River valleys on its way to the Mississippi River. When the immense weight of glacial ice left the Lake Superior region, the local landscape rebounded with a geologic upheaval. This caused the Brule River to reverse its course and flow north into Lake Superior, while the St. Croix River continued to flow south towards the Mississippi.

Several Native American cultures have lived along the Brule River over time. When French-Canadian explorers and fur traders paddled down the river in the 17th century, they found the both Ojibwa (Chippewa) and the Dakota Sioux inhabiting the region.

Ranger Trivia

The Ojibwa called the Brule River *Wiisaakode-ziibi*, meaning "a river through half-burnt wood." The French shortened this into Bois Brule or the "Burnt Wood" River. Historians believe the river was named after the charred, blackened trees that were often seen along the dry, sandy pine barren areas of the river, which regularly caught fire from lightning strikes.

During Wisconsin's lumber boom in the 1890s, nearly all the pine forests along the Brule River were cut. In 1907, Frederick Weyerhaeuser, a well-known lumber baron, donated 4,320 acres of cutover land to the State of Wisconsin. Efforts to reforest this barren land began in the 1930s with the help of the Civilian Conservation Corps (CCC), a federal Depression-era work program for unemployed young men. CCC crews planted trees, worked on forest fire prevention, and restored trout habitat along the river. The Brule River State Forest was established in 1936 and now encompasses 80,000 acres (41,000 state-owned). The forest continues to provide employment for timber and paper industry workers and outdoor recreation for thousands of visitors annually.

BEACHES, PICNIC AREAS, AND SHELTERS

The mouth of the Brule River is located in the far north end of the forest along the sandy shoreline of Lake Superior. This is great place to have a picnic, beachcomb or enjoy a cold, refreshing dip in the chilly waters of the lake.

Other picnic spots include the **Bois Brule Picnic Area**, located adjacent to the **Bois Brule Campground** and the **Upper St. Croix Lake Picnic Area**, a scenic lakeside park in the south end of the forest.

HIKING, BIKING, AND NATURE TRAILS

The forest has 40 miles of hiking trails, including a 22-mile segment of the **North Country National Scenic Trail**, a 4,600-mile multi-state trail that stretches from New York State to North Dakota. www.northcountrytrail.org.

Lake Superior beach

The **Old Bayfield Road Hiking Trail** (2.25 miles) follows a rustic roadway built in 1870 to connect the cities of Bayfield and Superior. The trail leads through hemlock/maple forests and past abandoned copper mines. A hilltop overlook along this trail offers panoramic views of the forest.

The **Brule and St. Croix River Portage Trail** (2.0 miles) begins at the **Upper St. Croix Lake Picnic Area.** This historic portage was once an important link in the fur trade route between Lake Superior and the Mississippi River. The **Brule Bog Boardwalk Trail** (2.3 miles) can be accessed from the Portage Trail. This accessible boardwalk leads through a cedar swamp and bog with close-up views of ferns, mosses, orchids and other wetland plants.

Horseback riding and hiking are allowed on **26** miles of snowmobile and hunter walking trails during the summer season. The **After Hours Bike and Ski Trail,** located southwest of the village of Brule, offers **18** miles of off-road biking trails.

Ranger Trail Pick

The **Stoney Hill Nature Trail** (2.0 miles) ascends through a pine and hardwood forest to the top of Stoney Hill with a beautiful view of the Brule River Valley below. Interpretive signs along the trail describe the natural and cultural history of the forest. Access to this trail is along Ranger Road near the forest headquarters.

WILDLIFE VIEWING

The pine and hardwood forests, wetlands, and bogs of the Brule River area provide ideal habitat for a wide array of wildlife. The forest is home to black bear, gray wolves, bobcat, coyote, otter and fisher, white-tailed deer, red squirrels, porcupine, snowshoe hares, and chipmunks. More than 200 species of birds have been recorded in the forest, including many types of woodland birds, hawks, owls, shorebirds, and waterfowl. Birds from the far north, such as great gray owls, goshawks, merlin, crossbills, and black-backed woodpeckers are often spotted here as well.

The **Brule River Boreal Forest Natural Area** (652 acres) preserves forests of black and white spruce, balsam fir, and white pine along the Brule River. The **Brule Glacial Spillway Natural Area** (2,656 acres) encompasses an ancient valley with many natural springs and bogs that serve as the headwaters of the Brule River. This wide, steep-sided gorge also has some of the largest and oldest trees in the state.

INTERPRETIVE PROGRAMS AND FACILITIES

A nature center located at the Brule State Fish Hatchery highlights wildlife of the forest and explains the restoration work performed by Civilian Conservation Corps crews in the 1930s. A self-guided tour of the fish hatchery explains the process of rearing brown trout in the ponds and raceways of the site. The hatchery is located just south of the village of Brule along East Hatchery Road. www.dnr.wisconsin.gov/brulehatchery

Ranger Trivia

The Brule River is often referred to as the River of Presidents. Ulysses S. Grant and Grover Cleveland were the first U.S. presidents to fish the river. President Calvin Coolidge enjoyed fishing the Brule so much that he set up his summer White House here in 1928. Other presidential anglers to cast a line into the Brule included Herbert Hoover, Harry Truman, and Dwight D. Eisenhower.

BOATING AND FISHING

The Brule River is one of the best known and revered trout streams in the Midwest. More than 30,000 anglers travel to this legendary river each year to fish for brook, brown, or rainbow trout. Coho and Chinook salmon can be found in the Brule during their annual spawning runs from Lake Superior as well.

The **Upper St. Croix Lake** (855 acres) is located in the southwest corner of the forest and supports good populations of panfish, walleye, trout, northern pike, and bass. **Rush Lake** (15 acres) located along the eastern border of the forest, is a carry-in lake with panfish and large-mouth bass fishing. **Lake**

Nebagamon (986 acres) and Lake Minnesuing (432 acres) are known for their excellent panfish, northern pike, and bass fishing. Both lakes are about four miles west of the forest.

The Brule River (44 miles) is one of the top canoe and kayak destinations for paddlers in the Midwest. The southern section of the Brule has a slower flow, making it easier to paddle. The northern route has stronger currents, especially the last 19 miles of the Brule, which drops 328 feet before emptying into Lake Superior. Forest staff maintain ten canoe landings along the river. Canoe and kayak rental outlets are available at several locations, including Brule River Canoe Rental located along Highway 2 in the village of Brule. www.brulerivercanoerental.com.

ACCESSIBLE FACILITIES AND TRAILS

The forest headquarters and most picnic areas and restrooms are accessible for mobility-impaired visitors. Both forest campgrounds have wheelchair-accessible campsites but do not offer electrical service. The Brule Bog Boardwalk Nature Trail located along the north side of the Upper St. Croix Lake, is wheelchair-accessible.

CAMPING

The Bois Brule Campground (22 campsites, 2 walk-ins) is located along the Brule River near the forest headquarters. A nearby picnic area has a canoe launch and a fishing pier.

The Copper Range Campground (17 campsites, 3 walk-ins), situated along a stretch of the Brule River, is known for its excellent fishing in spring and fall. The campground is located along County Road H.

Ranger Note

None of the forest campgrounds have electric service, showers or flush toilets. The Brule River Campground, located along Highway 2 in the nearby village of Brule, has electric sites, fee showers, and a RV sanitary/ water fill station. Lucius Woods County Park, located southwest of the forest along Upper St. Croix Lake, has modern amenities for campers as well. (See Area Attractions)

WINTER ACTIVITIES

The After Hours Ski Trail (18 miles), located west of the village of Brule, is groomed for both classic and skate skiing. Other popular off-season activities include winter camping, hiking, wildlife watching, snowshoeing, snowmobiling, and ice fishing.

AREA ATTRACTIONS

Douglas County is an outdoor recreation wonderland with its many resorts, lakes, and 280,000 acres of public forests and wildlife areas. www.douglascountywi.org

Lucius Woods County Park has 29 campsites (23 electric/water hookups). The park is along the western shoreline of Upper St. Croix Lake and offers hiking, picnicking, and swimming beach. Lucius Woods is located about four miles south of the forest along East Marian Avenue near the community of Solon Springs.
www.douglascountywi.org/luciuswoods

> ### Ranger Trivia
>
> Lucius Woods was established in 1950 as a Wisconsin state park. The park was named in honor of Nick Lucius, who purchased the property in 1891. Lucius operated a popular fishing and hunting resort here and preserved the towering, old-growth white pine trees found here. Lucius Woods remained a state park for 40 years before the state deeded it to Douglas County in 1990.

DIRECTIONS TO THE FOREST

The Brule River State Forest is located in northeastern Douglas County midway between the cities of Superior and Ashland. The forest headquarters is located near the village of Brule about a mile southwest of US Highway 2.

Brule River State Forest
6250 South Ranger Road
Brule, WI 54820
(715) 372-5678

60. BRUNET ISLAND STATE PARK

Chippewa River Island—A Scenic Gem

Brunet Island beach and shoreline

PARK SNAPSHOT

Brunet Island State Park is located along the Cornell Flowage, a backwater lake of the Chippewa and Fisher rivers. This scenic 1,300-acre park offers camping, swimming, picnicking, nature study, and miles of hiking trails that lead through a natural wonderland of river channels, woods, and wildlife.

GEOLOGY AND HISTORY

Brunet Island was shaped by the advance and retreat of several glaciers over the course of millions of years. The last great ice sheet, the Chippewa Lobe of the Wisconsin Glacier, pushed through this region about 30,000 years ago but came to an abrupt stop about 50 miles south of Brunet Island. This marked the extent of the Driftless Area, a region of southwestern Wisconsin that remained unglaciated during the last Ice Age.

As glaciers retreated north back to the Canadian arctic about 12,000 years ago, they left behind a landscape of kettle lakes, moraine hills, deep deposits of glacial drift, and re-routed the Chippewa River.

Several Native American cultures have hunted, fished, and lived along the Chippewa River for thousands of years. Around 1745, a band of Lake Superior Ojibwa (Chippewa) people migrated to this area and built permanent settlements near the headwaters of the Chippewa River. French-Canadian fur traders arrived some years later to set up trade pacts with Ojibwa hunters and trappers. French, English and eventually American companies all had fur trading posts along the Chippewa River at one time or another.

As part of the Treaty of LaPointe signed on Madeline Island in 1842, the Ojibwa were forced cede all their remaining lands to the U.S. government. This paved the way for lumber barrons to purchase large tracts of virgin pine throughout northern Wisconsin, including the forests of the Chippewa River Valley. Throughout the early 19th century and into the 1920s, the logging industry supported nearly all of the Chippewa River region's economy and development.

Brunet Island was named after Jean Brunet, a French nobleman and businessman who immigrated to America in 1818. Brunet (pronounced "brew-net") made his living as a teacher, missionary, and engineer. He became an Indian affairs agent, judge, and a legislator for the Wisconsin Territory. Brunet built the first dam and sawmill across the Chippewa River near the city of Chippewa Falls. He later moved upriver just south of Brunet Island State Park where he opened a fur trading post and a popular inn for travelers in the 1830s. Over time, his small settlement along the banks of the Chippewa River became known as Brunet Falls, now the city of Cornell.

In 1936, the Northern States Power Company donated the 179-acre Brunet Island to the State of Wisconsin to develop a new state park. The initial development of the park was completed by Civilian Conservation Corps (CCC) work crews. Some of the original CCC structures are still in use today, such as the attractive stone/log shelter building in the main picnic area. The state later acquired 1,124 onshore acres to add to the park.

> **Ranger Trivia**
>
> A regular guest at Jean Brunet's inn was Ezra Cornell, a wealthy businessman and founder of the Western Union Telegraph Company from Ithaca, New York. Cornell is best known for co-founding Cornell University, a private Ivy League school, in 1865. Cornell convinced New York State legislators to purchase nearly a half million acres of unclaimed government lands in Wisconsin's Chippewa River Valley as an investment to help fund Cornell University. New York paid less than one dollar per acre for this land on which timber was harvested and later sold for $5 million dollars. Cornell insisted that a parcel of this property be donated to the citizens of Brunet Falls. The town was eventually renamed Cornell in his honor.

60. Brunet Island State Park

BEACHES, PICNIC AREAS, AND SHELTERS

A wide, sandy swimming beach is located along the Chippewa River at the southern tip of Brunet Island. The nearby picnic area has changing stalls, restrooms, and a log shelter featuring a stone fireplace. The shelter has electrical service and can be reserved in advance.

HIKING, BIKING AND NATURE TRAILS

The park has ten miles of hiking trails to explore, including a two-mile section of the **Chippewa Moraine Unit** of the Ice Age Trail. The **Nordic Ski and Hiking Trail** (5.0 miles) begins near the park entrance station and leads through forested hills and valleys with scenic views of the Fisher River. The **Timber Trail** (0.75 miles) travels through the center of the island past impressive stands of hemlock

---

Ranger Trail Pick

The **Jean Brunet Nature Trail** (0.8 miles) is located on the north side of the island and follows backwater river channels along the Fisher and Chippewa Rivers. The trail travels beneath dense canopies of ancient hemlock, oak, and maple trees. Interpretive signs along the route describe the rich history and the unique flora and fauna of the park. Photographers will find many opportunities to capture stunning mirror-like waterfront photos along this scenic riverside trail.

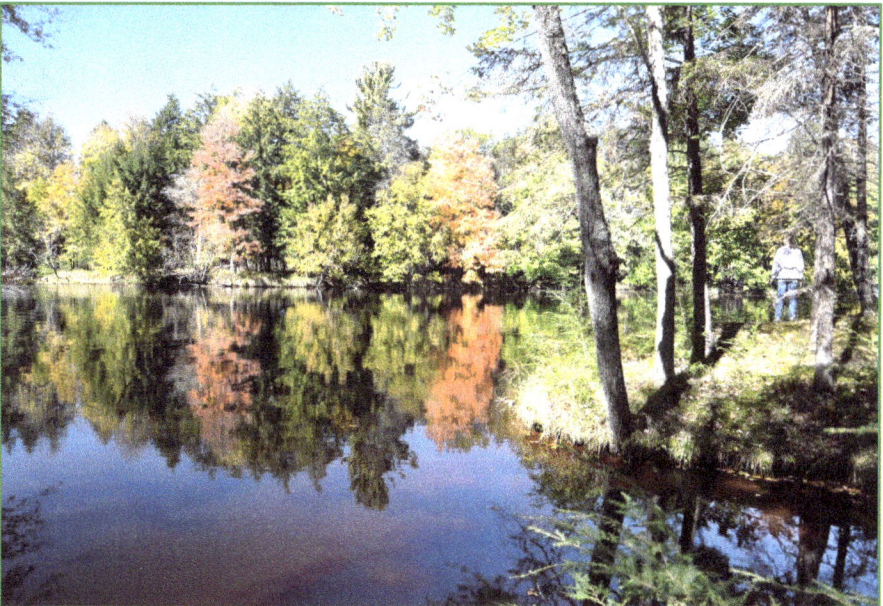

Jean Brunet Nature Trail

and aspen trees. This trail also serves to link the north campground to the beach and picnic areas.

The **Pine Trail** (0.5 miles) begins near the boat launch area where it ascends north through a red pine plantation and links up with the Timber Trail and the Jean Brunet Nature Trail. The **Spruce Trail** (0.50 miles) also starts at the boat launch and follows the scenic river shoreline south to the picnic and beach areas.

Brunet Island has two miles of designated bike routes along park roadways. The **Old Abe Link Trail** (1.0 miles) is a connector route that leads to the nearby city of Cornell and the trailhead for the **Old Abe State Trail**. (See Area Attractions)

WILDLIFE VIEWING

The park's forests, backwater river areas, and the Cornell Flowage attract many species of waterfowl, shore birds, owls, hawks, and woodland songbirds. Bald eagles and ospreys are often seen soaring above the river area. The park is also home to white-tailed deer, squirrels, chipmunks, porcupine, raccoon, coyote, beaver, otter, muskrat, and mink.

The **Jean Brunet Woods State Natural Area** (387 acres), located in the northern section of the park, has impressive old-growth forests of hemlock, white and red oak, sugar maple, and paper birch. Wetland areas and ephemeral (seasonal) ponds in this area harbor several species of frogs, salamanders, insects, and rare ferns and club-mosses.

INTERPRETIVE PROGRAMS AND FACILITIES

Interpretive panels describing the history of Brunet Island and the diverse flora and fauna found here are located along the Jean Brunet Nature Trail and at other locations in the park.

BOATING AND FISHING

Cornell Flowage (897 acres) is a backwater lake of the Chippewa and Fisher rivers and is known for its excellent boating and fishing opportunities. Anglers can expect to catch walleye, northern pike, smallmouth bass, perch, catfish, and panfish. The park has a paved boat launch with access to the flowage and Chippewa River.

The quiet backwater bays and river channels of the park are ideal for paddlers. Canoe and kayak rentals are available from local vendors. The Cornell Canoe and Kayak outlet is located in nearby city of Cornell and offers free delivery and pickup of their rentals to park guests.

ACCESSIBLE FACILITIES AND TRAILS

Most of the park's restrooms, shelters, and picnic areas are accessible for mobility-impaired visitors. Campsite 16 is wheelchair-accessible and has electrical service. A surfaced path provides access to a restroom/shower facility. A wheelchair-accessible fishing pier is located on a lagoon adjacent to the main road near the North Campground.

An asphalt-paved segment of the Jean Brunet Nature Trail is wheelchair-accessible. The walkway follows a scenic backwater channel and leads to a wheelchair-accessible picnic table and fishing platform.

CAMPING

The **North Campground** has 45 (non-electric) campsites. Most of these sites are small in size but very popular because many are located along the Chippewa River and backwater flowages. The **South Campground** has 24 electric-service campsites located within a wooded area north of the swimming beach. Showers/restroom facilities are located in this campground. An RV water fill and sanitary station is available on the mainland region of the park near the visitor entrance station.

WINTER ACTIVITIES

Hiking, snowshoeing, winter camping, and wildlife watching are popular off-season activities. Park staff groom five miles of cross-country ski trails for both classic and inline skiing. Many anglers enjoy ice fishing on the Cornell Flowage and backwater areas of the Chippewa and Fisher Rivers.

AREA ATTRACTIONS

Chippewa County and the city of Cornell have many fine resorts, parks, bike trails, historic sites and outdoor recreation areas to explore.
www.cityofcornell.com
www.chippewacounty.com.

The **Old Abe State Trail** (20 miles) is one of the state's most popular biking/hiking trails. This asphalt-paved trail begins in the nearby city of Cornell and travels south through scenic forests and farms along the Chippewa River to **Lake Wissota State Park** (See Chapter 72) near Chippewa Falls.
www.co.chippewa.wi.us/old-abe-state-trails

DIRECTIONS TO THE PARK

Brunet Island is located about 25 miles northeast of Chippewa Falls and the Eau Claire area. Take Highway 178 or Highway 27 north to the city of Cornell. From Bridge Street in Cornell, exit north onto Park Road to the visitor entrance station.

Brunet Island State Park
23125 255th Street
Cornell, WI 54732
(715) 239-6888

61. CHIPPEWA FLOWAGE

The "Big Chip"

Chippewa Flowage

PROPERTY SNAPSHOT

The Chippewa Flowage, located in Sawyer County, is the largest wilderness lake of its kind in the state. This expansive 15,300-acre reservoir has 233 miles of undeveloped shoreline bordered by 10,000 acres of state-owned forest. The flowage, also known as the "Big Chip," is renowned for its world-class fishing, boating, canoeing, and kayaking opportunities. The property has hiking trails and boat-in campsites on many of the 200 islands that dot the flowage.

GEOLOGY AND HISTORY

The original landscape of the Chippewa River Valley was a mix of rolling moraine hills, kettle lakes, streams, and wetlands created by several glaciers that had advanced through this region over the past two million years. As the last Ice Age came to an end about 12,000 years ago, grassy tundra meadows and pine forests reclaimed the landscape and wild rice regained its foothold in local streams and lakes. This lush green landscape attracted herds of big game animals and waterfowl, which in turn, enticed several Native American cultures to enter the area. The most recent were the Santee Sioux, Ottawa, and Ojibwa, also known as the Lake Superior Chippewa or the *Anisihnaabe*, "true or original people."

In 1660, French-Canadian explorers arrived to set up fur trading with Ottawa hunters and trappers and later with the Chippewa. Many fur traders spent the winter camped along the shores of Lake Lac Courte Oreilles, a 5,000-acre lake located west of the Chippewa Flowage.

> **Ranger Note**
>
> The Chippewa people's homeland is believed to have been in the Hudson Bay area. Over time, they expanded their range into the Great Lakes region and entered northern Wisconsin in 1745, where they built a permanent settlement near the headwaters of the Chippewa River.

As the fur trade came to an end, lumber from the expansive pine forests of northern Wisconsin became the most sought-after commodity in the state. In 1837, the Chippewa agreed to sell most of their land in north central Wisconsin to the U.S. government in exchange for annual cash payments and the right to hunt, fish, and gather rice on ceded lands. Later, Chippewa leaders gained ownership of three reservations in Wisconsin, including the 76,465-acre Lac Courte Oreilles Reservation located on the south side of the Chippewa Flowage. The virgin forests of the Chippewa River Valley held some of the finest stands of

> **Ranger Trivia**
>
> The French named Lac Courte Oreilles or "Lake of Short Ears" after a local band of Ottawa people living there, who were known for cutting off a portion of their ears for personal decoration. Ottawa is derived from the Algonquin word, "*Adawe*," which means "to trade."

timber in the state. During the timber industry's peak years in the late 19th century, more than 50 logging camps were located in Sawyer County alone. Lumbermen from the East Coast and immigrants from Germany, Norway, Poland, and Ireland provided labor for logging camps and sawmills.

The Chippewa Flowage was created in 1924 when the Northern State Power Company (NSP) built a dam across the Chippewa River near the town of Winter. The dam was originally intended only to create a water-holding reservoir to supply a

> **Ranger Note**
>
> The construction of the dam and flowage was strongly opposed by the Chippewa because its waters would submerge many of their sacred sites, cemeteries, churches, and homes in the village of Post. The power company did relocate some structures and eventually built a new town called "New Post" on higher ground, but the conflict dragged on for decades.

steady flow of water for hydroelectric power plants downstream, but it was later repurposed to produce electricity.

In 1988, the State of Wisconsin purchased most of the NSP Company's land, primarily to ensure public access to the Chippewa Flowage. The U.S. Forest Service also acquired a section of NSP land as an addition to the adjoining Chequamegon-Nicolet National Forest. The Lake Superior Chippewa were given 4,500 acres of NSP property, including several islands, to expand their Lac Courte Oreilles Reservation. The Chippewa also assumed ownership of the controversial Chippewa River Dam, which they continue to operate for electrical power. Today, the Chippewa Flowage is managed jointly by the State of Wisconsin, the National Forest, and the Chippewa and has become one of the most treasured recreational lakes in the state.

BEACHES, PICNIC AREAS AND SHELTERS
There are no developed swimming beaches or picnic areas in the Chippewa Flowage area. Both can be found at **Hayward City Park and Beach** located about 15 miles northwest of the flowage along South 2nd Street in Hayward.

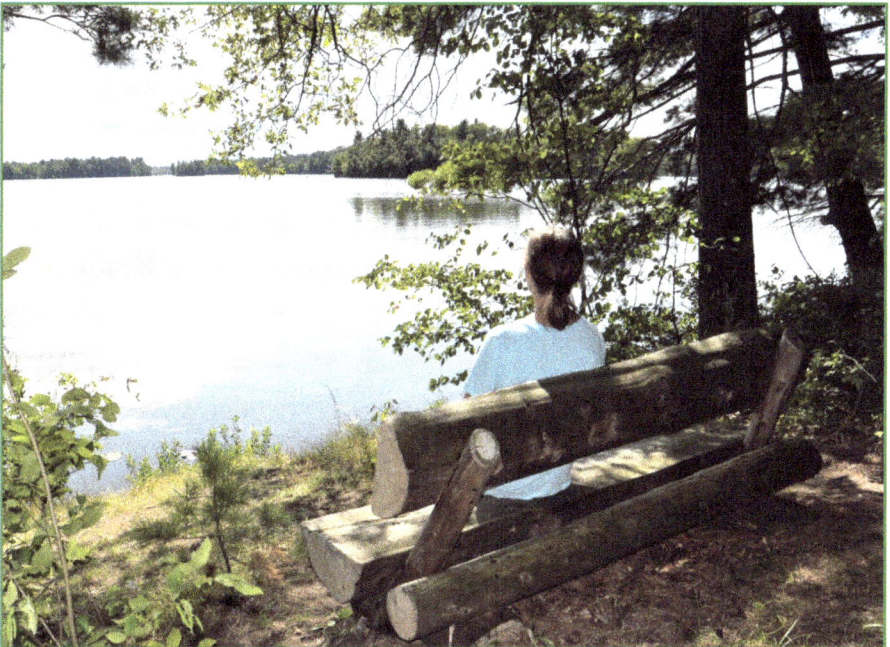

Chippewa Flowage overlook

HIKING, BIKING, AND NATURE TRAILS

The **Hay Creek Trail** (2.0 miles) follows the path of logging roads through an attractive forest area between Moss Bay Creek and Hay Creek. The trail is not signed so hikers are encouraged to bring along a trail map and compass. Trail parking is located on the north side of the flowage off County Highway B.

The **Tuscobia State Trail** (74 miles) follows the historic Omaha Railway line built in 1899 through sections of the **Flambeau River State Forest** and the **Chequamegon-Nicolet National Forest**. This multi-use route is open for hiking, horseback riding, ATV use, and off-road biking. A trailhead parking area is located 12 miles south of the flowage at **Ojibwa Park** along Highway 70.

Ranger Trail Pick

The *Mino-giizhigad Trail* (1.5 miles) leads through a mature hardwood and pine forest with scenic overlooks of the Chippewa Flowage. The term *Mino-giizhigad* is an Ojibwa phrase meaning "a good day." The trail is located east of the **CC North Boat Launch** adjacent to County Road CC.

WILDLIFE VIEWING

The open-water bays, islands, and forests of the Chippewa Flowage provide ideal habitat for wildlife. Over a 130 species of birds have been identified here, including many types of waterfowl and woodland songbirds. Many birdwatchers travel here to view the numerous ospreys, bald eagles, and common loons that nest along the flowage.

Ranger Note

Elk were reintroduced in the nearby Chequamegon National Forest near Clam Lake in 1995. A popular "elk-viewing area" is located along County Highway B on the north side of the Chippewa Flowage.

White-tailed deer, porcupine, rabbits, squirrels, raccoon, beaver, otter, and fox are often seen throughout the flowage. More secretive animals such as coyotes, fishers, black bears, gray wolves, and even elk are occasionally spotted here as well.

Ranger Trivia

The all-time world record muskellunge was caught in the Chippewa Flowage in 1949. This colossal fish measured over five feet in length and weighed 69 pounds, 11 ounces.

INTERPRETIVE PROGAMS AND FACILITIES

The **Namekagon River Visitor Center**, located southwest of Hayward along Highway 63, offers exhibits featuring river wildlife and has a 2.8 mile nature trail along the Namekagon River, a tributary of the St.

Croix River. The center is operated by the National Park Service. www.nps.gov/sacn/namekagonriver

BOATING AND FISHING

The Chippewa Flowage is probably best known for its outstanding fishing opportunities. Many anglers ply these waters in search of a trophy muskie but the flowage offers great fishing for walleye, bass, northern pike, yellow perch, and panfish as well.

The flowage has several boat launch and parking area facilities. The **CC South Landing**, located along Highway CC near the center of the property, is the largest launch area. It has restrooms and offers the only drinking water available on the flowage. The **CC North Landing** has a fishing pier and restrooms. Due to the size and irregular shoreline and bays of the Chippewa Flowage, boaters are advised to bring along lake maps or GPS units. Boaters need to be aware of submerged stumps, sandbars, and shallow rocky areas found throughout the flowage.

ACCESSIBLE FACILITIES AND TRAILS

A wheelchair-accessible fishing pier is available adjacent to the **CC North Boat Launch** area along County Highway CC. The **Cedar Tops West Campsite** is a boat-in accessible island camping area located on the west side of the flowage. The site can be reserved in advance by contacting the Hayward DNR office (715) 634-7433.

CAMPING

There are **18** (boat-in-only) campsites, all located on state-owned islands in the northwestern section of the flowage. Each site has a fire ring, picnic table, and an open-air box latrine. Camping is free and sites are available on a first-come, first-served basis. The Ojibwa Lac Courte Oreilles Conservation Department maintains six island campsites along the eastern edge of the flowage. Campsite reservations and payment of fees are required for these sites. (715) 634-0102.

Campers looking for drive-in campsites will find them at **Ojibwa Park**, a former state park, located along Highway 70 twelve miles south of the flowage. The park, situated along the Chippewa River, has 16 campsites (12 electric), a shower building, and an RV water fill/ dump station. **Lake Chippewa Campground** is a private camping resort located along County Road CC near the center of the flowage. The resort has 210 full hook-up campsites, a swimming beach, and watercraft rentals. www.lakechip.com.

WINTER ACTIVITIES

The **Hay Creek Cross-country Ski Trail** (2.0 miles) is located on the north side of the flowage along County Highway B. Ice fishing for crappie, bluegills, walleye, and northern pike is a popular winter activity. Several snowmobile routes pass through the flowage with links to the Sawyer County Snowmobile and ATV trail system

AREA ATTRACTIONS

Sawyer County is well known for its many fishing lakes, resorts, historic sites, and numerous outdoor recreational opportunities.
www.sawyercountygov.org, www.haywardlakes.com

The **Fresh Water Fishing Hall of Fame and Museum** in the nearby city of Hayward is often referred to as "Wisconsin's Shrine to Anglers." The museum houses antique fishing gear, outboard motors, trophy fish mounts, and many other displays. The site is best known for its 4-story-high and 143-foot-long muskellunge statue, the largest of its kind in the world. A staircase inside this gigantic fiberglass fish leads to an overlook inside the gaping, tooth-filled mouth of the muskie. The museum is located about 15 miles northwest of the Chippewa Flowage on Hall of Fame Drive in Hayward. www.freshwater-fishing.org.

DIRECTIONS TO THE FLOWAGE

The Chippewa Flowage is located in central Sawyer County. From U.S. Highway 53 near Spooner, exit northeast on Highway 63 towards Hayward. The flowage is located about 15 miles southeast of Hayward. It can also be accessed via Highway 27 just north of Highway 8 near Ladysmith.

Chippewa Flowage
(Property Office Address)
10220 State Road 27
Hayward, WI 54843
(715) 634-7433

62. Chippewa Moraine Ice Age State Recreation Area

A Glacial Treasure Trove

Ice Age Interpretive Center

RECREATION AREA SNAPSHOT

The Chippewa Moraine Ice Age Recreation Area is located within the scenic rolling hills of Chippewa County. The 4,177-acre property showcases a wealth of glacial formations. including moraine knolls, outwash plains, esker ridges, and over 70 small pothole lakes and ponds. A modern nature center here interprets the flora, fauna, human history, and unique glacial geology of the area. Chippewa Moraine offers walk-in campsites, nature study, fishing, hunting, and boating opportunities, along with some of the most scenic hiking trails in the state.

GEOLOGY AND HISTORY

The landscape of the Chippewa Moraine was formed during Wisconsin's last Ice Age which began about 30,000 years ago. Geologists believe at that time a large field of glacial ice stalled about a mile south of where the nature center stands today. Because of this the front edge of the glacier began to melt at about the same rate as the ice sheet was moving forward. This led to large amounts of sand, gravel, rocks, and other glacial debris to be deposited at the leading edge of the glacier, creating a massive ten-mile-wide elevated hill or "end moraine."

Archeological discoveries of spear and arrow points found locally suggest that indigenous people have occupied the Chippewa Moraine area for thousands of years.

The first European explorers to this area were French-Canadian fur traders who arrived in the middle of the 17th century to trade with the Ottawa people and later with Ojibway (Chippewa) hunters and trappers. When the fur trade came to an end in the early 1800s, the vast pine forests of became the most valuable commodity in the state. Logging and supplying lumber to the growing cities to the south became the most important industry in northern Wisconsin. In response to timber company interests, Chippewa leaders were compelled to sell their lands to the U.S. government in 1837 in exchange for annual cash payments and eventual ownership of three large reservation properties.

> **Ranger Note**
>
> The Chippewa Moraine Interpretive Center is situated on a hill overlooking a large grassy open plain below. Geologists surmise that both of these glacial landforms were formed when ice fields in this region became pot-marked with deep open holes. Meltwater from receding glaciers flowed into these holes, along with sand, silt, and rocks, creating small lakes surrounded by walls of ice. As the Ice Age came to an end about 12,000 years ago, the ice walls of these lakes collapsed, sending a flood of meltwater along with lighter glacial debris into washout plains. Heavier rocks and gravel remained behind, forming tall, flat-topped moraines like the hill at the Interpretive Center.

> **Ranger Note**
>
> One of the most influential leaders in creating and promoting the National Scientific Ice Age Reserve was U.S. Representative David R. Obey of Wausau, Wisconsin. In 2011, the State recognized the congressman's support of the Ice Age Reserve by renaming the Chippewa Moraine Nature Center the David R. Obey Ice Age Interpretive Center in his honor.

Immigrants from Germany, Ireland, and Norway began to arrive in the local area in the mid-1800s to set up dairy farms and raise crops on cutover forest lands. Today, dairy farms still dot the local landscape and second-growth forests once again provide timber for the lumber and paper industry.

In 1962, the U.S. Congress acknowledged the national significance of the many unique landforms created by the glaciers in Wisconsin. This led to the creation of the **Ice Age National Scientific Reserve** and the establishment of the **Chippewa Moraine Ice Age State Recreation Area** in 1971. The property is one of the nine units of the Ice

Age National Scientific Reserve and is managed by the State in conjunction with the National Park Service.

BEACHES, PICNIC AREAS, AND SHELTERS

There are no picnic areas or swimming areas at the Chippewa Moraine. A few picnic tables are available adjacent to the interpretive center, however. A swimming beach and picnic area is available at **Morris-Erickson County Park** located about five miles north of the Chippewa Moraine Recreation Area. (See "Camping")

HIKING, BIKING, AND NATURE TRAILS

There are ten miles of hiking trails at the Chippewa Moraine. The **Mammoth Nature Trail** (0.7 miles) has interpretive signs that describe how glacial plains and kettle lakes were formed here and the history of the local logging industry. The **Circle Trail** (4.5 miles) begins northwest of the interpretive center. The trail loops through a forested area and follows a boardwalk between **North** and **South Shattuck Lake**. The trail crosses over County Highway M and follows the south side of South Shattuck Lake atop a rare glacial esker.

> **Ranger Note**
>
> An "esker" is an elevated serpentine moraine ridge. This unique twisting Ice Age formation was formed when sand, gravel, and other debris was deposited by a river that once flowed beneath a glacier.

Kettle lake in autumn color

A six-mile section of the Ice Age Trail passes through the center of the Chippewa Moraine Recreation Area. The trail leads through oak and maple forests adjacent to several of the 21 scenic wilderness lakes and ponds found here. The 1,000-mile **Ice Age National Scenic Trail** follows the southern extent of the Wisconsin glacier during the last Ice Age.

Ranger Trail Pick

The **Dry Lake Trail** (1.8 miles) leads east of the interpretive center along the edge of a hilltop that was once the shoreline of an extinct Ice Age lake. The trail loops through a forested area past scenic pothole lakes. These lakes were formed when massive blocks of ice from retreating glaciers became buried under glacial deposits near the end of the Ice Age. When the underground ice melted, "pot-like" open basins or potholes emerged. Some remained "dry," while deeper ones filled with ground water, resulting pothole lakes. Dry Lake Trail merges with the **Ice Age Trail, Circle Trail,** and **Mammoth Trail,** all of which can all be explored along the way.

WILDIFE VIEWING

The lakes, ponds, wetlands, and hardwood forests of the recreation area provide ideal habitat for wildlife to thrive here. White-tailed deer, squirrels, rabbits, raccoon, muskrats, and beaver are regularly seen in the area. More secretive animals such as coyote, fox, and black bear are occasionally spotted here as well.

Birdwatching can be exceptional during the fall and spring migration periods. Several species of waterfowl, hawks, owls, and sandhill cranes, great blue herons, and an occasional loon can be spotted on local lakes and ponds. The bird feeders located at the interpretive center attract cardinals, goldfinches, red-headed woodpeckers, chickadees, blue jays, and nuthatches throughout the year.

The Chippewa Moraine has three natural areas that provide sanctuaries for several rare animal and plant species. The **Chippewa Moraine Lakes State Natural Area** (306 acres) preserves a concentration of nine beautiful glacial lakes. The **North Shattuck Lake State Natural Area** (297 acres), **Town Line Lake,** and **Woods State Natural Area** (635 acres) protect rugged glacial topography and old-growth oak and maple forests.

INTERPRETIVE PROGRAMS AND FACILITIES

The **David R. Obey Ice Age Interpretive Center** is one of the finest nature centers in the state. The center houses several museum-quality displays that highlight Wisconsin's glacial period. Rare archeological artifacts are also on display following the human history of the area from prehistoric Paleo-Indians to the arrival of European immigrants. The center is open every day from 8:30 a.m.

to 4:00 p.m. except on Mondays. Interpretive programs, hikes, and special events are scheduled throughout the year.

BOATING AND FISHING

The Chippewa Moraine area has seepage-type lakes, which are fed by groundwater springs and rainwater. Smaller lakes can only be accessed on foot, but most larger lakes have boat launch facilities, including **Town Line, Horseshoe, Knickerbocker,** and **Plummer** lakes, and both **North** and **South Shattuck lakes.** Anglers can expect to catch panfish, northern pike, and large-mouthed bass in most lakes. **Horseshoe Lake** (27 acres) is occasionally stocked with walleye.

ACCESSIBLE FACILITIES AND TRAILS

The Ice Age Interpretive Center and restrooms facilities are wheelchair-accessible. Due to the steep and sometimes rocky moraine topography of the recreation area, many of the hiking trails here may be challenging for visitors with walking disabilities.

CAMPING

The recreation area has two hike-in campsites located along the Ice Age Trail west of the Ice Age Interpretive Center, plus a shoreline campsite accessible by watercraft only along Town Line Lake. Each campsite has a fire ring and an open-air latrine. Campers need to pack in their own drinking water. All campsites are first-come, first-serve and can be checked out at the interpretive center.

Campers who prefer more modern amenities will find them at **Morris-Erickson County Park,** which has 28 campsites (all electric) with restrooms and a swimming beach on **Long Lake.** The park is located about five miles north of the Chippewa Moraine along Highway 40 northeast of New Auburn. www.chipewacounty.com/parks.

WINTER ACTIVITIES

Hiking, snowshoeing, wildlife watching, and cross-country skiing are popular off-season activities. None of the trails are groomed. Anglers enjoy ice fishing on many of the areas lakes for northern pike, large-mouth bass, and panfish.

AREA ATTRACTIONS

Chippewa County is a popular vacation destination with a variety of outdoor activities, such as fishing, hunting, boating, hiking and camping. www.gochippewacounty.com

The **Chippewa County Forest** system has more than 34,000 acres of pristine northern woodlands and dozens of secluded lakes to explore. The forest offers camping, hiking, wildlife watching, berry-picking, horseback riding, and ATV trails. An 18-mile segment of the Ice Age Trail leads through county forests as well. www.co.chippewa.wi.us

DIRECTIONS TO THE RECREATION AREA

The Chippewa Moraine Recreations Area is located in Chippewa County about 38 miles north of Eau Claire. From Highway 53, exit onto County Highway M West to New Auburn. Follow County Highway M/SS north out of New Auburn and then turn east on County Highway M to the entrance to the Interpretive Center.

Chippewa Moraine Ice Age State Recreation Area
13394 County Highway M
New Auburn, WI 54757
(715) 967-2800

63. COPPER FALLS STATE PARK

Waterfalls Park—A State Treasure

Copper Falls

PARK SNAPSHOT

Copper Falls State Park is a true Wisconsin treasure. People have been traveling to this 3,496-acre park to view the picturesque waterfalls along the Tyler Forks and the Bad River for generations. Copper Falls offers camping, picnicking, swimming, nature study, and miles of hiking trails, yet the park seems to have retained its wilderness aura over time. The thunder of the waterfalls echoing through the river gorge adds a primordial mystique to this treasured park.

GEOLOGY AND HISTORY

The forested hills, waterfalls, and rocky gorges of Copper Falls offer subtle clues to the fiery volcanos, ancient oceans, glacial ice, and turbulent rivers that formed them. The story of how these dramatic geologic events occurred is one of the most complex episodes of Wisconsin's northern landscape. Ancient inland seas that once submerged this region eons ago deposited thick layers of sand, gravel, and clay. Over time, these sediments were transformed into sedimentary rock such as sandstone, shale, and iron ore. About 2.2 billion years ago, an immense continental tectonic force uplifted this region of the state, creating the nearby the **Penokee Mountain Range**, a 100-mile ridge of iron-rich hills.

Much of the exposed rock at Copper Falls was formed by volcanic activity that occurred here and throughout the Lake Superior basin about 1.1 billion years ago. Molten lava from deep beneath the earth's crust oozed onto the landscape over a period of millions of years. The fiery lava hardened into the dark-colored red and brown basalt rock seen within the parks waterfalls today.

The final geologic event to shape the park's landscape was a series of massive glaciers that pushed through this region from the Canadian Arctic beginning about two million years ago. Each of these powerful glaciers excavated the Lake Superior deeper and wider to the point that its southern shoreline once reached the edge of Copper Falls. Glaciers also deposited layers of sand, gravel, boulders, and red-colored clay here as they retreated northward about 12,000 years ago. Geologists believe the **Bad River** and **Tyler Fork's River** have been eroding the rocky gorge at Copper Falls for nearly 200 million years. Today, these ancient rivers continue to cut through layers glacial deposits, clay, and sandstone and shale rock, exposing the harder volcanic basalt rock of the waterfalls.

The Copper Culture people were one of earliest tribes to inhabit this area about 5,000 years ago. They were the first to mine copper and fashion it into fish hooks, ornaments, arrow heads, axes, and other tools. When French-Canadian explorers reached this area in the mid-17th century, the Sioux and later the Ojibway (Chippewa) people controlled this region. The Ojibway referred to the Bad River as the *Mashkiziibe* or "Swampy River" due to the brownish color of its water. The French translated the term to *mauvais* or the "Bad" River.

Attempts to mine copper in this area began in the mid-1800s, primarily by Welch and Cornish miners. Several vertical and horizontal mineshafts were dug within the park but none were known to produce much copper. Eventually, iron-ore deposits found in the nearby Penokee Mountain Range, not copper, would draw mining companies to this region. In 1886, the nearby town of Iron City was established to serve the area's booming iron ore and lumber industry of the time. New railway lines were built to transport logs to sawmills and haul supplies to iron-mining towns. The railroad was so important to the citizens of Iron City they eventually renamed their town Mellen in honor William Mellen, the general manager of the Wisconsin Central Railroad.

> **Ranger Note**
>
> The Bad River Band of the Lake Superior Chippewa first entered northern Wisconsin in 1745 from the eastern Great Lakes region. Today more than 7,000 Chippewa people make their home on the 156,000-acre Bad River Reservation located along the shores of Lake Superior north of Copper Falls.

In the early 1900s, immigrant farmers attempted to grow crops on cutover forest land but most failed due to the poor soils and short growing seasons of northern Wisconsin. With no other industry to fall back on, the area's once vibrant economy began to falter. Fortunately, interest in Northwoods tourism was just getting started. Within a few years, thousands of urban visitors arriving by train to resorts, cabins, and campgrounds to hunt, fish, or just enjoy the beauty of the area. Local business leaders recognized the potential of the Copper Falls in attracting tourists to Mellen area. They petitioned state legislators to establish Copper Falls State Park in 1929.

BEACHES, PICNIC AREAS, AND SHELTERS
The **Bad River Picnic Area** is located downstream of the park's main waterfalls. This day-use area features an attractive stone and timber shelter/concession facility, which was built by Civilian Conservation Corps (CCC) craftsmen in the 1930s. A wide, sandy swimming beach is located on **Loon Lake** on the far south side of the property.

Shelter/concession building

HIKING, BIKING, AND NATURE TRAILS
Copper Falls has 17 miles of hiking trails to explore, including a section of the **North Country National Scenic Trail**, a multi-state 4,800-mile-long walking

route that leads from Vermont to North Dakota. The **Red Granite Falls Trail** (2.5 miles) loops through the southwest section of the park with views of the lower **Bad River Gorge.**

The **Pipeline Trail** (5.0 miles), the longest trail in the park, stretches through the center of the park. The **Vahttera Trail** (1.7 miles) and the **Takesson Trail** (2.5 miles) loop through mature hardwood forest areas with scenic views of the upper Bad River Gorge. Off-road bike riding is allowed on both of these trails.

> **Ranger Note**
>
> The Doughboy's Trail was named in honor of World War I soldiers, nicknamed "Doughboys." Local veteran doughboys from the nearby community of Mellen built the first trails and bridges in the park in 1920.

The **CCC 692 Trail** (1.0 miles) leads up a stone staircase to the park's 65-foot-tall observation tower. On a clear day, Lake Superior and the Penokee Mountain Range can be seen from the tower. This trail was named after the Civilian Conservation Corps (CCC) Camp 692, which was based at Copper Falls. The CCC was a federal Depression-era work program for young unemployed men prior to World War II. Many of the park's trails, bridges, shelters, and the original tower were built by CCC workers in the 1930s.

> **Ranger Trail Pick**
>
> The **Doughboy's Waterfall Nature Trail** (1.7 miles) is often described as the "most scenic trail in Wisconsin." The trail begins at the **Bad River Picnic Area,** where it crosses a small stream and ascends about 100 feet to the crest of the Bad River Gorge. Overlook areas along the gorge offer great views the park's namesake, the iconic 29-foot-high Copper Falls. The trail then descends to a bridge spanning the Bad River with a great view of Devil's Gate, a narrow, door-like rock opening of the river gorge. Beyond the bridge is **Brownstone Falls,** 30-foot-high thundering waterfall of the **Tyler Forks River.**

WILDLIFE VIEWING

The forests, rivers, bluffs, and wetlands of the park provide ideal habitat for a variety wildlife species. White-tailed deer, gray squirrels, raccoons, porcupine, chipmunks, northern ravens, chickadees, ruffed grouse, turkey vultures, bald eagles, and many of species of songbirds and waterfowl make their home here. Black bear, fisher, coyote, and gray wolves are occasionally seen in the park as well.

INTERPRETIVE PROGRAMS AND FACILITIES

Evening programs and special events are held at the Bad River Picnic Area shelter building. Nearby, a small log cabin serves as the park's nature center and houses interpretive displays and historical artifacts. Nature hikes are offered throughout the use season.

BOATING AND FISHING

Tyler Forks and the Bad River are both rated as Class II trout streams with populations of brook, brown, and rainbow trout. The park's **Loon Lake** (32 acres) has a boat launch (electric motors only) on the west side of the lake. Anglers can expect to catch a variety of panfish, large-mouthed bass, and northern pike in this lake.

ACCESSIBLE FACILITIES AND TRAILS

The park's office, picnic areas, shelters, and most restrooms are accessible for mobility-impaired visitors. A handicap-only vehicle access road leads from the Bad River Picnic Area to the top of the river gorge. A paved wheelchair-accessible walkway here provides a great view of Copper Falls. Campsite 7 in the North Campground is wheelchair-accessible and has electrical service. A rustic cabin for campers with physical disabilities is located in the South Campground. The cabin has electrical service and access to the nearby restroom/shower facility.

CAMPING

The park's **North Campground** has 33 campsites sites (25 electric). The **South Campground** has 18 (non-electric) campsites and four walk-in sites. A restroom/ shower building and the park's RV water fill/sanitary station are both located along the entrance road to South Campground. A group campground (tents only) located west of the South Campground can accommodate up to 40 people. The park also has a backpack campsite located along the Bad River in the far north side of the park.

WINTER ACTIVITIES

Many people visit Copper Falls in winter just to view the fanciful ice sculptures that form along the Bad River Gorge and the park's frozen waterfalls. The park is located within the **Lake Superior Snow Belt**, a region that often receives heavy snow cover in winter. Park staff groom 17 miles of cross-country ski trails for both classic and skate skiing. Other off-season activities include hiking, wildlife watching, snowshoeing, winter camping, and ice fishing on Loon Lake for panfish and northern pike.

AREA ATTRACTIONS

Ashland County is known for its many pristine lakes and streams and friendly historic towns to explore. www.travelashlandcounty.com   www.mellenwi.org

Morgan Falls and St. Peter's Dome Recreation Area is a popular wilderness area of the Chequamegon-Nicolet National Forest, located about 17 miles west of Copper Falls. A short (0.75 mile) trail leads to Morgan Falls, a beautiful 70-foot cascading waterfall surrounded by old-growth forest. A one-mile trail here ascends to the summit of St. Peter's Dome, a massive 1,000-foot-high red granite rock bluff. On a clear day, Lake Superior can be seen on the horizon 20 miles to the north.

DIRECTIONS TO THE PARK
Copper Falls State Park is located in Ashland County about two miles north of the city of Mellen. From U.S. Highway 13, exit northeast onto Highway 169 about two miles to the park entrance along Copper Falls Road.

Copper Falls State Park
36764 Copper Falls Road
Mellen, WI 54546
(715) 274-5123

64. COUNCIL GROUNDS STATE PARK

Wisconsin River Park—a Scenic Gem

Wisconsin River shoreline

PARK SNAPSHOT

Council Grounds is the only state park that has a continuous shoreline along both a major river and a lake. The property has more than 12,000 feet of frontage bordering the Wisconsin River and Alexander Lake making it one of the most a popular boating and fishing destinations in central Wisconsin. Council Grounds is known for its attractive campground, picnic areas, swimming beach, and scenic hiking trails, but the park also has a rich history as a peaceful gathering spot for Native American tribes in the past.

GEOLOGY AND HISTORY

Council Grounds State Park is located along the southern edge of a large uplifted area of the state geologists refer to as the Northern Highlands. This dome-like region of northern Wisconsin is part of the Canadian Shield, a massive formation that has some of the oldest granite and sedimentary rock on the American continent. Outcrops of this ancient rock can be seen along the banks of the Wisconsin River in the northwest section of the park.

The forested hills of Council Grounds are composed of glacial deposits of sand, gravel, rocks, and other glacial debris that was deposited here by retreating glaciers near the end of the last Ice Age about 12,000 years ago.

The park was named after a former Native American gathering place or council grounds believed to have been located within the park boundaries. According to early written accounts of the area, a forested riverbank along a bend in the Wisconsin River in this location was an important meeting spot for various Native American tribes to trade, barter, socialize, and hold council to discuss important self-governing issues. Several Native American cultures are known to have lived within the Wisconsin River Valley region for generations. One of the most recent was the Ojibway, or Chippewa, people. In 1660, Father René Ménard, a French Jesuit missionary and explorer, was in regular contact with the Chippewa, who had established semi-permanent village along the Wisconsin River in this area.

In 1843, a fur trading post was built at the mouth of the **Prairie River** about a mile east of Council Grounds near a Chippewa village of 800 people. A year earlier, the Chippewa had been forced to sell their lands to the U.S. government through the Treaty of 1842. The Chippewa retained the right to hunt, trap, fish, and gather wild plants on their ceded lands. Many chose to continue living in the local area until well into the 1850s.

> **Ranger Note**
>
> Father Menard tragically disappeared while making a canoe portage around a rapids area of the Wisconsin River about four miles north of the Council Grounds. He was attempting to reach a Huron encampment to help starving refugees sheltered here after fleeing from their enemy, the Iroquois. His body was never found. An historical monument to Father Menard is located along Highway 107 north of the park.

During the logging boom of the 19th century, untold millions of pine logs were floated down the Wisconsin River to riverside sorting stations, including one located along the shoreline of Council Grounds. After the logs were sorted by size and ownership, they were sent on to nearby sawmills. In 1846, the area's first sawmill was built along a river rapids area east of the park known as **Jenny Bull Falls**.

> **Ranger Trivia**
>
> Jenny Bull Falls was named after the French term, *taurea* or "bull" sections of rivers that had dangerous rapids or waterfalls. "Jenny" is believed to be name of a sweetheart of the local fur trader who lived along the river at the time.

The bustling lumber industry and the railroad lines that followed led to the establishment of a new community called Jenny Bull Falls in 1854. The town was renamed Merrill in 1881 in honor

Sherburn Merrill, the general manager of the Chicago, Milwaukee, St. Paul, and Pacific railroads. In 1938, Merrill community leaders sold its 278-acre **Wildwood City Park** to the State Highway Commission for use as a highway rest area. Park improvements such as parking areas, shelters, and tree planting were completed by crews from the Works Progress Administration (WPA), a federal work-relief program for young men in the 1930s. The park was re-designated as Council Grounds State Park in 1978 and is now 508 acres in size.

BEACHES, PICNIC AREAS, AND SHELTERS

Council Grounds has a wide, sandy beach along the shoreline of **Lake Alexander**. Adjacent to the beach is a picnic area with changing stalls, vault-type restrooms, and a large open air shelter facility.

An attractive log and stone indoor shelter building is located in the far southeast area of the park. The shelter has electrical service and features two stone fireplaces. The shelter can be reserved in advance for larger groups.

Picnic shelter building

HIKING, BIKING, AND NATURE TRAILS

There are five miles of hiking trails to explore at Council Grounds. All routes are marked by colored emblems. The **Northwest Trail** (1.5 miles) leads through stands of hemlock, red pine and spruce trees, and a northern hardwood forest.

The **Green Trail** (1.1 miles) is located in southern end of the park and travels through an attractive forest area and wetland area. A short side trail leads to the shoreline of the Wisconsin River. **The Red Trail** (0.4 miles) provides a link from the campground to the 2.5-mile **River Road.** Many visitors also enjoy walking River Road to enjoy scenic views of the Wisconsin River.

The **Brown Trail** (0.7 miles) travels parallel to Council Grounds Drive across from the park entrance and visitor station. A side trail east of the Brown Trail provides access to **River Bend Trail** (3.5 miles), an off-property trail that loops through the 96-acre **Merrill Area Recreation Complex.** A roadside trail from here leads to a paved biking/hiking trail along the Wisconsin River in the city of Merrill.

> **Ranger Trivia**
>
> Senator Clifford "Tiny" Krueger was once a sideshow circus performer with the Sheboygan-based Seils-Sterling Circus as the "Fat Boy." Krueger is said to have tipped the scales at 425 pounds during his circus days.

The **Blue Trail** (0.9 miles) begins near the park's log shelter building in the southeast corner of the park. The trail loops through the **Krueger Pines State Natural Area,** which preserves an old-growth white pine forest. The natural area was named in honor of former Wisconsin Senator Clifford "Tiny" Krueger, a leading conservationist and natural resource advocate who served in the Wisconsin State Senate for 34 years.

> **Ranger Trail Pick**
>
> The **Big Pines Nature Trail** (0.75 miles) leads through some of the largest white pines in the park. Many of these majestic trees are over two feet in diameter and more than 130 years old. Walking among these giants gives visitors an idea of what the virgin pine forests of northern Wisconsin may have looked like prior to logging in the mid-1880s. The trail has interpretive signs along its route that describe the flora, fauna, and history of the park.

WILDLIFE VIEWING

The forests, wetlands, and open grassland areas of the park attract a wide array of wildlife. White-tailed deer, fox, coyotes, raccoon, rabbits, squirrels, chipmunks and more than a hundred different species of songbirds have been recorded here. Bald eagles, osprey, otter, mink, beaver, and many species of waterfowl and shorebirds can be observed along the shoreline of the park, especially during the spring and fall migration periods. More secretive animals, including black bears and gray wolves, are occasionally spotted in the local area as well.

INTERPRETIVE PROGRAMS AND FACILITIES
Interpretive programs, special events and guided nature hikes are offered from time to time at Council Grounds throughout the use season.

BOATING AND FISHING
Alexander Lake is a 677-acre reservoir lake created in 1925 when a hydroelectric dam was built across the Wisconsin River to generate power. The lake is known for its outstanding boating, waterskiing, canoeing, kayaking, and fishing opportunities. Anglers can expect to catch bluegill, crappie, muskie, walleye, bass, and northern pike in the lake. The park has a paved boat launch along Alexander Lake. Park visitors can rent canoes, kayaks and paddleboards through the park office.

Fly-fishing the shallows, pools, and riffles of the Wisconsin River downstream from the dam can be productive at times. The **Prairie River,** which joins the Wisconsin River in the nearby city of Merrill, is considered one of the best brook and brown trout streams in the state. The river is also a well-known by paddlers for its great canoeing and kayaking routes.

ACCESSIBLE FACILITIES AND TRAILS
Most of the park's picnic areas, shelters, and restrooms are accessible for mobility -impaired visitors. Campsite 46 is wheelchair-accessible and has electrical service. An accessible path leads to a restroom/shower facility. A paved walkway adjacent to Alexander Lake provides access to the park's wheelchair-accessible fishing pier and several fishing platforms.

CAMPING
Council Grounds has 52 campsites (19 electric) located within a well-shaded forest area. The campground has a restroom/shower facility and a RV water fill and sanitary station located along the campground entrance road. The park's group campground has three sites with electrical service. Each group site can accommodate up to 32 campers.

WINTER ACTIVITIES
Hiking, wildlife watching, and snowshoeing are popular off-season activities. Park staff groom three miles of cross-country ski trails. Many anglers enjoy ice fishing on Alexander Lake for bluegill, crappie, and northern pike.

AREA ATTRACTIONS

The Merrill area has long been considered the gateway to Wisconsin's Northwoods. Lincoln County has 700 lakes, 100,000 acres of public land, and many historic towns, museums and other attractions to explore.
www.northcentralwisconsin.com/lincoln-county
www.merrillchamber.org/tourism

Grandfather Falls is the highest waterfall on the Wisconsin River with an 89-foot drop over a one-mile section of the river. Many smaller waterfalls and beautiful cascade areas can be viewed along a four-mile section of the Ice Age Trail that follows the shoreline of the river from **Camp New Wood County Park** to the falls area. Grandfather Falls is located ten miles northwest of Council Grounds along Highway 107.

DIRECTIONS TO THE PARK

Council Grounds is located about 25 miles north of the city of Wausau in Lincoln County. From U.S. Highway 51, take the Highway 64 (Main Street) Merrill exit and travel west about two miles through the downtown. Turn right (northwest) onto Highway 107 (Grand Avenue) about 1.8 miles and then left (south) onto Council Grounds Drive to the park entrance.

Council Grounds State Park
N 1895 Council Grounds Drive
Merrill, WI 54452
(715) 536-8773

65. Flambeau River State Forest

Wisconsin's Second-Largest State Forest—Second to None

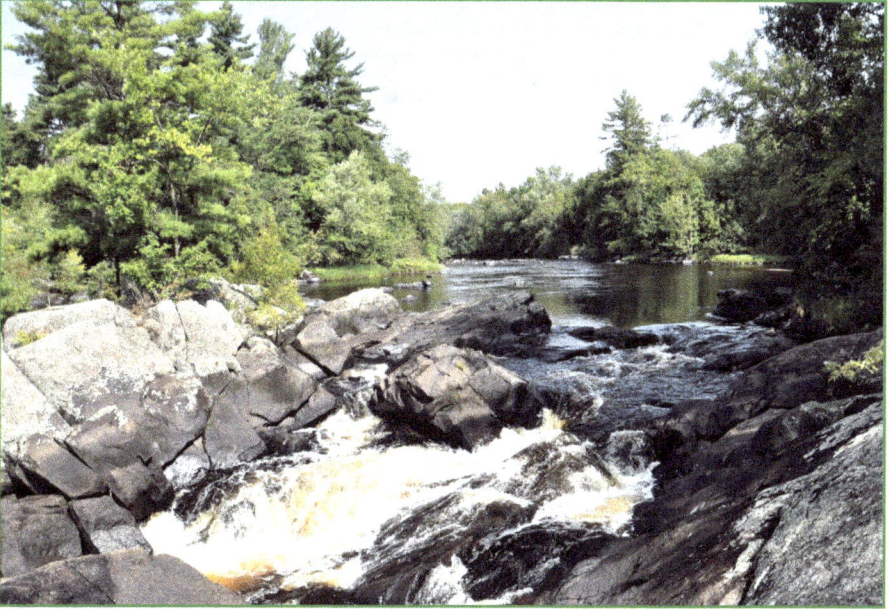

Flambeau River Falls

FOREST SNAPSHOT

The Flambeau River State Forest encompasses 91,000 acres within parts of Sawyer, Price, Rusk, Ashland, and Iron counties, making it the second-largest state forest in Wisconsin. The property contains more than 75 miles of the Flambeau River which is known for its outstanding whitewater canoeing and kayaking. In addition to being a paddler's paradise, the Flambeau River State Forest offers camping, hiking, picnicking, swimming, wildlife watching, fishing many other outdoor recreational opportunities.

GEOLOGY AND HISTORY

The silent forests of the Flambeau River give little hint as to the powerful forces that shaped its landscape in the past. More than 1.1 billion years ago volcanic eruptions formed the metamorphic granite rock seen along the river today. Much later, about 300 million years ago, shallow inland oceans submerged this area, leaving behind deep layers of sandstone. Then, about two million years ago, massive glaciers from the Canadian Arctic pushed through this region, depositing sand, gravel, rocks, and other glacial debris, creating a landscape of moraine hills and outwash plains. As the Ice Age came to an end about 12,000 years ago, retreating ice fields left behind kettle lakes, wetlands, and tremendous amounts glacial melt water that carved out the Flambeau River Valley.

Many Native American cultures have hunted, fished, and gathered wild rice along the Flambeau River over time. When French-Canadian fur traders explored this region in the 17th century, the Ojibwa (Chippewa) people had laid claim to most of the Flambeau River region.

In 1837, the Chippewa were forced to sell one million acres of their forest land to the U.S. government through a settlement agreement known as the "White Pine Treaty." The Chippewa wisely negotiated their right to hunt, fish, trap, gather wild rice, and continue to live on their former lands, which included the Flambeau River region.

By the mid-1800s, timber companies had logging camps throughout the Flambeau River Valley area. Lumbermen sent thousands of white and red pine logs down the Flambeau River to sawmills. The barren, cutover land they left behind was not suitable for farming due to the sandy, acidic soils and short growing season of the Northwoods. In 1920, a group of local citizens petitioned the State to purchase cutover land along the Flambeau River to establishment of Flambeau River State Forest, which opened in 1930.

> **Ranger Note**
>
> The French named the river the Flambeau or "flaming torch" after the Ojibwa's method using torchlight to spear fish in the river by canoe at night.

> **Ranger Trivia**
>
> In 1861, a Chippewa Indian captured a young bald eagle along the North Fork of the Flambeau River. He sold the bird to Daniel McCain, a local resident, in exchange for a bushel of corn. McCain resold the eagle for $2.50 to a member of the Eighth Regiment of the Wisconsin Volunteer Infantry. The eagle was named "Old Abe" and became the mascot for Company C. throughout the Civil War.

BEACHES, PICINIC AREAS AND SHELTERS

Connors Lake Recreation Area is located on the north shoreline of Connors Lake adjacent to County Highway W. The recreation area has picnic tables, flush restrooms, an open-air shelter, a wide, sandy swimming beach, and a boat launch.

HIKING, BIKING, AND NATURE TRAILS

Lakeside Nature Trail (0.4 miles) and the **Woodland Nature Loop** (0.8 miles) are located adjacent to the **Lake of the Pines Campground**. Both have interpretive signs describing the flora and fauna of the area. The **Flambeau Hills**

65. Flambeau River State Forest

Trail System offers 13 miles of trails through scenic pine and hardwood forests located above the Flambeau River Valley. Off-road bicycling is allowed on all these trails. Parking areas for Flambeau Hills arc located along County Highway W and Highway 70.

Bass Lake Access Trail (.25 miles) leads to stand of majestic old-growth pine and hemlock trees along the shoreline of Bass Lake. Surrounding this area is the **Bass Lake Peatlands State Natural Area** (921 acres), which preserves a black spruce wetland containing sphagnum moss, leather leaf, Labrador tea, and other bog plants. A parking area is located at the end of Bass Lake Road off of Tower Hill Road.

Slough Gundy Trail (0.3 miles) can be accessed just north of the Little Falls. This rustic trail follows the Flambeau River shoreline through a cedar/hemlock forest up to a massive granite outcrop overlooking the river. Along the trail are fanciful tree and rock formations, including **Elephant Rock**, a large boulder with a trunk-like tree root growing on its side, making it appear (somewhat) like its namesake.

Ranger Trail Pick

One of the most popular stops in the forest is **Little Falls Trail** (0.7 miles) located off County Highway M in the south end of the forest. This short walking trail leads to the picturesque cascading waterfall tumbling over massive granite boulders along the South Branch of the Flambeau River.

WILDLIFE VIEWING

The forest is home to gray wolves, black bear, and bobcat, as well as more common wildlife, including white-tailed deer, fox, coyote, bald eagles, river otter, porcupine, and squirrels. In 2017, about 30 elk were relocated to the Flambeau River State Forest, followed by another 48 elk from Kentucky in 2019. Today, a healthy and growing herd of these majestic animals continues to flourish here.

Ranger Note

A massive log from a tree known locally as *The Big White Pine* is on display near the old ranger station. This 130-foot-tall forest giant stood undisturbed in the forest for more than 300 years before it was maliciously cut down by vandals in 2000.

The rivers, wetlands, bogs, and woodlands of the forest are home to more than 200 species of birds. The **Sobieski Flowage** (268 acres), a former commercial cranberry bog, is a great place to observe waterfowl, shorebirds, and cranes.

INTERPRETIVE PROGRAMS AND FACILITIES
The Forest **Headquarters and Visitor Center** has several interpretive displays and a nature learning station for kids. Adjacent to the visitor center is the forest's original 1950s log ranger station, which is being repurposed as an interpretive center and logging-era museum. Interpretive programs, hikes, and special events are scheduled throughout the use season.

BOATING AND FISHING
The Flambeau River is one of the most popular canoeing and kayaking streams in Wisconsin. The **South Fork** section of this 75-mile river has many whitewater rapids, rocks, and portages along its route. The **North Fork** route is considered the easiest for beginners but still has some challenging rapids. An organization called Paddle the Flambeau River (www.paddleflambeau.com) is a good source for river maps, watercraft, rental advice, and whitewater ratings. Canoe route maps are also available online and at the park headquarters.

The forest has ten boat launch and canoe landings along the Flambeau River and at several inland lakes, including **Connors Lake** (410 acres), **Lake of the Pines** (223 acres), and **Mason Lake** (190 acres). Anglers can expect to catch walleye, musky, perch, crappie, bluegill, largemouth, and smallmouth bass in these waters.

Bass Wilderness Lake

Bass Lake (94 acres) is located within a designated wilderness area. Anglers can access the lake on foot along the **Bass Lake Access Trail** (.25 miles) to fish for panfish, smallmouth, and largemouth bass.

ACCESSIBLE FACILTIES AND TRAILS

The forest headquarters visitor center and attached restroom and showers are fully accessible for mobility-impaired visitors. The **Connors Lake Picnic Area** has paved walkways and accessible picnic tables and restrooms. The **Connors Lake Campground** has an accessible site with electrical service. The hilly topography of the forest and rocky, unsurfaced hiking trails can be challenging for some visitors with walking difficulties.

CAMPING

The forest has 60 campsites located within two campgrounds. Each campsite has a picnic table, fire ring, and access to drinking water and vault-type restrooms.

Connors Lake Campground has 29 (4 electric) well-shaded campsites located along south shore of Connors Lake. The campground is located south of the forest headquarters off County Highway M and has a small swimming beach for campers.

Lake of the Pines Campground offers 30 campsites (no electric) and has a small swimming beach and boat launch. The campground is about four miles northeast of the forest headquarters at the end of the Lake of the Pines Road.

> **Ranger Note**
>
> Flush toilets, showers, and an RV water fill and sanitary station are all available at the forest headquarters along County Highway W.

Overnight paddlers may camp at any of the 14 rustic canoe campsites located along the Flambeau River. Camping is free but limited to one night only. Backpack camping is allowed throughout the forest by permit, which can obtained by mail or from the forest headquarters.

WINTER ACTIVITIES

The **Flambeau Hills Ski Area** has 15 miles of cross-country ski trails groomed for both traditional and skate skiing. Other off-season activities include hiking, snowshoeing, wildlife watching, ice fishing, and hunting. Winter camping is available at the Connors Lake Campground. The forest has 50 miles of snow-mobile trails, which connect to county trail networks.

AREA ATTRACTIONS

Sawyer County has many fine resorts and campgrounds along many of its 200 lakes, rivers, and 400,000 acres of publicly owned forest land.
www.sawyercountygov.org

Ojibwa Park is a 366-acre park located along the Chippewa River. The property was originally established as Ojibwa State Park in 1932 and was operated by the State for 58 years until it was transferred to the Town of Ojibwa in 1990. The park has 19 campsites (16 electric), picnic areas, hiking trails, and a canoe landing on the Chippewa River. The property also has parking for access to the **Tuscobia State Trail** (74 miles). Ojibwa Park is located along State Highway 70 four miles west of the village of Winter.

DIRECTIONS TO THE FOREST

The Flambeau River State Forest is located in north-central Wisconsin within Sawyer, Price, Rush, Ashland, and Iron counties. From U.S. Highway 13 near the city of Phillips, exit onto County Highway W and head northwest for about 23 miles to the forest headquarters located along the Flambeau River in the town of Winter.

Flambeau River State Forest
W1613 County Highway W
Winter, WI 54896
(715) 332-5271

66. Governor Earl Peshtigo River State Forest

Smallest Northern Forest is Big in Beauty

Peshtigo River

FOREST SNAPSHOT

Governor Earl Peshtigo State Forest encompasses a pristine section of the beautiful Peshtigo River and several of its reservoir lakes. This narrow 9,200-acre forest stretches 25 miles along both sides of the river in Marinette and Oconto counties. The Peshtigo River provides some of the best whitewater canoeing, kayaking, and wild river tubing in the Midwest. The forest offers camping, picnicking, hiking, swimming, wildlife watching, cross-country skiing, fishing, and hunting.

GEOLOGY AND HISTORY

The rocky shoreline and boulder-strewn route of the Peshtigo River suggest powerful forces were at work here in the distant past. Volcanic eruptions and molten lava formed the metamorphic and igneous bedrock of the region. Evidence of the area's fiery geologic period can be seen at the many basalt rock outcrops and boulders within the Peshtigo River. Vast ancient seas also covered this region at one time and left behind deep deposits of layered sandstone rock. Later, several continental glaciers from the Canadian Arctic pushed through this area reshaping the entire landscape of the forest once again. As the ice fields began to retreat about 12,000 years ago, they left behind rolling moraine hills of sand and gravel deposits and outwash plains. Melt water from decomposing glaciers carved a new path for the Peshtigo River along its 136-mile route south to the waters of Green Bay.

Native American people, such as the Menominee, have lived within the Peshtigo River Valley for at least 5,000 years. The term *Menominee* was derived from an Ojibwe (Chippewa) word for "people of (wild) rice" but many tribe members prefer to be called their original name, *Mamaceqtaw*, "the people." French-Canadian explorers were the first Europeans to make contact with the Menominee in the 1780s, when they set up fur trading posts along the Peshtigo River.

In 1848, the Menominee reluctantly sold all their remaining lands to the U.S. government and agreed move to reservation lands in Minnesota. Several years later, tribal leaders appealed to then U.S. President Millard Filmore to allow the Menominee people to remain in Wisconsin. Their

> **Ranger Trivia**
>
> The origin of the word *Peshtigo*, originally pronounced *Pe-shet-i-go* by native people, is unclear. The Menominee used the term, which translated into the "Wild Goose" River. In the Ojibwa language, however, the same word meant "snapping turtle."

request was granted and they obtained ownership of a quarter-million-acre reservation (now Menominee County) northwest of Green Bay.

The Peshtigo River was a major transportation highway during Wisconsin's lumber boom in the late 1800s. Millions of pine logs were floated downriver each spring to sawmills in Green Bay. Cutover forest land was left strewn with tinder-dry tree tops from logging. In 1871, a horrific forest fire erupted near the community of Peshtigo and burned millions of acres of timber. The fire took over 2,000 lives in its wake.

> **Ranger Note**
>
> Where are all the "falls" today? Unfortunately, all the waterfall areas were incorporated into the construction of the dams and no longer exist except for a few rock ledges exposed at the base of the dams.

In the mid-1800s, newly arrived immigrant farmers attempted to raise crops on cutover forest land but most failed due to the unfertile sandy soils of the region. New employment arrived in the area from 1907 to 1959, when several hydroelectric dams were built across the Peshtigo River. The backwaters of these dams led to the formation of the **Sandstone Flowage** and the **Caldron Falls**, **High Falls**, and **Johnson Falls** reservoirs.

66. Governor Earl Peshtigo State Forest

In 2001, the State purchased most of flowage property from the Wisconsin Public Service Corporation to establish the Peshtigo River State Forest. The forest was renamed "Governor Earl" Peshtigo River State Forest in 2019 to honor of Tony Earl, who served as the head of the Department of Natural Resources for six years and as Wisconsin's governor from 1983 to 1986.

BEACHES, PICNIC AREAS, AND SHELTERS

Musky Point Picnic Area, located on the north side of **Caldron Falls Reservoir,** has a wide, sandy swimming beach, restrooms, charcoal grills, and tables. **Wood Lake Picnic Area,** located on the south side of the reservoir at **Governor Thompson State Park,** also has a swimming beach, picnic tables, and an attractive log shelter facility.

HIKING, BIKING, AND NATURE TRAILS

The forest has 12 miles of hiking trails to explore. The **Old Veteran's Lake Trail** (1.0 miles) loops around the perimeter of Veteran's Lake. **Sweet Fern Trail** (0.5 miles) leads to a scenic view of the **High Falls Flowage.** A parking area for both these trails is located near **Old Veteran's Lake Campground.**

Spring Rapids Hiking Trail System (7.0 miles) is located along the Peshtigo River in the far south end of the property. The **Big Oak Trail Loop** (2.25 miles) loops around the entire perimeter of the Spring Rapids Trail System with scenic views of the Peshtigo River. Nearby is the **White Cedar Riverside Trail** (1.0 mile), a somewhat rustic path through a cedar/ balsam fir lowland forest and eventually leads to the Peshtigo River.

> ### Ranger Note
>
> Both the **Spring Rapids** and **Seymour Rapids** areas are accessed by rustic roads marked with directional signs for fly-fishing anglers. Light 4x4 trucks or AWD vehicles are recommended for travel on these roadways. Both routes can be hiked or biked as well for visitors who may not want to drive their vehicles on dirt roads.

> ### Ranger Trail Pick
>
> The **Seymour Rapids Trail System** (3.5 miles) has several looped trail routes but the most popular is a well-worn shoreline trail along the Peshtigo River. Trail markers with camera emblems along this trail lead to several scenic rock outcrops along the river with outstanding views of the rapids. Seymour Rapids is located in the south end of the forest at the end of half-mile dirt road off of Bushman and Forest Road.

Peshtigo River

WILDLIFE VIEWING

The Peshtigo River's upland hardwood/pine forests and lowland cedar/spruce woodlands provide ideal habitat for wildlife including white-tailed deer, squirrels, rabbits, porcupines, coyote, fox, hawks, owls, and wild turkeys. Black bear and gray wolves are occasionally spotted here as well. More than 200 species of birds are known to nest or migrate through the Peshtigo River State Forest. The river wetlands and flowages attract many kinds of waterfowl, herons, cranes, and shorebirds, as well as bald eagles, osprey, river otter, beaver, and muskrats.

INTERPRETIVE PROGRAMS AND FACILITIES

There are no scheduled interpretive programs or self-guided nature trails in the forest.

BOATING AND FISHING

The 25-mile-long section of the Peshtigo River in the forest is the longest continuous whitewater rapids in the Midwest. Paddlers from around the country travel here to test their skills, especially along the challenging **Roaring Rapids Area** of the

> **Ranger Note**
>
> A 5-mile "fly-fishing-only" stretch of the Peshtigo River is located below the Johnson Falls Dam off of High Falls Road. This narrow section of the river is known for excellent brown trout fishing.

upper river area. Other sections of the river have more moderate rapids, making them ideal for less experienced paddlers. Canoe, kayak, and whitewater raft rentals and shuttle service are available from local outfitters.

Boaters and anglers have more than 3,200 acres of open water to explore. The **High Falls Reservoir** (1,498 acres) and **Caldron Falls Reservoir** (1,018 acres) have good populations of walleye, northern pike, perch, bass, and panfish in these waters. Caldron Falls is also a "Class-A" muskellunge lake. The smaller **Johnson Falls Flowage** (158 acres) and the **Potato Rapids (Bagley) Flowage** (281 acres), located 25 miles south of the main forest, offer walleye, northern pike, muskie, perch, panfish, and bass fishing.

ACCESSIBLE FACILITIES AND TRAILS
The **Muskie Point Picnic Area** on the north shore of the Caldron Falls Reservoir has wheelchair-accessible picnic tables, restrooms, and an accessible ramp to the swimming beach. The nearby boat landing features a paved walkway to a wheelchair accessible fishing pier.

CAMPING
Old Veteran's Lake Campground, located off Parkway Road in the northern section of the forest, has 16 non-electric campsites. The campground has vault-type restrooms, drinking water, and a small swimming beach. The forest also has seven backpack campsites along the Peshtigo River and three sites adjacent to the Johnson Falls Reservoir. These can be reached by watercraft or by foot via unmarked volunteer trails. Each site has a tent pad, fire ring, and a primitive toilet. A free camping permit for these sites can be obtained online or from the forest headquarters located at Governor Thompson State Park.

Several boat-in campsites are located at High Falls and Caldron Falls reservoirs. Each site has a tent pad, canoe rack, picnic table, and a primitive latrine. Campers must register and pay a camping fee prior to setting up for these sites at forest headquarters at Governor Thompson State Park. Campers who prefer more modern amenities will find them at Governor Thompson State Park, which has 103 campsites (16 electric) and shower/restroom facilities. (See Chapter 68)

WINTER ACTIVITIES
Forest staff groom five miles of cross-country ski trails for classic and skate skiing at the Spring Rapids Trail System. The Seymour Rapids Trail has three miles of marked snowshoe trails. Ice fishing for panfish, perch, walleye, and northern pike is popular on all the reservoir lakes. The forest has 25 miles of snowmobile trails

that connect to county trail systems. ATVs are allowed on snowmobile trails during the winter season.

AREA ATTRACTIONS
Marinette and Oconto counties have several historical sites and small towns to visit, plus 400 lakes, dozens of scenic waterfalls, and rivers to explore. www.therealnorth.com www.ocontocounty.org

The **Peshtigo Fire Museum** commemorates the deadliest wildfire ever to occur in the United States. On the night of October 8th, 1871, a raging wildfire near the community of Peshtigo killed nearly 2,500 people, destroyed 17 towns, and burned 1.3 million acres of forests in northeastern Wisconsin. A monument and mass grave containing 300 victims of the fire adjacent to the museum is listed on the National Register of Historic Places. The museum is located 20 miles southeast of Crivitz along Oconto Avenue in Peshtigo. www.peshtigofiremuseum.com

DIRECTIONS TO THE PARK
Peshtigo River State Forest is located in Marinette and Oconto counties. From Green Bay, take Highway 141 north for 57 miles and exit west onto County Highway X north of Crivitz. Follow Highway X about 12 miles to High Falls Reservoir.

Peshtigo River State Forest Headquarters
(Governor Thompson State Park)
N10008 Paust Lane
Crivitz, WI 54114
(715) 757-3965

67. GOVERNOR KNOWLES STATE FOREST

Wisconsin's Longest State Forest

St. Croix River

FOREST SNAPSHOT

Governor Knowles is the longest state forest in Wisconsin. The property stretches for 55 miles along the east bank of the St. Croix River on the Minnesota/ Wisconsin border. The forest is also one of the narrowest as well, averaging only two miles or less in width. Governor Knowles has campgrounds, backpack sites, a picnic area, hiking and equestrian trails, plus canoeing, kayaking, fishing, and hunting opportunities. The forest is managed jointly between the State of Wisconsin and the National Park Service as part of the **Saint Croix National Scenic Riverway.**

GEOLOGY AND HISTORY

The forest's rocky bluffs, hills, wetlands, and rivers are evidence of several geologic events that occurred here in the distant past. The basalt bedrock of the forest area was laid down by volcanic magma, which oozed from the earth's crust here 1.1 billion years ago. The exposed sandstone rock seen along the river corridor formed beneath an ancient sea that once covered this area 600 million years ago. Both of these ancient rock formations were buried beneath sand, gravel, and other glacial debris deposited by glaciers during the last Ice Age 30,000 years ago. As the earth's climate began to warm again about 15,000 years ago, massive amounts of glacial meltwater from the Lake Superior Basin thundered down the St. Croix River toward the Mississippi. The powerful river currents cut through glacial deposits and sandstone, sculpturing the scenic bluffs and rock outcrops seen along the river in the forest today.

The abundance of game animals, fish, and wild rice along the St. Croix River drew several Native American tribes to this area. When the first Canadian-French fur traders and Jesuit missionaries arrived here in 1660, both the Dakota Sioux and Ojibwa (Chippewa) people had laid claim to the region. When the fur trading began to decline in the early 1800s, commercial logging interests took over. In 1837, Chippewa and Sioux leaders were induced to sell all of their land that drained to the Mississippi to the U.S. government. Within a few years, logging became the largest industry in the St. Croix River area and continued on for the next 75 years. The nearby village of Grantsburg was built primarily to serve logging and railroad interests. Several grist, shingle, and sawmills were built along the Wood River near the community of Grantsburg.

> ### Ranger Trivia
> "Grants"burg was established as the Burnett County seat in 1886. The village was named after General Ulysses S. Grant's Civil War victory at "Vicks"burg, Mississippi.

In 1968, the Wild and Scenic Rivers Act was approved by the U.S. Congress, which resulted in the establishment of the **Saint Croix National Scenic Riverway**. In 1970, the State of Wisconsin created the **St. Croix State Forest** along the eastern shore of the St. Croix River. The property was later renamed **Governor Knowles State Forest** in honor of former Governor Warren P. Knowles for his lifelong efforts to protect Wisconsin's natural resources.

BEACHES, PICNIC AREAS, AND SHELTERS
The **St. Croix Picnic Area** is located along the St. Croix River south of the Highway 70 Bridge. The day-use area has picnic tables, grills, and vault-type restrooms. Swimming in the St. Croix River is not recommended due to strong river currents. A swimming beach is available at **Clear Lake Park**, located about 15 miles east of the forest along Highway 35 in the village of Siren.

Sandrock Cliffs Trail

HIKING, BIKING, AND NATURE TRAILS

Governor Knowles State Forest has over 40 miles of hiking trails to explore. Off-road bicycles are allowed on most trails in the forest. The **Brandt Pines Interpretive Trail** (2.5 miles) leads through towering 130-year-old red and white pines above the Brandt Brook ravine. The **Cedar Interpretive Trail** (0.4miles) follows a section of the Iron Creek through old-growth white cedar within the **Norway Point Bottomlands State Natural Area** in the north area of the forest. Both the Cedar and Brandt Pines trails are part of the 23-mile-long **St. Croix River North Trail Route**.

The **Sandrock Cliffs Trail** (5.0 miles) follows the crest of a forested sandstone bluff with great panoramic views of the St. Croix River below. Trail parking is available adjacent to the Highway 70 Bridge or at the Sandrock Cliffs parking area located north of Highway 70 off Benson Road.

The **Trade River Equestrian Trail System** (35 miles) is located in the southern region of the forest. Access to these horse trails is available from a day-use parking area or from the nearby horse-rider campground off of Evergreen Avenue.

> **Ranger Trail Pick**
>
> The **Wood River Interpretive Trail** (1.0 miles) begins at the **St. Croix Campground** and loops through a mature upland forest adjacent to the Wood River. The trail has interpretive panels that describe the flora and fauna of the forest and an overlook platform to view the scenic Wood River below. A side trail off of the Wood River Trail leads downhill to bridge over the Wood River with access to the **Raspberry Trail** (2.2 miles). Both trails are part of the 17-mile-long **St. Croix River South Trail**, which includes the **Benson Brook** and **Lagoo Creek** trails.

WILDLIFE VIEWING

The rivers, streams, and wetlands of the forest provide habitat for osprey, waterfowl, shorebirds, swans, cranes, herons, mink, beaver, and muskrat. Forested areas are home to white-tailed deer, rabbit, raccoon, squirrels, opossum, fox, coyote, and occasionally black bear. Wild turkey, ruffed grouse, woodpeckers, bald eagles and more than 100 species of songbirds, owls and hawks, can be seen in the forest as well.

There are six state natural areas within the property that protect old-growth maple, white cedar, and rare short-grass openings on bluff areas called "goat prairies." Restored prairie areas provide vital habitat for the **Karner Blue Butterfly**, a national endangered species.

INTERPRETIVE PROGRAMS AND FACILITIES

The **Brandt Pine** and **Wood River** nature trails both have interpretive signs that explain the unique geology and wildlife of the forest. A kiosk highlighting the history of the forest is located at St. Croix Picnic Area. Nearby is a monument to former Governor Warren P. Knowles, the namesake for the forest.

BOATING AND FISHING

The forest contains a 55-mile stretch of the St. Croix River and a short section of the 47-mile Wood River. Both waterways are highly regarded by canoers, tubers, and kayakers for the near-wilderness paddling experience they provide. A river map indicating small craft launch sites and other information is available online or can be picked up at the forest headquarters in Grantsburg. Canoe and kayak rental outlets are available locally including Wild River Outfitters in Grantsburg. www.wildriverpaddling.com

The St. Croix River is known for its excellent populations of smallmouth bass, walleye, catfish, muskie, and sturgeon. The forest has access points along twelve Class II trout streams, including the popular **Trade River** and **Wolf Creek**.

Wood Lake (520 acres), located about five miles southeast of Grantsburg, offers largemouth bass, northern pike, bluegill, sunfish, crappie, and perch fishing. A public boat launch for this lake is located within **Thoreson American Legion Park**.

ACCESSIBLE FACILITIES AND TRAILS

The St. Croix Picnic Area along Highway 70 has picnic tables, grills, and a vault-type restroom that are accessible mobility-impaired visitors. The nearby **St. Croix Campground** offers two wheelchair-accessible campsites with electrical service.

An accessible section of the Wood River Interpretive Trail leads from the campground area to an overlook platform above the Wood River gorge. Most other hiking trails in the forest are very rustic and some have steep climbs and staircases, which may be challenging for some visitors with walking difficulties.

CAMPING

The St. Croix Campground has 31 semi-rustic campsites (5 electric). The campground, located south of the Highway 70 Bridge, has vault-type restrooms, drinking water, and an RV water fill and sanitary dump station.

The **Sioux Portage Group Camp** (tents only) is located in the far north end of the forest. The campground has drinking water, vault-type restrooms and can accommodate up to 60 campers. A canoe landing with access to the St. Croix River is located below the camp area.

> **Ranger Note**
>
> There are no flush toilets or showers at Governor Knowles State Forest. Campers who prefer more modern amenities can find them at the nearby **James McNally Campground,** which has 38 campsites, all with electric, water and sewer hookup. The campground is located along the Wood River off West Olson Drive in Grantsburg.

The **Trade River Equestrian Campground** is located along Evergreen Drive in the south end of the forest. The horse camp has 40 campsites (18 electric) with vault-type restrooms and an RV water fill station. The campground provides direct access to more than 40 miles of equestrian trails.

The forest has nine backpack campsites with fire rings and picnic tables. Several are located along streams so campers can purify their own drinking water. There is no camping fee but a backpack permit is required from the forest headquarters at least two days in advance.

WINTER ACTIVITIES

Forest staff groom nine miles of cross-country ski trails at the **Brandt Pines Ski Area** for both traditional and skate skiing. Other winter activities include hiking, snowshoeing, fat-tire biking, and hunting. The forest maintains 33 miles of snowmobile/ATV trails.

AREA ATTRACTIONS

Burnett and Polk counties contain hundreds of lakes, fine resorts, and thousands of acres of public land to explore. www.co.polk.wi.us/tourism www.burnettcountyfun.com

Crex Meadows State Wildlife Area (30,000 acres) is the largest marsh of its kind in Wisconsin. The area's expansive wetlands and brush prairies can be viewed by vehicle or explored on foot along hiking trails, boardwalks, and overlook platforms. The **Wildlife Education and Visitor Center** has an auditorium and interpretive displays featuring the wildlife of the area. Crex Meadows is home to more than 280 species of birds, including sharp-tailed grouse, osprey, trumpeter swans, sandhill cranes, and thousands of ducks, geese, and shoreline birds. Crex Meadows is located just north of Grantsburg off East Crex Avenue (Highway D). www.crexmeadows.org.

DIRECTIONS TO THE FOREST

Governor Knowles State Forest is located along the St. Croix River in Burnett and Polk counties about five miles west of the village of Grantsburg. From U.S. Highway 53 near Spooner, exit onto Highway 70 west 40 miles west to Grantsburg. Stop at the DNR Ranger Station along Highway 70 in Grantsburg to pick up brochures and maps of the forest.

Governor Knowles State Forest
Ranger Station
325 State Road 70
Grantsburg, WI 54840
(715) 463-2898

68. GOVERNOR THOMPSON STATE PARK

Wisconsin's Centennial State Park

Woods Lake

PARK SNAPSHOT

Governor Thompson State Park was established at the start of the new millennium in 2000 to commemorate the 100th anniversary of the Wisconsin State Park System. The property is located along the sprawling Caldron Falls Reservoir of the Peshtigo River in Marinette County. This scenic 3,000-acre park offers camping, picnicking, swimming, hiking, wildlife watching, nature study, boating, and fishing.

GEOLOGY AND HISTORY

The landscape of the park was shaped by glaciers that pushed through this area during the last Ice Age about 30,000 years ago. As global temperatures rose and the ice fields began to retreat northward about 12,000 years ago, an intricate maze of sand and gravel moraine hills, outwash plains, kettle lakes, and wetlands were left behind. Melting glaciers also released vast amounts of water, creating a new 130-mile route for the Peshtigo River south to Green Bay.

Beneath the glacial deposits are layers of sandstone rock, which formed beneath an ancient sea that once submerged this area 600 million years ago. Deeper still is bedrock of red granite, an igneous rock formed from molten lava released from the earth's crust more than a billion years ago.

Native American people have hunted, fished, and gathered wild plants in the Peshtigo River area for generations. When the first French-Canadian fur traders canoed down the Peshtigo River in the 17th century, they met and began to trade with the Menominee or "wild rice people." As the fur trade industry came to a close in the 1820s, it was replaced powerful logging company interests. The Menominee Nation reluctantly sold most of their forest land in Wisconsin to the U.S. government in 1831. Tribal leaders successfully negotiated the right for the Menomonie people to hunt, fish and gather wild rice on all ceded territory. The tribe later gained ownership of a quarter-million acres of land northeast of Green Bay, now Menominee County.

> **Ranger Note**
>
> Watch for colorful red granite rock outcrops throughout Governor Thompson State Park. Many of these massive rock domes were exposed and polished smooth by the grinding force of moving glaciers during the last Ice Age.

> **Ranger Note**
>
> In 1871, wildfires ignited near the village of Peshtigo and other areas of northeastern Wisconsin. These devastating fires burned over a million acres of forest land, destroyed 17 towns, and killed 2,500 people in a single night. The Peshtigo Fire remains the deadliest wildfire ever to occur in the United States.

Lumber companies had begun to cut timber in the region even before the treaty settlement with the Menominee was finalized. Within a few decades, loggers had leveled nearly all the virgin pine forests of the Peshtigo River area, leaving behind thousands of acres strewn with dry brush piles and tree tops.

Immigrants, mostly from Poland and Germany, arrived in the 1880s to farm the cutover forestland in the Peshtigo area, but the dry sandy soils made growing crops difficult. Local employment got a boost in the early 20th century when several hydroelectric dams were built across the Peshtigo River, including the nearby Caldron Falls Dam built in 1925. Today, the 1,063-acre **Caldron Falls Flowage** borders the northwest section of Governor Thompson State Park.

To commemorate the 100th anniversary of the Wisconsin State Parks, the State purchased the **Paust Woods Lake Resort** (1,987 acres) and 200 acres from the Wisconsin Public Service Corporation to establish Governor Tommy Thompson Centennial State Park during the millennium year in 2000. The park is named in honor of former Governor Tommy Thompson, Wisconsin's longest-serving governor.

BEACHES, PICNIC AREAS, AND SHELTERS

A picnic area and sandy swimming beach are located along **Woods Lake** near the park entrance. This day-use area has vault-type restrooms, changing stalls, and an indoor log shelter with a stone fireplace. A small hike-in picnic site along the north shoreline of Woods Lake can be accessed via short side trail from the **Pine View Hiking Trail.**

HIKING, BIKING, AND NATURE TRAILS

The park has 16 miles of hiking trails to explore. The **Pine View Trail** (1.0 miles) and **Otter Trail** (1.0 miles) both begin at the **Woods Lake Picnic Area.** The **Forest View Trail** (3.5 miles) can also be accessed near the picnic area. This trail leads through several wooded and grassy field areas of the park. A short side trail leads to **Popular Rock**, a massive 150-foot-tall red granite outcrop, the tallest in the park.

The **Flowage Trail** (1.9 miles) connects the park's boat launch on Caldron Falls Reservoir to **Huber Lake** on the western edge of the park. The trail also leads to the **Sunset Trail** (1.4 miles), which follows the eastern shoreline of Huber Lake and **North Trail** (1.0 miles), a route that leads to three canoe campsites along Caldron Falls Reservoir.

The **Thunder Mountain Trail** (2.0 miles) travels through the center of the property past a large granite rock outcrop. The **Starlight Trail** (1.3 miles) is a one-way trail that ends along Ranch Road in the southwest section of the park. Along this trail is **Big Buck Ravine**, a towering flat rock bluff overlooking a deep forested ravine.

> **Ranger Note**
>
> **Forest View Trail** leads past an historic 100-foot-high fire tower built by Civilian Conservation Corps (CCC) crews in 1934. The lookout tower was originally located near town of Penbine and used by the State before being decommissioned in the 1980s. The tower was purchased and moved to its present location by the former owner of the Paust Woods Lake Resort.

> **Ranger Trail Pick**
>
> The **Granite Path Trail** (2.1 miles) travels through scenic upland forest areas adjacent to some of the most outstanding red granite dome outcrops in the park. A side trail leads to beautiful **Vista Rock**. There are several large granite domes throughout the park, each with its own unique plant life and miniature ecosystem. Most do not have marked paths but well-worn side trails allow access to these impressive rock outcrops.

WILDLIFE VIEWING

The park's upland woods, open fields, rock outcrops, wetlands, and lakes provides habit for a variety of wildlife, including white-tailed deer, raccoon, rabbits,

Red granite rock outcrop

squirrels, chipmunks, fox, coyote, mink, muskrat, river otter, ruffed grouse, wild turkey, and other gamebirds. Black bear and gray wolves are occasionally spotted in the area but rarely encountered.

Many species of waterfowl and songbirds, sandhill cranes, herons, and egrets can be viewed here, especially during spring and fall migration periods. Red-tailed hawks, turkey vultures, bald eagles, and osprey are often seen flying above the park area.

INTERPRETIVE PROGRAMS AND FACILITIES
Interpretive panels are placed at various locations throughout the park. Many visitors enjoy a day trip to see the **Cathedral of the Pine State Natural Area.** This site preserves some of the largest and oldest white pine, hemlock, and red pine found anywhere in the Midwest, some trees dating back to the year 1735. The site has hiking trails and interpretive panels. The natural area is located west of the park off of National Forest Road 2121 near the town of Lakewood. www.fs.usda.gov/cathedralofthepines

BOATING AND FISHING
The park has **6.5 miles** of shoreline along **Caldron Falls Reservoir** (1,063 acres). A paved boat launch, boarding dock, and a fishing pier provide access to this

man-made lake. Motorboats, personal watercraft, and waterskiing are allowed on the reservoir. Anglers can expect to catch walleye, northern pike, rock bass, crappie, perch, smallmouth and largemouth bass. Caldron Falls Reservoir is also the only Class-A muskellunge water in Marinette County. **Handsaw Creek** (6 miles) located in the southwest corner of the park, is a cold-water trout stream that supports native brook trout. Marinette County has 191 trout streams and rivers for anglers to explore.

Woods Lake (48 acres), located near the park entrance, has good populations of bluegill, large-mouthed bass, and northern pike. Small watercraft (no motors) can be launched from the picnic area or beach. Kayak and paddleboard rentals are available through the park office. **Huber Lake** (30 acres) located near the western boundary of the park, has limited populations of bass, bluegill, and northern pike

ACCESSBILE FACILITIES AND TRAILS

The park's entrance station, picnic area, and indoor shelter building are accessible for mobility-impaired visitors. A wheelchair-accessible path provides access to the swimming beach along Woods Lake. An accessible fishing pier is located near the boat launch along the Caldron Falls Reservoir. Campsites 29 and 30 are wheelchair-accessible and offer electrical service. Paved walkways provide access to the nearby restroom and shower facility. Most hiking trails in the park are grass-covered and gently rolling. None are surfaced for wheelchair use.

CAMPING

The park has 103 campsites (16 electric). A restroom/shower facility is centrally located within the campground and an RV water fill/sanitary station is available along Campground Road. The park has three rustic campsites along the shoreline of the Caldron Falls Reservoir and can be accessed by watercraft or on foot. Each site has a canoe rack, picnic table, fire ring, a bear-proof food locker, and a primitive outdoor toilet.

WINTER ACTIVITIES

Park staff groom six miles of cross-country ski trails for both traditional and inline skiing. The park's indoor shelter serves as a warming house for skiers in winter. Hiking, wildlife watching, and snowshoeing are also popular off-season activities. Winter camping is available at the hike-in sites along the Caldron Falls Reservoir. Many anglers enjoy ice fishing on Woods Lake and the Caldron Falls Reservoir.

AREA ATTRACTIONS

Marinette County has hundreds of lakes, streams, rivers, and waterfalls to explore and more than 230,000 acres of county forest land.

www.therealnorth.com www.crivitz.com

The **Marinette County Waterfall Tour** is one of the most popular road trips in Wisconsin. The route includes 14 scenic waterfalls, most within Marinette county parks. Waterfalls near the state park include **Veteran's Memorial Falls** located seven miles to the south and **Strong Falls** at Goodman Park about 21 miles to the north. A guidebook, *Marinette County Waterfalls*, is available online at www.marinette.com

DIRECTIONS TO THE PARK

Governor Thompson State Park is located in southern Marinette County about 70 miles north of Green Bay. From Highway 141 north of Green Bay, exit onto County Highway A northwest of Crivitz and turn left (west) on County Highway X. After crossing the High Falls Reservoir Bridge, turn right (north) on Parkway Road and then left (west) on Ranch Road about one mile. Turn right (north) on Paust Lane to the entrance of the park.

Governor Thompson State Park
N10008 Paust Lane
Crivitz, WI 54114
(715) 757-3979

69. Hoffman Hills State Recreation Area

A Hilltop Beauty

Park view

PARK SNAPSHOT

Hoffman Hills State Recreation Area is nestled atop some of the highest hills in Dunn County. The property is known for its natural beauty and scenic views of the surrounding rural countryside. The recreation area has several picnic areas and miles of hiking trails that lead through upland hardwood forests, pine plantations, wetlands, restored prairies, and oak savannahs.

GEOLOGY AND HISTORY

Hoffman Hills is located just north a geologic transition zone between the Driftless Area, the unglaciated region of southwestern Wisconsin and the rest of the state, which experienced the full impact of continental glaciers. As the last Ice Age came to an end about 12,000 years ago, retreating glaciers left behind a rolling landscape of washout plains, wetlands, and towering moraine mounds such as Hoffman Hills. Glacial meltwater carved out numerous streams in the area, including the nearby Chippewa and Red Cedar rivers.

Native American people have been drawn to this region of the state for thousands of years. Archeologists have unearthed Clovis points (stone spearheads) used by pre-historic Paleo-Indian hunters who entered this area in search of big game animals about 10,000 years ago. Late Woodland Indians, known as the Effigy Mound Builders, occupied this region from about 550 B.C. to 1,000 A.D. These early inhabitants built earthen burial mounds in the geometric shapes or in the outline of spiritual animals. Some of these ancient effigy mounds can still be seen near the Red Cedar State Trail and at Wakanda Park in the nearby city of Menomonie.

> **Ranger Note**
>
> Unlike the Kettle Moraine and Chippewa Moraine areas, which have hundreds of pothole lakes and ponds, most of the pre-glacial lakes in the Hoffman Hills region were "filled in" with deep deposits of glacial fill and erased from the landscape by glaciers. As a result, there are very few natural lakes in this part of the state.

When French fur traders arrived in the region in the 17th century, the Santee Dakota (Sioux) dominated the area. Later, the Ojibewa (Chippewa) entered the area to hunt and trap as well, which led to many bloody conflicts between the two tribes. In 1831, Henry Schoolcraft, an Indian agent for the U. S. government, arrived by canoe on the Red Cedar River in attempt to negotiate a peace treaty. The two tribes agreed on a dividing line through Dunn County but warfare did not end until the Dakota Sioux relocated to the western plains. The Chippewa tribe's claim of land ownership was brief, however. Within a few years, Chippewa leaders were induced to sign the White Pine Treaty of 1837, which required the tribe to relinquish their northern forest lands to the U.S. government in exchange for annual payments, reservation land, and the right to hunt, fish, and gather wild plants on ceded lands.

> **Ranger Note**
>
> In 1767, Johnathan Carver, an English explorer, described the prairies surrounding Hoffman Hills as "fine meadows where droves of buffaloes (American bison) and elk were feeding." Although the bison and elk are no longer here, the prairies they helped maintain by grazing for thousands of years became the productive farm fields seen throughout the local area today.

Logging the great northern pine forests became the predominant industry of the area throughout the 19th century. By 1844, the nearby community of Menomonie was home to the largest lumber company in the world. European immigrant farmers also moved to the area and found the deep, rich soils of the prairies ideal to raise wheat and other crops.

In 1980, retired Boston Store Company executive Richard Hoffman and his wife Marian donated their 280-acre hilltop property to the State of Wisconsin to establish "The People's Park," now called Hoffman Hills State Recreation Area.

BEACHES, PICNIC AREAS, AND SHELTERS

The **Entrance Picnic Area** has picnic tables, grills, a hand water pump, and a vault-type restroom. The **Shelter Picnic Area** is located about a half mile east of the entrance area along the **Upper Pines Trail**. This area has picnic tables and an attractive shelter facility with a stone block fireplace. The **Tower Picnic Area** is located near the 60-foot-high observation tower along the **Hawk Ridge Trail** in the northern section of property.

There are no swimming areas at Hoffman Hills but a public swimming beach is available at **Wahanda Park** along Lake Menomin. The park is about ten miles southwest of Hoffman Hills in the city of Menomonie.

HIKING, BIKING, AND NATURE TRAILS

Hoffman Hills has over 11 miles of hiking trails within 16 named trail routes. Many trail intersections have directional signage but a property map is useful for navigating more complex trail loops. The **Buck Ridge Trail** (0.9 miles), **Skyline Trail** (1.1 miles), and **Spring Run Trail** (0.4 miles) are all attractive forested routes but have steep inclines.

The **Memorial Wetland Trail** (1.1 miles) travels through wetland areas and circles two small, scenic ponds in lower area of the property. The **West Savannah** (1.0 miles) and **East Savannah** (0.8 miles) trails loop through a restored

Memorial Wetland Trail pond

15-acre long-grass prairie and oak savannah. Interpretive panels along this route describe the plants,

The nearby **Red Cedar River State Trail** (15.0 miles) is one of the most scenic biking trails in the state. The trail follows the forested riverbank of the Red Cedar River, where it skirts sandstone rock outcrops and crosses eleven bridges. The trail can be accessed from the **Depot Visitor Center** along Brickyard Road in the nearby community of Menomonie. www.redcedarhoffman.org.

Ranger Trail Pick

Hoffman Hills' observation tower is located along the **Hawk Ridge Trail** (1.2 miles) on one of the highest hills in Dunn County. The **Lower Pines Trail** (1.3 miles) and the **Plantation Trail** (0.3 miles) both lead uphill towards the tower from the entrance parking area. These trails connect to other short trails with directional signs to the observation tower. All the trails to the tower have an uphill climb but most have modest inclines. The 60-foot-high observation tower offers beautiful views of the surrounding landscape.

WILDLIFE VIEWING

Hoffman Hills' landscape of upland oak and maple forests, pine plantations, wetlands, and restored prairies provide ideal habitat for wildlife. White-tailed deer, rabbits, chipmunks, raccoons, fox, coyote and four kinds of squirrels can be found here. The property is known as one of the top birding spots the area. More than 200 species of birds either nest or migrate through the area, including many species of woodland songbirds, hawks and owls, plus turkey vultures, osprey, and bald eagles. A small creek and two man-made ponds in the western section of the property attract waterfowl, herons, cranes, and shorebirds.

INTERPRETIVE PROGRAMS AND FACILITIES

The **Tower Nature Trail** (2.3 miles) begins along the **Hawk Ridge Trail** near the observation tower. The trail has interpretive signs that highlight the flora and fauna of the area. Interpretive panels located along the **Savanah Hiking Trails** describe the unique plants and animals found in prairie regions.

BOATING AND FISHING

Two man-made ponds along the **Memorial Wetland Trail** have modest populations of bluegills and large-mouthed bass. **Tainter Lake** (1,605 acres) located northwest of Hoffman Hills and **Lake Menomin** (1,009 acres) adjacent to the city of Menomonie, offer boating, waterskiing, swimming, and fishing. Both are reservoir lakes of the Red Cedar River and harbor walleye, northern pike, bass, yellow perch, crappie, and other panfish.

The nearby Red Cedar River is a popular canoeing and kayaking destination. Many visitors take a combined paddling/biking outing along the Red Cedar. The route begins with paddling downstream (south) from the city of Menomonie to the **Dunnville Wildlife Area** and then biking back north along the **Red Cedar State Trail**. Watercraft and bike rentals outlets are available in the city of Menomonie.

ACCESSIBLE FACILITIES AND TRAILS
A wheelchair-accessible picnic table and grill are available at the **Entrance Picnic Area**.

The **Memorial Wetland Trail** (1.1 miles) was designed specifically to accommodate most mobility-impaired visitors with access from the entrance parking area. Most other trails in the park have steep terrain and natural tread surfaces, which may them difficult for visitors with walking disabilities.

CAMPING
Hoffman Hills is a day-use-only property so there are no camping facilities. **Myron Township Park** located along the Red Cedar River, has **45** campsites (all electric). The park is about **26** miles north of Hoffman Hills along Highway I south of the city of Chetek. www.co.dunn.wi.us/parksdivision. **Lake Wissota State Park** about **36** miles east of Hoffman Hills has **118** campsites (60 electric) with restroom/shower facilities. (See Chapter 72)

WINTER ACTIVITIES
Property staff groom nine miles of cross-country ski trails for both traditional and skate skiing. Hiking, snowshoeing, and wildlife watching are also popular off-season activities at Hoffman Hills.

AREA ATTRACTIONS
Dunn County is known for its beautiful rural scenery, small friendly towns and its many outdoor recreational opportunities. www.co.dunn.wi.us/visitors

Devil's Punchbowl Nature Preserve, located 15 miles southwest of Hoffman Hills, features a water-carved sandstone canyon and a scenic waterfall that cascades over a rocky ledge. Devil's Punchbowl can be accessed near the Red Cedar River State Trail from Paradise Valley Road southwest of the city of Menomonie.

DIRECTIONS TO THE PARK

Hoffman Hills is located 11 miles northeast of the city of Menomonie in Dunn County, about 23 miles northwest of Eau Claire. From Interstate I-94, exit onto Highway12/29 West about three miles and then turn north on 730th Street. After about two miles, turn west onto 650th Avenue and then north on County Highway E for about one mile before turning onto 740th Street to the entrance to the recreation area.

Ranger Note

Hoffman Hills is open from 7:00 a.m. to 9:00 p.m. daily. There is no visitor entrance station and admission fees are not required at this recreation area.

Hoffman Hills State Recreation Area
Red Cedar River Trail Office
N7284 740th Street
Menomonie, WI 54751
(715) 232-1242

70. INTERSTATE STATE PARK

Wisconsin's First State Park is Still "Number One"

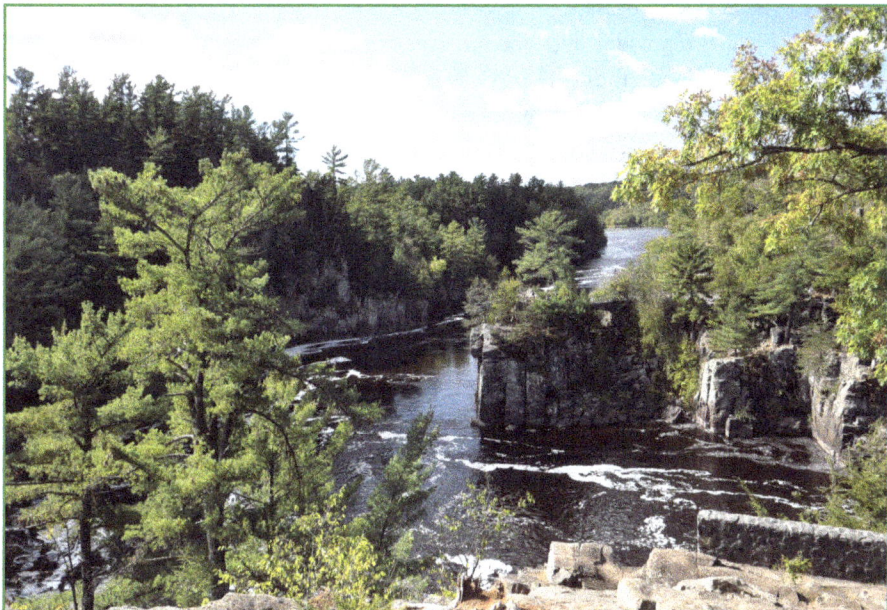

Dalles of the St. Croix River

PARK SNAPSHOT

Interstate was established as Wisconsin's first state park in 1900. This scenic property is perched above The Dalles of the St. Croix River within the St. Croix National Scenic Riverway, one of the least developed wild river systems in the country. Generations of visitors have flocked to this popular park over the years to camp, hike, swim, canoe, or just enjoy the stunningly beautiful view from its rocky bluffs.

GEOLOGY AND HISTORY

Few sites showcase the powerful geologic forces that shaped northern Wisconsin better than Interstate. The park's towering basalt rock bluffs along the St. Croix River were molded from volcanic lava that oozed from the earth's crust 1.1 billion years ago. Sandstone and shale rock at Interstate were formed from sand and clay deposits beneath a shallow inland sea that once covered this region 600 million years ago.

The park's landscape was also shaped by several continental glaciers that pushed through this region of Wisconsin over the past two million years. The earth-crushing weight of these massive glaciers compressed the nearby Lake Superior Basin ever wider and deeper. Near the end of the Ice Age about 11,000 years ago, retreating glaciers to the north blocked the flow of meltwater creating Glacial Lake Duluth. When the ice dam finally collapsed, enormous amounts of water thundered down the St. Croix River, cutting through the sandstone and basalt rock of the park on its way to the Mississippi. Left in its wake were many small waterfalls, rapids, and the 200-foot-high bluffs seen along The Dalles of the St. Croix today.

Ranger Trivia

The word "Dalles" is a French term for a waterfall or rapids area within steep rock-walled gorge. The nearby city of St. Croix Falls and its sister city Taylor Falls in Minnesota were both named after the many rapids along St. Croix River. Unfortunately most of these rapids and waterfalls were submerged when a hydroelectric dam was built between the two cities in 1906.

One of the most unusual rock formations along the St. Croix River are potholes. Potholes were excavated by powerful underwater whirlpools that caused sand and stones to spin violently, carving holes into solid rock over time. Most potholes are small but a few are 25 feet in diameter and 60 feet deep. The largest and deepest potholes in the world can be seen at Minnesota's Interstate State Park on the west side of the river.

Ranger Trivia

A three-mile stretch of the St. Croix River north of Interstate Park has the peculiar distinction of having the world's largest log jam ever recorded. In 1886, an estimated 150 million board feet of logs became wedged in massive piles in the river. It took 175 men working around the clock over a period of six months to break it up.

Several Native American cultures have lived, hunted, and fished within the St. Croix River Valley for thousands of years. Archeological excavations within the park revealed skeletal remains of extinct bison along with spear points used by Paleo Indian hunters more than 10,000 years ago. When French-Canadian fur traders canoed down the St. Croix River in the 17th century, they encountered the Ojibwe (Chippewa) and Dakota Sioux people. At first, indigenous people welcomed the opportunity to trade with Europeans for metal knifes, axes, and wool blankets but this was short-lived.

The onset of logging and dam building by timber companies in the early 19th century led to ownership disputes with native people. In 1837, Chippewa leaders were forced to sell most of their forest land to the U.S. government in return for annual cash payments, reservation land, and the right to hunt fish and gather plants on ceded lands. Soon after, lumber camps began sending millions of pine logs down the St. Croix River to sawmills further south.

In the late 1800s, Wisconsin and Minnesota citizens lobbied their respective state representatives to preserve The Dalles of the St. Croix River from development and mining. As a result, the first "interstate" system of parks in the nation was created with Minnesota's Interstate Park opening in 1895, followed by Wisconsin's Interstate Park in 1900.

BEACHES, PICNIC AREAS, AND SHELTERS

Swimming in the St. Croix River is not recommended because of its dangerous river currents and deep drop-off areas. A swimming beach and bathhouse is located along the north shore of *Lake O' The Dalles*, an inland lake near the center of the park.

The **River Bottoms Picnic Area** on the west side of Lake O' The Dalles has grills, picnic tables, and two open-air shelters that can be reserved in advance. **Camp Interstate Picnic Area** south of Lake O' The Dalles has a large solar-powered shelter facility that can accommodate up to 100 people. Other shelter facilities are located along the **Horizon Rock Trail** (0.5 miles) north of the Ice Age Center and at the **Meadow Valley Picnic Area** east of Lake O' the Dalles.

HIKING, BIKING, AND NATURE TRAILS

Interstate State Park has nine miles of hiking trails to explore. The park also serves as the western terminus of **Ice Age National Scenic Trail**, a 1,200-mile-long hiking route that follows the southern extent of glaciation across Wisconsin to **Potawatomi State Park** in Door County. (See Chapter 23)

The **Skyline Nature Trail** (0.75 miles) has interpretive panels that describe various plant communities found in the park. The trail is part of the **Skyline Trail** (1.6 miles) and begins near the **Ice Age Nature Center**. The trail has a few steep and rocky sections as it ascends to the rim of the St. Croix Valley.

The **Echo Canyon Trail** (0.7 miles) follows an extinct river valley with steep canyon walls that often stay cool even on warm summer days. This trail connects to **Summit Rock Trail** (0.5 miles), which leads to a lookout atop a 200-foot-high river bluff with great views of the St. Croix River Valley.

The **Eagle Peak Trail** (0.8 miles) leads through a pine forest before ascending a stone staircase to the highest inland bluff in the park. Nearby is **Silverbrook Trail** (1.2 miles), which follows an abandoned roadway to the south end of the park. Along this route are the ruins of an

> **Ranger Note**
>
> The face profile of the park's iconic Old Man of the Dalles rock formation can be seen on a bluff above the river from the **Summit Rock Trail** observation area.

abandoned copper mine structure and a scenic 18-foot-high waterfall. Adjacent to the waterfall area is the former site of **Silverbrook Mansion**, an elaborate summer home built in 1895 by a copper mine businessman.

The nearby **Gandy Dancer State Trail** (47 miles) can be accessed from **Interlink Trail** (1.9 miles) from the Interstate Park. This trail leads past the **Polk County Information Center** along Highway 35 and travels beneath Highway 8 with connections to both the Gandy Dancer State Trail and the **Woolly Mountain Bike Trail.** www.woollybikeclub.com.

Ranger Trail Pick

The **Pothole Trail** (0.4 miles) is the most popular trail in the park and the easiest to access. This bluff-top trail features interesting rock formations, including dozens of potholes. An observation platform along the trail offers panoramic views of the St. Croix River below and colorful rock bluffs.

WILDLIFE VIEWING

The pine and hardwood forests, bluffs, prairies, lakes, and rivers of the area provide exceptional habitat for wildlife. More than 200 species of birds are known to live or migrate through the park, including many woodland songbirds, waterfowl, shorebirds, turkey vultures, hawks, owls, osprey, and bald eagles. White-tailed deer, fox, raccoon, rabbits, squirrels, and chipmunks are often seen in the park. More secretive animals such as coyote, gray wolves, and black bear are occasionally spotted as well.

INTERPRETIVE PROGRAMS AND FACILITIES

The **Interstate Ice Age Interpretive Center** is one of the finest nature centers in the state. The center has an auditorium and exhibits featuring Wisconsin's glacial past. Nature hikes and evening programs are scheduled throughout the season. Some are held at the outdoor amphitheater in the **Meadow Valley Picnic Area.**

The nearby **Saint Croix National Riverway Visitor Center** has exhibits highlighting the wildlife and ecology of the riverway. The center is located along North Hamilton Street in St. Croix Falls. www.nps.gov/saintcroix

BOATING AND FISHING

The Dalles of the St. Croix River is one of the most popular paddling routes in the Midwest. A small watercraft launch is available at the **River Bottoms Picnic Area**. Canoe and kayak rentals are available in the local area.

The St. Croix River offers excellent walleye, muskie, northern pike, and smallmouth bass fishing. The Lake O' The Dalles (25 acres) has a fishing pier and small boat launch (non-motorized) along the west side of the lake. Anglers can expect to catch bluegill, largemouth bass, and northern pike in this lake.

ACCESSIBLE FACILTIES AND TRAILS

The Ice Age Interpretive Center and most picnic areas, restrooms and shelters are accessible for mobility-impaired visitors. Campsite 32 in the North Campground is wheelchair-accessible and has electrical service. A paved walkway provides access to a restroom/shower facility. A wheelchair-accessible fishing pier on the St. Croix River is located at the River Bottoms Picnic Area.

CAMPING

The park has 84 campsites (23 electric). The **North Campground** has electric service sites and a restroom/shower facility. The **South Campground** has non-electric campsites and vault-type restrooms. An RV water fill/sanitary station is located along the entrance road near park office. The **Pines Group Campground** (tent-only) has two campsites with a combined capacity of 60 people. Each site has a fire ring and picnic tables and access to vault-type restrooms and drinking water.

WINTER ACTIVITIES

Winter camping, hiking, snowshoeing, and wildlife watching are popular off-season activities at Interstate. Park staff groom 3 miles of cross-country ski trails for inline skiing. Ice fishing is available on Lake O' The Dalles.

AREA ATTRACTIONS

Polk County has more than 500 lakes, 200 miles of rivers, 98 miles of trout streams, and thousands of acres of public land to explore. www.co.polk.wi.us/tourism. Many park visitors enjoy taking the **Taylor Falls Scenic Boat Tour**

Boat cruise on the St. Croix River

down the St. Croix River aboard an authentic paddlewheel boat. The cruise offers close-up views of the towering bluffs along both the Wisconsin and Minnesota sides of the river. Boat tour docks are located in Taylor Falls. www.taylorfallsboat.com

DIRECTIONS TO THE PARK
Interstate State Park is on the western edge of Polk County. From Interstate I-94 near the St. Paul/Minneapolis area, exit north onto Highway 35 about 40 miles to the park entrance just outside the city of St. Croix Falls.

Interstate State Park
1275 State Highway 35
St. Croix Falls, WI 54024
(715) 483-3747

71. KINNICKINNIC STATE PARK

A Hiking and Boating Utopia

Confluence of the Kinnickinnic and St. Croix rivers

PARK SNAPSHOT

Kinnickinnic State Park is located along the scenic St. Croix River in Pierce County. The park was named after the Kinnickinnic River, which flows through the center of the property before joining the St. Croix River. Both rivers offer great canoeing and kayaking opportunities for paddlers. A sand delta at the mouth of the Kinnickinnic River serves as a popular overnight mooring site for boat camping. The park has a swimming beach, picnic areas, and hiking trails and offers nature study, especially birdwatching.

GEOLOGY AND HISTORY

The upper region of Kinnickinnic State Park is located along a 180-foot-high bluff area divided by deep, wooded ravines. The lower section of the property has river-bottom forests and a rock-lined gorge bordering the Kinnickinnic River. The sandstone and dolomite limestone rock of the park was formed beneath a shallow inland sea that once covered this area 600 million years ago from deposits of sand, silt and clay. During the last Ice Age, a massive sheet of ice known as the

Superior Lobe of the Wisconsin Glacier pushed through most of the Northwoods but stopped short of reaching the Kinnickinnic area. Although the park's landscape escaped the rock-crushing force of glacial ice, meltwater from decaying ice fields within the Lake Superior Basin to the north thundered through this area, carving out a new path the Kinnickinnic and St. Croix rivers on its way to the Mississippi.

Many Native American cultures, including the Ho-Chunk (Winnebago), Santee Dakota (Sioux) and the Ojibwe (Chippewa), have lived, hunted, and fished within the St. Croix and Kinnickinnic River area for generations.

In the 17th century, French-Canadian explorers established fur trading posts up and down the St. Croix River to trade with Native American hunters and trappers. Fur trading continued for many years with French, English, and eventually American companies until the mid-1800s when logging interests took precedence. In 1837, U.S. government agents induced Ho-Chunk, Sioux, and Chippewa leaders to sign treaties and sell most of their land to the federal government. In return, tribal members were to receive annual cash payments and the right to hunt, fish, and collect native plants on all ceded lands. Soon after, dozens logging camps, sawmills, grist (flour) mills, and villages were built along rivers throughout the area to accommodate the logging industry and influx of immigrant wheat farmers to the area.

The first American settler in the Kinnickinnic area was Joel Foster, who arrived here during the winter of 1848-49. Foster spent his first year living in a walled-in overhang cave along the banks of the Kinnickinnic River before erecting a log home adjacent to a waterfall. Foster later helped other immigrants, including two of his brothers, to settle in what would become the nearby city of River Falls.

The effort to develop a state park along the St. Croix River was championed by local citizen groups such as the Save Our St. Croix Association. Several civic-minded land owners offered to donate their land to protect the river shoreline

---

Ranger Trivia

The term *Kinnickinnic* was adapted from the Ojibwe word, *Giniginige*, which means "what is mixed." This refers to the Native American practice of combining the shredded leaves and bark of sumac, dogwood, and other native plants with pipe tobacco. Smoking Kinnickinnic was most often used for social occasions but was also used for medicinal and spiritual purposes.

Ranger Note

In the 1850s, a small town called Clifton Hollow was built along a waterfall area of the Kinnickinnic River within what would become the state park. The town had a flour mill, sawmill and several homes, but the site was abandoned in the 1870s when the railroad bypassed Clifton Hollow in favor of the city of River Falls.

St. Croix River swimming beach

from commercial development. Their efforts paid off with the establishment of the 1,239-acre Kinnickinnic State Park in 1972.

BEACHES, PICNIC AREAS, AND SHELTERS
The **St. Croix Picnic Area** is perched on a scenic wooded bluff above the junction of the St. Croix and the Kinnickinnic rivers. The day-use area has picnic tables, restrooms, and an open-air shelter. Below the picnic area is a sandy swimming beach along the St. Croix River.

HIKING, BIKING, AND NATURE TRAILS
There are seven miles of hiking trails at Kinnickinnic to explore. Most hiking trails are located in the upland region of the park and are marked by color-coded emblems. The **Yellow Trail** (1.2 miles) and **Green Trail** (1.1 miles) loop through attractive hardwood forests and restored prairies. The **Pink Trail** (1.6 miles) loops around the park entrance area through grassland areas and provides access to **Vulture's Peak**, an elevated region of the park. The **Blue Trail** (0.7 miles) parallels the entrance road with several steep areas as it traverses the hilly landscape of the park. The **Brown Trail** (0.1 mile) leads to a scenic overlook of the Kinnickinnic River Valley below.

The **Red Trail** (2.9 miles) is the longest trail in the park and leads to a scenic rocky gorge along the Kinnickinnic River. The trail has interpretive signs highlighting the flora and fauna of the park. Access to the trail is off 770th Avenue along the southern boundary of the park.

Many bike riders enjoy peddling the **White Kinnickinnic Pathway**, a two-mile paved bike path located in the nearby city of River Falls. This community participates in the **Blue Bike Program**, which offers free use of city-owned bikes.

Ranger Trail Pick

The **Purple Trail** (1.0 miles) is the most popular route in the park. The trail begins in the St. Croix Picnic Area and follows the edge of a wooded bluff to an overlook platform with great views of the confluence of the St. Croix and Kinnickinnic rivers below. The trail connects to the **Yellow Trail** loop, which loops back to picnic area. Hikers can also opt to take the **Orange Trail** (0.5 miles), which is a bit more challenging and also connects to the **Yellow Trail.**

WILDLIFE VIEWING

The park's forested bluffs, prairies, wetlands, and rivers are home to more than 30 species of mammals including white-tailed deer, fox, coyote, rabbits, squirrels, mink, and raccoon. At least 140 different types of birds are known to nest in the park making it one of the top birding areas in Wisconsin. The Kinnickinnic and St. Croix rivers are both within the **Greater Mississippi River Migration Route** used by hundreds of thousands of migrating waterfowl, cranes, herons, songbirds, and hawks.

The **Kinnickinnic River Gorge** and **Delta State Natural Area** (88 acres) provide protection for many unique plant and animal species, including several rare ferns and mosses.

INTERPRETIVE PROGRAMS AND FACILTIES

Interpretive panels are located along the Red Trail and at bluff overlook areas. The **Great River Road Visitor and Learning Center** is located about ten miles south of Kinnickinnic. The center is perched atop a 100-foot-high sandstone bluff overlooking the confluence of the St. Croix and Mississippi rivers. Displays here explain the unique history and aquatic ecosystem of the rivers. The Learning Center is located at **Freedom Park** along Monroe Street in city of Prescott. www.freedomparkwi.org

Ranger Note

The St. Croix River can get very busy on warm summer weekends. High traffic from powerboats, pontoons, and jet skis can make travel for smaller craft unpleasant at times. Paddlers will find less river traffic early in the morning and during midweek days.

BOATING AND FISHING

The lower 22-mile route of the Kinnickinnic River is a Class I trout stream with good populations of brook trout. An angler's parking area is located within the park off of County Highway F. The **Kinnickinnic State Fishery Area** (893 acres) located upstream from the park has one of the highest densities of naturally reproducing brown trout populations in the state. **Lake St. Croix** (7,696 acres) is a wide section of the St. Croix River north of the park. Anglers can expect to catch walleye, northern, muskie, panfish, white bass, and catfish in these waters.

A carry-in canoe/kayak launch area near the swimming beach can be accessed along the paved **Black Trail** from the picnic area. Motorboat launch sites with access to the St. Croix and Mississippi rivers are available in Prescott, about ten miles south of the park.

An eight-mile stretch of the Kinnickinnic River from River Falls to the St. Croix River is a popular kayaking route. Paddlers are advised that the shallow water, boulders, and rapids of this section of the river can be challenging. There are several kayak rental and shuttle services in the local area, including **Kinni Creek Outfitters** located in the nearby city of River Falls. www.guide@kinnicreek.com

ACCESSIBLE FACILITIES AND TRAILS

The **St. Croix Picnic Area** has paved walkways that lead to wheelchair-accessible picnic tables, restrooms, and an overlook platform with a panoramic view of the junction of the Kinnickinnic and the St. Croix rivers. The park's steep topography and rocky, unsurfaced trails can be challenging for visitors with walking disabilities.

CAMPING

Camping at Kinnickinnic is limited to overnight boat mooring along a 70-acre sand delta at the junction of the St. Croix and Kinnickinnic rivers. Boat-in campers are required sleep onboard and must have self-contained restrooms. A self-registration station is located onshore. More traditional camping is available at **Hoffman Park**, which has 20 campsites (15 electric), restrooms, and showers. Hoffman Park is located nine miles northeast of Kinnickinnic along Hanson Drive near River Falls. www.rfcity.org/298/camping

WINTER ACTIVITIES
Park staff groom seven miles of cross-country ski trails for in-line skiing. A popular sledding hill can be accessed from the Brown Trail parking area. The Red Trail (2.9 miles) located in the south area of the park is a designated snowshoe trail.

AREA ATTRACTIONS
Pierce County is known for its small, welcoming towns, historical attractions, and many outdoor recreation opportunities. www.pcedc.com.

Crystal Cave, a multi-level cavern located more than 70 feet below the surface, is a popular tourist attraction. Guided one-hour tours showcase a variety of cave formations, including stalagmites, stalactites, helicities, and cave pearls. Crystal Cave is located 29 miles west of the park along Highway 29 near Spring Valley. www.acoolcave.org

DIRECTIONS TO THE PARK
Kinnickinnic is located in Pierce County near the city of River Falls. From Interstate I-94 in Hudson, exist onto County Highway F south about eight miles and then turn right onto Cedar View Road (820th St.) to the park entrance.

Kinnickinnic State Park
W11983 820th Avenue
River Falls, WI 54022
(715) 425-1129

72. LAKE WISSOTA STATE PARK

Lakeshore Park—A Natural Beauty

Lake Wissota

PARK SNAPSHOT

Lake Wissota State Park is located along the shoreline of Lake Wissota, a 6,000-acre reservoir flowage of the Chippewa River. Many visitors enjoy boating, fishing, canoeing, kayaking, and other water-based activities on this beautiful lake. The state park encompasses 1,062-acres of scenic forests, restored prairies and wetland areas. Lake Wissota has one of the most popular campgrounds in the state, plus picnic areas, miles of hiking trails and a wide sandy swimming beach.

GEOLOGY AND HISTORY

The quiet backwater bays, wetlands, and forested hills of Lake Wissota were shaped by both man-made and natural forces. The park's bedrock is composed of deep layers of Cambrian sandstone formed from sand and silt deposits that settled beneath an ancient sea that submerged this area about 600 million years ago. Sandstone outcrops can be seen along much of the shoreline of the lake. During the last Ice Age 15,000 years ago, the advance of the **Chippewa Lobe** of the Wisconsin Glacier was stopped short just six miles northeast of the park. Although not directly altered by glacial ice the park's landscape was shaped indirectly by glaciers none the less. Decomposing ice fields far to the north sent tremendous amounts of glacial melt water surging down the Chippewa River basin. These massive floodwaters deposited a 12-mile-wide outwash plain of boulders, gravel and deep layers of sand and other glacial debris throughout the Lake Wissota area.

Several Native American cultures have hunted, fished and lived within the Chippewa River Valley for thousands of years. The Ojibwe (Chippewa) people migrated to this part of Wisconsin from the Lake Superior region in 1745. Sometime later, French-Canadian explorers set up fur trading posts up and down the Chippewa River to trade with Chippewa and other Native American trappers and hunters. The fur trade continued with the English and eventually American fur companies well into the 19th century.

The upper Chippewa River region had some of the finest white pine forests in Wisconsin. Several logging companies began cutting timber here as early as 1822. This led to land and timber ownership disagreements with the Chippewa people. After years of conflict, Chippewa and Dakota Sioux leaders were induced to sign a treaty in 1837 giving ownership of all their land that drained southwest to the Mississippi River to the U.S. government.

In 1836, Jean Brunet, the namesake of **Brunet Island** State Park (See Chapter 60), built the first sawmill dam across the Chippewa River near what would become the nearby city of Chippewa Falls. The arrival of rail lines to the city in 1875 allowed logs to be shipped by train to sawmills instead of floating them down river. As a result, Chippewa Falls became a major lumber town and the site of the largest sawmill in the world at that time.

As the logging boom came to an end around 1911, local factories turned to shoe-making and other industries. Immigrant farmers attempted to farm cutover forest land but found it difficult to grow crops in the sandy soils of the area. Beginning in the mid-1950s, the **U.S. Federal Soil Bank** program paid local farmers to plant trees on their sandy fields instead of crops. Many of these trees can still be seen on former farm fields that are now part of the state park. Local employment was enhanced in 1917 when a hydroelectric dam was built across the Chippewa River. The dam created a 6,024-acre reservoir called **Lake Wissota**.

The State of Wisconsin began to acquire land along the northeast shoreline of Lake Wissota in 1961 to establish Lake Wissota State Park, which opened in 1972.

Ranger Trivia

Lake Wissota was named by Louis Arnold, an engineer who helped plan and build the Chippewa River Hydroelectric Dam for the Wisconsin-Minnesota Power and Light Company. The "Wis" part of Wissota stands for Wisconsin. The last part, "sota," comes from the last four letters of Minnesota, despite the fact that the lake is nearly 100 miles from the Minnesota state line.

BEACHES, PICNIC AREAS, AND SHELTERS

A 285-foot-long sand swimming beach and bathhouse are located along Lake Wissota. Nearby is a large picnic area, playground, ball field, and an open-air shelter facility. Smaller picnic areas and additional shelters are located along shoreline as well.

HIKING, BIKING, AND NATURE TRAILS

Lake Wissota has 18 miles of hiking trails to explore. The **Lake Trail** (1.4 miles) follows the shoreline of the lake and provides a walking path to the swimming beach from the campground area. The **Plantation Trail** (0.7 miles), **Red Pine Trail** (1.5 miles), and **Jack Pine Trail** (0.75 miles) all lead through forested areas, including second-growth conifers planted in former crop lands during the 1950s.

The **Eagle Prairie Trail** (0.5 miles) and the **Prairie Wildflower Nature Trail** (0.5 miles), as their names suggest, loop through restored prairie areas. In late summer, the prairies come alive with colorful wildflowers, tall grasses and butterflies.

Off-road biking and horseback riding are allowed on nine miles of multi-use trails. The nearby **Old Abe State Trail** (20 miles) is a popular asphalt-paved hiking/biking trail that connects Chippewa Falls to the city of Cornell and **Brunet Island State Park** (see Chapter 60) to the north. A parking area for this trail is off County Highway O about four miles north Lake Wissota.

Ranger Trail Pick

The **Beaver Meadow Nature Trail** (1.0 miles) is a self-guided interpretive trail that loops through several different ecosystems, including hardwood forests, a beaver pond wetland area, and natural springs. The trail crosses over several small creeks and offers scenic views of the quiet, backwater bays of Lake Wissota.

WILDLIFE VIEWING

Lake Wissota is located within an ecological and vegetative transition zone. The park has both northern forest trees, such as white pine, spruce, balsam fir, and sugar maple and southern woodland species, including oak, hickory, and jack pine.

Canada geese on Lake Wissota

The Chippewa River and Lake Wissota are important migration stops for thousands of ducks, geese, and shorebirds in spring and fall. More than 200 bird species of birds are known to live or migrate through this area, including dozens of waterfowl and woodland songbirds, cranes, egrets, herons, owls, hawks, osprey, and bald eagles.

White-tailed deer, squirrels, rabbits, raccoon, chipmunks, fox, and coyote are the most common mammals of the park. Beaver, muskrat, and river otter are occasionally spotted in the backwater bay areas of the park. Black bear and gray wolf sightings have also occurred in the Chippewa Falls area in recent years.

INTERPRETIVE PROGRAMS AND FACILITIES

The park has interpretive kiosks and two self-guided nature trails that explain the history of the Lake Wissota and the flora and fauna of the area. Guided nature hikes, evening programs, and special events are scheduled throughout the year. The **Karen Lea Nature Center**, located in south end of the park, has exhibits featuring the park's wildlife, geology, and history. The center is open periodically on weekends during the summer months.

BOATING AND FISHING

Lake Wissota (6,024-acres) offers fishing, sailing, motor boating, waterskiing, canoeing, and kayaking. A boat launch and two fishing piers are located along a backwater channel of Lake Wissota in the south end of the park. Anglers can expect to catch walleye, musky, northern pike, catfish, crappie, and a variety of panfish. Canoe, kayak, paddleboard, and rowboat rentals are available near the boat launch.

> **Ranger Trivia**
>
> In the 1997 Hollywood movie *Titanic*, Leonardo DiCaprio's character Jack Dawson mentions that he often went ice fishing on Lake Wissota in Wisconsin as a boy. In reality this would have been impossible since Lake Wissota did not exist until the hydroelectric dam was built across the Chippewa River in 1917, five years "after" the *Titanic* sank in the Atlantic Ocean in 1912.

ACCESSIBLE FACILITIES AND TRAILS

The park's visitor entrance station, restrooms, picnic areas, shelters, and bathhouse are accessible for mobility-impaired visitors. Campsites 14 and 91 are wheelchair-accessible and have electrical service. Paved access trails provide access to restroom/shower facilities. Accessible fishing piers are available along a paved walkway between the swimming beach and the boat launch area. Most hiking trails have natural sand and grass tread surfaces, which can make them challenging for visitors with walking impairments.

CAMPING

Lake Wissota has 116 campsites (60 electric). Most sites are well-shaded and several are located along the shoreline of Lake Wissota. The campground has two restroom/shower facilities. An RV water fill/sanitary station is located between the park office and shop area. The park has two large group campsites (tent-only) that can accommodate up to 100 campers each. The group area has vault-type restrooms and an open-air shelter.

WINTER ACTIVITIES

Park staff groom eight miles of cross-country ski trails for both traditional and skate skiing. A marked ten-mile hiking/snowshoe trail is also available. Ice fishing for walleye, northern pike, crappie, and panfish on Lake Wissota is also a popular winter activity.

AREA ATTRACTIONS

Chippewa County is popular tourist destinations known for its natural beauty, hiking and biking trails and historic towns.

www.gochipewafalls.com

or www.chippewacounty.com

The **Leinenkugel Brewing Company** is one of the oldest breweries in Wisconsin. The brewery was established by German immigrant Jacob Leinenkugel in 1867 and continues to be one of the most recognized brands in the beer industry. The Wisconsin Historical Society has recognized the brewery's spring house, horse stable, and three-story malt house as historic 19th-century structures. Brewery tours, beer samples, and merchandise are available at the company's **Leinie Lodge Visitor Center** located along Duncan Creek on East Elm Street in Chippewa Falls. www.leinie.com

DIRECTIONS TO THE PARK

Lake Wissota State Park is located on the northeast side of Lake Wissota near the city of Chippewa Falls in Chippewa County. From the Eau Claire area, take Highway 53 north to Highway 29 east and then Highway 178 north. Exit onto County Highway S and head east crossing the bridge over the Chippewa River. Turn onto County Highway O and travel about two miles east to the park entrance.

Lake Wissota State Park
18127 County Highway O
Chippewa Falls, WI 54729
(715) 382-4574

73. MENOMINEE RIVER STATE RECREATION AREA

A Wild River Beauty

Menominee River

RECREATION AREA SNAPSHOT

The Menominee River State Recreation Area is located in Marinette County along Wisconsin's northeastern border with Upper Michigan. The 4,817-acre property follows a 17-mile-stretch of the Menominee River known for its outstanding kayaking and tubing opportunities. The recreation area is a semi-wilderness property so there are no visitor station, picnic areas, or modern camping and restroom facilities. In addition to its outstanding paddling opportunities, the property offers canoe camping, hiking trails, nature study, birdwatching, hunting, and fishing.

The Menominee River Recreation Area is one of the few public lands in the nation that shares a common boundary and is co-managed by two states. The Wisconsin section of the property is located west of the river and the State of Michigan side is to the east. The combined recreation area of both states encompasses nearly 10,000 acres of northern forestland.

GEOLOGY AND HISTORY

The rugged beauty of the Menominee River is enhanced by its rock-lined shoreline and cascading waterfalls. Granite boulders and exposed bedrock along the river are volcanic in origin and over one billion years old. Outcrops of iron-rich sedimentary rock, which formed beneath shallow seas that once covered this region 600,000 years ago, can also be seen here.

The streams, wetlands, and moraine hills of gravel and sand of the Menominee River region were shaped by continental glaciers that advanced and retreated through this area over a period of two million years. As the last Ice Age came to an end about 12,000 years ago, several Native American cultures migrated to the Menominee River Valley in search of big game animals, fish, and wild plants. The most successful of these early inhabitants were the Menominee, the only Native American tribe in the state whose original homeland was northeastern Wisconsin.

Menominee warriors fought with British troops against American militia during the War of 1812, but later supported the United States Army during the Black Hawk War of 1832. Despite this turn-around in patriotism, the Menominee were still forced to sell all their remaining land in north-central Wisconsin to the federal government in 1848. During the treaty negotiations, tribal leaders fought for the right of all Menominee people to hunt, fish, and gather wild plants on their ceded property. The tribe later gained ownership of 236,000 acres of reservation land northwest of Green Bay, now Menominee County.

> ### Ranger Trivia
>
> The Menominee River was named after an Algonkian word *Manomin* or "rice," which refers to the tribe's traditional use of wild rice as an important food source. During the 17th century, French fur traders and Jesuit Missionary priests referred to the Menominee people as the *Folles Avoines* or "Wild Oats People." Today, many tribal members prefer to be called *Mamacequta*, "The People."

During Wisconsin's lumber boom in the late 19th century, millions of white pine logs were sent down the Menominee River to lumber and paper mills in Green Bay. At the same time, mining operations began in the nearby iron-rich region known as the **Menominee Range**. By the early 20th century, most of pine forests had been leveled and iron mines began to close, leaving the local economy struggling. Fortunately, passenger railroad service and new highways brought waves of tourists from southern cities to camp, hike, hunt, fish, or just enjoy the pristine beauty of the Menominee River region.

Most of the land within the recreation area was once owned by the Wisconsin Public Service (WPS), who intended to build a hydroelectric dam on the Menominee River. The project stalled in the 1970s due to new environmental regulations, which made construction of the dam impractical. The State of Wisconsin eventually purchased most of the WPS lands to establish the Menominee River State Recreation Area in 2010.

BEACHES, PICNIC AREAS, AND SHELTERS

Quiver Falls Overlook Day-Use Area is located along County Highway R and Pemene Dam Road. An accessible foot path here leads to an observation area with great views of the waterfalls. Historic dam and cribbing remnants from the 19th-century logging era can still be seen along the river. Swimming is not recommended in the turbulent, fast-moving Menominee River. A public swimming beach is located at nearby Morgan County Park. (See Camping section)

The **Pemene Falls Day-Use Area** is located along State Road W 2 on the Michigan side of the recreation area. A short path from a parking area leads to an overlook area above the falls. The **Piers Gorge Scenic Hiking Trail** (2.6-miles), also on the Michigan side, follows the shoreline of the Menominee River past several sets of pier rapids (perpendicular rock ledges) and ends near the **Sand Portage Waterfall**. This section of the river is a great place to watch kayakers and rafters make their way through the surging whitewater rapids of the gorge.

HIKING, BIKING, AND NATURE TRAILS

Sand Portage Falls Trail (0.5 miles) leads to outstanding views waterfalls and the **Piers Gorge** in the northern part of the property. A rustic path leads to the base of the falls. Parking for this trail is located along Highway 141.

Ranger Trail Pick

The **Pemene Falls Trail System** (0.5 to 1.9 miles) is located in the lower section of the property along the east side of the Menominee River. A small parking lot is located at the end of Verheyen Lane, a half-mile dirt road off of Highway Z. A short unmarked trail from the parking area leads to a massive rock outcrop above the river with great views of Pemene Falls. Visitors can also park along Highway Z and walk down Verheyen Road to the falls area.

WILDIFE VIEWING

The recreation area is known as one of the best places in Wisconsin to view both common and rare woodland songbirds, woodpeckers, waterfowl, grouse, owls, and hawks. Several bald eagles nest within the Menominee River Valley and can often be seen catching fish along nearby streams and flowages. In winter, a seasonal influx of bald eagles and osprey arrive here to catch fish in ice-free sections of the river, such as waterfall areas.

Bald eagle perched along Pemene Falls

White-tailed deer, squirrels, chipmunks, raccoon, snowshoe hare, coyote, and river otter are commonly seen here. More secretive animals, such as black bear, fisher, bobcat, and gray wolf live within the recreation area as well. Four species of bats are drawn to this area in fall to spend the winter months hibernating in nearby abandoned iron mines.

The forests of the recreation area are composed of towering white pine, oak, and maple trees and also birch, aspen, white cedar, hemlock, and black spruce. Most of the larger trees date back to Wisconsin's logging era more than a hundred years ago but several areas of the forest were never cut and are much older.

INTERPRETIVE PROGRAMS AND FACILITIES

There are no signed nature trails or interpretive centers within the recreation area but ongoing development of the property may lead to these amenities in the future. The nearby **Iron Mountain Iron Mine** is great place to learn about Wisconsin's iron mining history. Underground mining began here in 1877 and remained active well into to the 1940s. Visitors can ride a mine train 400 feet below the surface to view iron ore rock formations and antique mining equipment. Iron Mountain Mine is located northeast of Penbine along Highway 2 near Vulcan, Michigan. www.ironmountainironmine.wix.site.com

73. Menominee River State Recreation Area

BOATING AND FISHING

The headwaters of the Menominee River are located in northern Wisconsin and Upper Michigan. The river flows south for 116 miles before emptying into Green Bay. For most of its course, the Menominee River is wide and smooth except for the 17-mile stretch that flows through the state recreation area. The river here turns violent with many powerful whitewater rapids as it thunders through narrow, rocky areas. Many kayakers and rafters consider the Menominee River to be the most challenging and beautiful river gorge in the Midwest. Local kayak and rafting outfitters in the area offer rental and pickup services. Small craft landings are located along both sides of the river.

The Menominee River is known for its excellent smallmouth bass fishing. Anglers can also catch walleye, northern pike, rock bass, and panfish in the river. Several side streams of the Menominee River harbor brown, rainbow, and brook trout. Marinette County has over 500 miles of trout streams, plus 242 named and 200 un-named lakes for anglers to fish in as well.

ACCESSIBLE FACILITIES AND TRAILS

The **Quiver Falls** overlook trail off of Highway R is reasonably accessible for most visitors. Most other trails are unsurfaced and often have rocky terrain, which make them challenging for mobility-impaired visitors.

CAMPING

There are four canoe campsites along the Menomonie River on Wisconsin side of the recreation area. Each site has a fire ring and a primitive latrine. Campsite reservations are required and can be made by calling **888-947-2757**. Campers looking for more traditional campsites will find them at **Morgan County Park** located west of the recreation area along Timms Lake. The park has **45** campsites (all electric), restrooms, showers, hiking trails and a swimming beach. www.marinettecouinty.com/parks/camping

WINTER ACTIVITIES

Snowshoeing, cross-country skiing, and wildlife watching are popular off-season activities in the recreation area. Many visitors enjoy hiking to ice-sculptured waterfalls in winter. Bald eagles can often be spotted catching fish in the open waters below waterfall areas. A ten-mile section of the Marinette County snowmobile/ATV trail passes through the recreation area.

AREA ATTRACTIONS

Marinette County is known as the "Waterfall Capital of Wisconsin." There are 14 scenic waterfalls located within the county park system and 230,000 acres of forest land to explore. www.therealnorth.com

Dave's Falls County Park has several small but spectacular waterfalls. The Pike River, a tributary of the Menominee River, cascades through a rocky gorge here. Dave's Falls is located about nine miles south of the recreation area adjacent to Highway 141 near the village of Amber.

DIRECTIONS TO THE RECREATION AREA

The Menominee River State Recreation Area is located in northeastern Marinette County about eleven miles west of the village of Penbine. From Green Bay, take Highway 141 north and exit onto County Highway R east.

Menominee River State Recreation Area
Mailing/Contact Address: Governor Thompson State Park
N100008 Paust Lane
Crivitz, WI 54114
(715) 757-393

Ranger Note

Vehicle parking fees are not required on the Wisconsin side of the recreation area, but a daily or annual recreation pass is required in Michigan. Passes can be purchased at self- registration stations at parking areas.

---

74. NORTHERN HIGHLAND-AMERICAN LEGION STATE FOREST

Wisconsin's Largest State Forest

Loon on Firefly Lake

FOREST SNAPSHOT

The Northern Highland-American Legion State Forest is the largest state-owned property in Wisconsin. This sprawling forest encompasses **236,282** acres within Vilas, Oneida, and Iron counties. With over 900 pristine lakes and more than 300 miles of streams to explore, the forest offers some of the best fishing and boating opportunities in the state.

Nearly two million people visit the forest each year to enjoy its many outdoor activities such as camping, hiking, swimming, hunting, wildlife viewing, and biking or just to relax and enjoy the beauty of this Northwoods icon.

GEOLOGY AND HISTORY

Northern Highland State Forest's name gives a hint as to its location and origin. This "northern" state forest is situated atop a geologic "highland" or raised expanse of the state known as the **Wisconsin Dome**, a massive underground structure of volcanic igneous stone, the oldest rock on the continent. Much of northcentral Wisconsin was uplifted about 1.1 billion years ago due to continental drift collisions. This left the Northern Highlands region above the waters of a shallow inland sea that flooded most of the state 600 million years ago. The current landscape of the forest was shaped by action of several glaciers that pushed through most of Wisconsin over a period of two million years. Near the end of the last Ice Age about 12,000 years ago, retreating ice fields left behind a labyrinth of streams, wetlands, bogs, moraine hills, sandy plains, and more than 3,000 kettle lakes throughout northern Wisconsin.

Several Native American cultures have hunted, fished, and lived throughout the Northern Highland area for generations. The Ojibwe (Chippewa) entered northern Wisconsin from the Lake Superior region in 1745. They traded with French-Canadians, English, and Americans until the early 19th century, when the fur trade market began to sag and logging interests took precedence in the Northwoods. The demand for lumber to build the great cities to the south led to "White Pine Treaty" of 1842, in which the Chippewa agreed to sell 24,400 square miles of their land to the U.S. government. The Chippewa retained their right to hunt, fish, and gather wild plants on all ceded lands and later gained ownership of the **Lac du Flambeau Reservation** northwest of Woodruff. Today more than a thousand Chippewa descendants continue to live on reservation lands.

By the late 19th century, northern Wisconsin was the center of the lumber industry. The first railroad line reached Minocqua in 1887, which allowed logs to be transported to sawmills by rail much faster than floating them down river. When the logging boom came to an end, these same rail lines were used to transport anglers, hunters, and resort guests to the Northwoods. Since then, the tourism industry has continued to drive the local economy of the local area.

The Northern Highland State Forest was established in 1925 to safeguard the headwaters of the Manitowish, Flambeau/Chippewa, and Wisconsin rivers. In 1927, the U.S. Veterans American Legion donated land to create the American Legion State Forest to the south. Today, the Northern Highland and American Legion state forests are managed as one unit and encompass more than a quarter million acres.

BEACHES, PICNIC AREAS, AND SHELTERS

There are nine picnic areas located throughout the forest. Each has a swimming beach, picnic tables, shelters, and restrooms. The two largest are **Crystal Lake** west of Sayner and **Clear Lake** southwest of Woodruff. Other popular picnic areas are along **Carol Lake, Tomahawk Lake,** and **Big Arbor Vitae Lake.**

HIKING, BIKING, AND NATURE TRAILS

The forest has 78 miles of hiking trails to explore. The **Raven Nature Trail** (1.4 miles), located north of the **Clear Lake Campground**, loops through old age hardwood and hemlock forests to the shoreline of **Hemlock Lake**. The **North Trout Lake Nature Trail** (1.0 miles) follows a boardwalk through a black spruce bog with great views of Trout Lake. Bog plants such as insect-eating pitcher plants and lady slipper orchids can be seen here in summer

Star Lake Nature Trail (2.5 miles), located off County Road K in the northeast section of the forest, follows the edge of a peninsula that juts out into **Star Lake**. The trail leads past an experimental pine plantation established by state foresters in 1913 and over a boardwalk through a black spruce bog.

The **Van Vliet Hemlock Trails** (3.2 miles) northwest of Boulder Junction offers several routes through a mix of lakes, ponds, bogs, and rolling glacial topography. The **Escanaba Trail System** (2.4 to 8.5 miles), located east of the **North Trout Lake Campground**, provides access to five lakes surrounded by balsam, aspen, and maple forests.

The forest offers 50 miles of paved bicycle trails and 58 miles of off-road trails. The **Sayner-to-Boulder Junction Trail** (14.0 miles) is a popular asphalt-paved trail that connects the **Crystal**, **Musky**, and **Firefly campgrounds**. The trail is part of the **Heart of Vilas County Trail** (52 miles), which connects several small towns along its route. https://biketheheart.org

The **Bearskin State Trail** (21.5 miles) follows an historic logging railroad grade through upland forests and black spruce bog areas. This popular trail

Fallison Lake

crosses a 375-foot wooden train trestle bridge in Minocqua and 13 other bridges along its route.

WILDLIFE VIEWING

The near-wilderness forests, streams, and lakes of the Northern Highland region provides habitat for wildlife to thrive here. More than 50 species of mammals can be found here including whitetail deer, squirrels, chipmunks, porcupines, beaver, otter, bears, coyotes, bobcats, pine martins, fishers, and gray wolves.

> **Ranger Note**
>
> The Boulder Junction area is home to the largest herd of albino (white-colored) deer in the Midwest. Watch for them along forest edges when driving through this area for a rare encounter with these enchanting animals.

Hundreds of species of waterfowl, woodland songbirds, shorebirds, game birds, owls, osprey and bald eagles live in the forest area including the North-wood's most iconic bird, the common loon. About 100 pairs of loons nest on forest lakes each year.

INTERPRETIVE PROGRAMS AND FACILITIES

Nature programs and hikes are presented at various locations during the use season. The **Crystal Lake Nature Center** has interpretive displays, animal mounts, and nature activities for kids. The **Northern Lakeland Discovery**

Center, located on state forest land adjacent to **Statehouse Lake** north of Manitowish Waters, offers 12 miles of nature trails to explore. www.discoverycenter.net.

BOATING AND FISHING

There are 930 named lakes within the forest, the highest concentration of lakes in the state. Boat launch access is provided along most lakes. Carry-in watercrafts (electric motors only) are allowed on many wilderness lakes. Anglers can expect to catch a variety of fish on most lakes including musky, northern pike, walleye, bass, perch, crappie, and panfish. **Trout Lake** (3,864 acres), one of the largest lakes in forest, harbors nearly every fish species in Wisconsin.

ACCESSIBLE FACILITIES AND TRAILS

Visitor entrance stations and most developed picnic areas are accessible for mobility-impaired visitors. **Crystal Lake**, **Firefly**, and **Clear Lake** campgrounds have designated accessible campsites with electric service and paved access to restroom/ shower facilities.

The **Crystal Lake Nature Center** and the nearby asphalt-paved **Tom Roberts Nature Trail** (0.5 miles) are both wheelchair-accessible.

CAMPING

The forest has 967 campsites located within 18 campground areas. The most popular campgrounds are located along **Crystal Lake, Muskellunge Lake**, and **Firefly Lake** along County Highway N in the northern area of the forest. The **Clear Lake Campground** located off County Highway J southeast of Woodruff is also a favorite camping area. All of these campgrounds have restrooms/ shower facilities. RV water fill and sanitary stations are available at Clear Lake, Crystal Lake, and North Trout Lake Campgrounds

> **Ranger Note**
>
> There is no electrical service at any of the campsites in the state forest (except certain handicap sites). Portable generators may be operated from 10 a.m. to 5 p.m. in most campgrounds except those designated as quiet zones. Campers should check with forest staff regarding current policies before using a generator.

The forest has two group campgrounds located along **Jag Lake** and **North Muskellunge Lake**. Each group site can accommodate up to 80 campers. Backpack camping is allowed along marked snowmobile routes and within the Lumberjack Trail area. Backpack camping permits are available free of charge at any forest office.

There are **78** primitive canoe campsites within the forest. Camping is free but stays are limited to one night only. The most popular canoe trail in the forest is along the Manitowish River, which flows 44 miles from High Lake to the Flambeau Flowage. Canoe route maps and portage information are available online or at forest offices.

WINTER ACTIVITIES
Forest staff groom 41 miles of cross-country ski trails. The **Raven, Escanaba, Madeline,** and **McNaughton** trails are the most popular ski trails. Snowshoeing and big-tire biking are allowed on any trail that's not groomed for skiing. Other winter activities include ice fishing, hunting, hiking, and wildlife watching. Winter camping is available at the Clear Lake Campground. The forest has 486 miles of snowmobile trails that all connect to county trail systems.

AREA ATTRACTIONS
The nearby communities of Minocqua, Woodruff, Boulder Junction, and many other small towns have been popular vacation destinations for generations of visitors. www.minocqua.org

The **Northwoods Wildlife Center** is a wildlife rehabilitation center that cares for sick and injured wildlife from throughout northern Wisconsin. The center offers nature-based programs and daily guided tours of its facility including live bird displays of bald eagles, hawks, falcons, and owls. The center is located along South Blumenstein Road in Minocqua. www.northwoodswildlifecenter.org

DIRECTIONS TO THE FOREST
The Northern Highland-American Legion State Forest is located within Vilas, Oneida and Iron Counties. The forest is about 68 miles north of Wausau along State Highway 51.

Northern Highland-American Legion State Forest
4125 Forestry Headquarters Road
Boulder Junction, WI 54512
(715) 356-3668

75. Pattison State Park

Home of Wisconsin's Highest Waterfall

Big Manitou Falls

PARK SNAPSHOT

Pattison State Park is one of the most scenic parks in Wisconsin. The Black River here drops 31 feet over Little Manitou Falls before tumbling 164 feet down Big Manitou Falls, the tallest waterfall in Wisconsin and fourth highest east of the Rocky Mountains. The park offers many outdoor recreation activities, including camping, picnicking, wildlife viewing, nature study, and miles of hiking trails. Interfalls Lake, located between the two waterfalls, is a favorite spot for swimming and fishing.

GEOLOGY AND HISITORY

Pattison has an amazing geologic story to tell, most of which can be seen along its beautiful waterfalls and rocky gorges. The brownish basalt rock of the waterfalls was formed 1.1 billion years ago from molten lava that oozed to the surface here from deep fissures within the earth's crust. About 600 million years ago, a shallow inland sea covered this area. Deep deposits of sand, silt, clay, and other debris that settled beneath this ancient ocean became the sandstone and shale rock seen in the park today. Then, about 500 million years ago, a cataclysmic collision of continents caused several massive breaks or faults in the earth's crust. One of these breaks, the Douglas Fault, caused a massive uplift of the bedrock along a line stretching from the Ashland area all the way to the Twin Cities in Minnesota. This powerful earthquake event split the rocky landscape of the park and created Pattison's beautiful waterfalls.

During the last Ice Age, advancing glaciers crushed layers of sandstone and deposited glacial till, moraine hills, wetlands, and outwash plains throughout the park area. Near the end of the Ice Age about 10,000 years ago, glacial ice fields south of Pattison disintegrated rapidly, sending enormous amounts of melt water raging down the Black River. This massive surge of water cut through hundreds of feet of glacial deposits and sandstone, exposing the basalt rock ledges of the waterfalls.

Native American people have lived in this area for thousands of years. One of the first, the Copper Culture people, lived here as far back as 1,500 BC. These early inhabitants mined for copper throughout the region to fashion tools, knives, arrow points, and decorative items as trade goods. Later in the early 1600s, the Ojibwe (Chippewa) people were among the first to meet and trade with French-Canadian fur traders. The French set up a local fur trading post along the Black River near Big Manitou Falls. The Chippewa called the river *Mucudewa Sebee* or "Black River" because of its dark color. The brownish tinge of the water is caused by tannin leaching from decaying vegetation and upstream bogs.

> **Ranger Note**
>
> The Chippewa regarded Big Manitou Falls as sacred place and referred to it as *Gichi Monido*, home of the "Great Spirit." They believed that the Monido were supernatural deities that lived in all animals, plants, and natural objects. Many felt they could hear a spirit's voice in the thundering sound of the waterfall. French explorers transcribed the Ojibwe term Monido into *Manitou*, a name the waterfall retains to this day.

The first settlers arrived in this area around 1847. Some were miners searching for copper in the park area but none found much success here. In 1879, Martin Pattison, a lumberman from upper Michigan, arrived to set up a logging camp along the Black River. Pattison became a wealthy man thanks to his successful lumber business and iron ore mining interests. In 1917, Pattison learned of a plan to build a hydroelectric dam on the Black River that would have destroyed Big Manitou Falls. To stop the dam project, Pattison secretly purchased 660 acres along the river and then donated the property to the State to develop a new state park in 1920. The initial development of the park was completed by the Civilian Conservation Corps (CCC), a Depression-era federal government employment program for young men. The CCC had a work camp at Pattison throughout the 1930s. Pattison State Park celebrated its 100-year anniversary in 2020.

BEACHES, PICNIC AREAS, AND SHELTERS

A popular swimming beach and bathhouse is located along **Interfalls Lake**. Adjacent to the beach is a large picnic area and an historic indoor shelter, which was built with local stone and timber by CCC craftsmen in 1937. Other picnic areas are located near Big Manitou Falls adjacent to County Highway B and along Highway 35 at Little Manitou Falls.

HIKING, BIKING, AND NATURE TRAILS

The park has seven miles of hiking trails. **Big Manitou Falls River Trail** (0.5 miles) is located on the west side of the Black River. This is one of the best routes to see and hear the roar of Big Manitou Falls and the only trail that leads down to rapids area of the Black River below the waterfall.

The **Beaver Nature Trail** (2.0 miles) encircles Interfalls Lake and leads over a bridge spanning the Black River built by CCC workers in the 1930s. The trail has interpretive signs describing flora and fauna of the park. The **Overlook Trail** (1.2 miles) and the **Oak Ridge Trail** (0.8 miles) lead through upland forest areas on the east side of the park.

Both trails connect to the **River View Trail** (2.0 miles), which follows the Black River for much of its course.

Little Manitou Falls is a 31-foot-tall "twin" waterfall located about a mile upstream of Big Manitou Falls. The Black River here tumbles over a split basalt rock ledge into a shimmering pool below. This beautiful waterfall can be accessed via the **Little Manitou Falls Hiking Trail** (0.5 miles) or by driving south on Highway 35 to a parking area adjacent to the falls.

The **Osaugie Bike and Hiking Trail** (5.0 miles) is located in the nearby city of Superior. This paved trail follows the Lake Superior shoreline with great views of the Superior Bay harbor, one of the busiest ports on the Great Lakes.

> **Ranger Trivia**
> The Ojibwe (Chippewa) people believed that they could hear the voice of the Great Spirt, *Gitchi Monido*, within the heart of Big Manitou Falls and that of *Bohiwum Sasigewon* or "Laughing Rapids Spirit" in the churning river below the waterfalls. Today, the park's thundering waterfalls continue to inspire visitors, some of whom still listen for the voice of the spirit *Gichi Monido*.

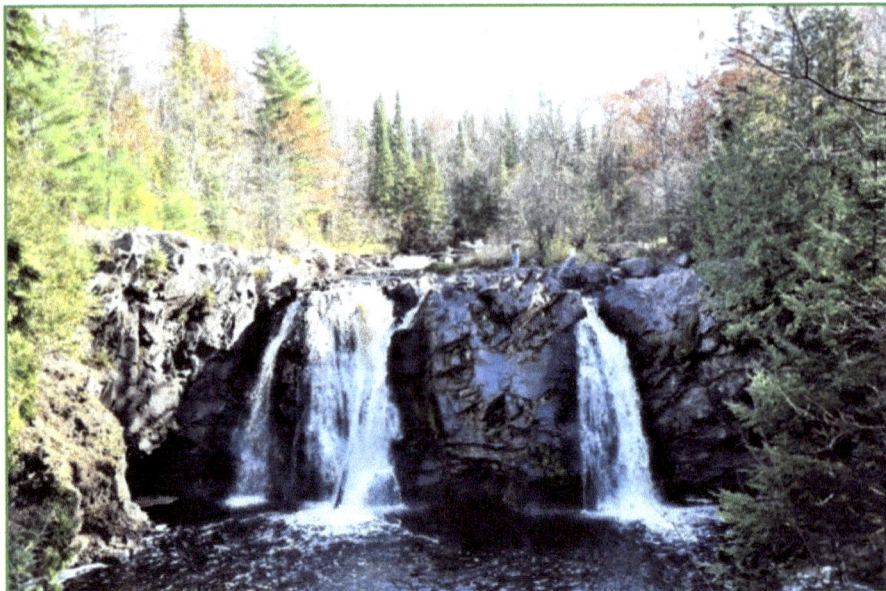

Little Manitou Falls

> **Ranger Trail Pick**
>
> The Big Manitou Falls Overlook Trails (0.5 miles) are short, asphalt-paved walkways that lead to overlook platforms on both sides of the river. The overlook areas offer outstanding views of Wisconsin's tallest waterfall and allow visitors to hear the deafening roar of the Black River as it plunges over the 165-foot-high rock ledge into the gorge below.

WILDLIFE VIEWING

Pattison's diverse landscape of rocky gorges, waterfalls, lowland spruce swamps, and pine and oak woodlands provides ideal habitat for wildlife. More than 200 species of birds are known to nest or migrate through the park, including dozens of species of woodland songbirds, waterfowl, shorebirds, owls, bald eagles, and ospreys. The park is home to more than 50 species of mammals, including white-tailed deer, porcupine, rabbits, red squirrels, and beaver. More secretive mammals such as black bear, fisher, and gray wolves are also spotted in the park occasionally.

INTERPRETIVE PROGRAMS AND FACILITIES

The **Gitche Gumee Nature Center** has several interesting displays highlighting the flora, fauna, human history and unique geology of the park. The center is located inside the park's indoor shelter building and is open every day in summer. Interpretive hikes and evening programs are scheduled throughout the use season.

Interpretive panels that explain the unique volcanic rock and sedimentary sandstone geology of the park are located at waterfall overlook areas. A free guide entitled "Big Manitou Geology Walk" has illustrations and descriptions of the rock formations of both Big Manitou and Little Manitou falls. The guidebook can be obtained from the park office.

BOATING AND FISHING

Interfalls Lake (23 acres) offers shoreline fishing opportunities for panfish, bass, and northern pike. The shallow, turbulent waters of the Black River within the park area are not suitable for fishing but the lower ten miles of the river near the city of Superior is popular with anglers for bass, catfish, walleye, and smallmouth bass fishing. **Amnicon Lake** (390 acres), **Dowling Lake** (141 acres), and **Lyman Lake** (370 acres) are all located about seven miles southeast of the park. All three lakes have a boat launch and good populations of musky, walleye, bass, and a variety of panfish.

ACCESSIBLE FACILITIES AND TRAILS

The park's nature center, indoor shelter, and most picnic areas are accessible for mobility-impaired visitors. Campsite **36** is wheelchair-accessible with electrical service and access to a flush toilet/shower facility. An asphalt-paved walkway leads to an overlook platform with an outstanding view of Big Manitou Falls. Most other hiking trails in the park have natural turf, steps, or steep grades, making them challenging for visitors with walking impairments.

CAMPING

The park has **62** campsites (18 electric) located in the northern area of the park. Most are pull-through sites for easy access. The campground has a central flush toilet/shower facility. Three backpack campsites can be accessed from the **Oak Ridge Trail** near the Little Manitou Falls area. Each site has a fire ring, picnic table, and access to vault-type restrooms and a nearby shelter.

WINTER ACTIVITIES

Pattison's waterfall areas become even more beautiful in winter when cloaked in deep snow and lined with fanciful ice sculptures. Off-season activities at the park include hiking, wildlife watching, snowshoeing, and winter camping. Park staff groom four miles of cross-country ski trails for inline skiing. Ski routes are color-coded and rated from easy to advanced skill levels. A warming shelter is located along the **Oak Ridge** (Orange) trail in the south end of the park.

AREA ATTRACTIONS

Douglas County and the Superior area are known for their outstanding natural beauty, historical sites, and many outdoor recreational opportunities. www.visitsuperior.com www.visitdouglascounty.com

The **Fairlawn Mansion and Museum** has a direct historical tie to Pattison State Park. The 42-room Queen Anne Victorian mansion was built in 1891 as the family home of Martin Pattison, the namesake for the park. The mansion features a towering, four-story circular turret overlooking Lake Superior. The museum is located along East 2nd Street in the nearby city of Superior. www.superiorpublicmuseums.org/fairlawn

DIRECTIONS TO THE PARK

Pattison State Park is located in northwestern Wisconsin within Douglas County about 13 miles south of the city of Superior. From U.S. Highway 53, exit onto County Road B west to Highway 35 and then turn south for a quarter mile to the park entrance.

Pattison State Park
6294 South State Road 35
Superior, WI 54880
(715) 399-3111

76. RIB MOUNTAIN STATE PARK

Highest Peak in Wisconsin—A Scenic Wonder

Rib Mountain Observation Tower

PARK SNAPSHOT

Rib Mountain State Park is perched atop the highest peak in Wisconsin. At 1,942 feet above sea level, the park's rocky summit is considered the tallest in the state due to its steep vertical rise above the Wisconsin River near Wausau. Many people visit Rib Mountain just to enjoy the panoramic view of the surrounding countryside below. The park has picnic areas, shelters, an observation tower, and miles of hiking trails to explore.

GEOLOGY AND HISTORY

Rib Mountain was originally known as Rib Hill but was renamed Rib "Mountain" sometime in the early 1940s by local promoters as a marketing ploy to increase publicity of the downhill ski resort built on its slope in 1937. Renaming this towering landmark a mountain may not have been that far off the mark since geologists have yet to agree on a universal definition of what separates a mountain from a hill.

Rib Mountain, along with nearby **Mosinee Hill** and **Harwood Hill**, are among the oldest geologic formations found on earth. They were formed about 1.7 billion years ago by intense heat and fusion, which transformed sandstone bedrock into quartzite, a very hard, metamorphic stone. Geologists refer to Rib Mountain as a **monadnock**, a single, colossal block of quartzite that has been eroded by wind, water, and ice over millions of years.

The massive glaciers that covered most of the state during the last Ice Age stopped short about 15 miles east of Rib Mountain. As a result, Paleo-Indian hunters may have entered this area much sooner than other areas of the state, perhaps even as the last glacial ice fields were melting about 12,000 years ago. By the time French-Canadian fur traders canoed down the Wisconsin River in the mid-1600s, the Ojibwe (Chippewa) people had been living along the river at what now is the city of Wausau for generations.

In the early 19th century, most of northern Wisconsin was one vast, unbroken forest of mostly old-growth white pine. Lumber companies pressed U.S. government officials to open these valuable timber areas for logging. Their efforts paid off in 1874 when Chippewa leaders were forced to sell most of their forest land to government agents. Soon after, hundreds of logging camps and dozens of sawmills were built throughout the Northwoods. When the railroad line reached Wausau in 1874, transportation of logs and lumber by rail became much more efficient than floating them down rivers. This

greatly increased the pace of logging and made Wausau the heart of the lumber industry.

Rib Mountain State Park was established in 1927 thanks to the donation of 160 acres of land to the State by the Wausau Kiwanis Club. The early development of the park, including many of the existing trails, roads, and the first downhill ski run, were built by the Civilian Conservation Corps (CCC), a federal employment program for young men in the 1930s. Today, Rib Mountain encompasses 1,627 acres of parkland and continues to attract thousands of visitors to enjoy the outstanding views from Wisconsin's highest peak.

BEACHES, PICNIC AREAS, AND SHELTERS

The **Upper Picnic Area** has open-air shelters, a concession stand, and access to the park's 60-foot lookout tower. A modern indoor shelter with restrooms, a fireplace, and a gathering room with tables and chairs is also located here. The **Lower Picnic Area** has a playground, an amphitheater, and bluff-side overlooks.

There are no lakes or streams on Rib Mountain but the city of Wausau has two outdoor swimming pools: the **Weston Aquatic Center** along Alta Verde Street and the **Kaiser Pool** on East Bridge Street.

> **Ranger Note**
>
> The term "talus" refers to expansive fields of broken rock fragments created by natural erosion from ice and frost action over the course of millions of years.

HIKING, BIKING, AND NATURE TRAILS

There are more than 14 miles of color-coded trails to explore in the park. The **Yellow Trail** (3.79 miles) has three looped routes and descends through the forested south slope of the park. This trail is the longest and most challenging route in the park. The **Red Trail** (1.14 mile) located west of the picnic area, loops through a hardwood/ pine forest and section of the **Rib Mountain Talus Forest State Natural Area.**

The **Quarry Trail** (0.5 miles) loops around the upper perimeter of an abandoned quartzite quarry. This trail provides access to **Turkey Vulture Trail** (1.28 miles), **Homestead Loop** (1.26 miles), and **Dynamite Trail** (0.25 miles), all of which are located in the northwest section of the park.

The **Walking Path** (1.5miles) is a paved hiking/biking trail that parallels the park entrance road to the top of Rib Mountain. The nearby **Mountain-Bay State Bike Trail** (83 miles) is the longest Rails-to-Trails bike route in Wisconsin. The trail connects Wausau to the Green Bay area. Parking for this trail is located just west of Wausau at the Weston Municipal Center. www.co.marathon

> **Ranger Trail Pick**
>
> The **Gray Trail** (1.2 miles), **Blue Trail** (0.61 miles) and **Green Trail** (0.34 miles) lead to some of the most interesting quartzite rock formations and scenic overlook areas of the park. All three trails provide access to the Rib Mountain's iconic 60-foot observation tower, which offers a panoramic view of the surrounding forests and hills of the Wausau area. These trails begin in the Upper Picnic Area.

View from Rib Mountain

WILDLIFE VIEWING

The hardwood forests and rocky slopes of the park provide habitat for a variety of wildlife. White-tailed deer, raccoon, squirrels, and porcupine are often seen here. More secretive animals such as fox, coyote, black bear, and gray wolves are occasionally spotted in the area as well. More than 200 species of birds are known to either nest or migrate through the park area, including many different types of woodland songbirds, woodpeckers, owls and hawks, bald eagles, and turkey vultures.

INTERPRETIVE PROGRAMS AND FACILITIES

Interpretive programs and guided hikes are scheduled throughout the use season. Evening band concerts are occasionally held at the park amphitheater. Interpretive panels describing the geology, history, flora, and fauna of Rib Mountain are located along trails, overlooks, and at the observation tower.

BOATING AND FISHING

The nearby Wisconsin River is a favorite destination for boaters and paddlers. The city of Wausau maintains a world-class whitewater canoe/kayak competition training run along a quarter-mile stretch of the Wisconsin River. The run is located at **Whitewater Park** and is available for public use for a small fee. www.wausauwhitewater.org.

76. Rib Mountain State Park

Lake Wausau (1,851 acres) and **Wausau Dam Lake** (152 acres) are Wisconsin River reservoirs and located about seven miles northwest of the park. Both lakes have a boat launch and good populations of panfish, northern pike, muskie, walleye, and bass. The **Big Rib River State Fishery Area** in northwestern Marathon County offers shoreline fishing for smallmouth bass, brook, and brown trout.

ACCESSIBLE FACILITIES AND TRAILS

The visitor entrance station, amphitheater, indoor shelter, and most restrooms, picnic areas, and overlook areas are wheelchair-accessible. Due to the steep topography and rocky terrain of the park, some hiking trails may be challenging for visitors with walking disabilities.

CAMPING

Rib Mountain is a day-use park so there are no camping facilities here. **Marathon Park**, located in the city of Wausau, has 35 campsites (all electrical) with restrooms and showers. This urban camping area is located within a grove of towering white pine trees adjacent to Highway 52 (Stewart Avenue) about seven miles north of Rib Mountain.

The **Dells of the Eau Claire County Park** has 27 campsites (16 electric) with vault-type restrooms. (See Area Attractions) **Council Grounds State Park**, located about 22 miles north of Rib Mountain near the city of Merrill, has 52 campsites (19 electric) with restrooms and showers. (See Chapter 64)

WINTER ACTIVITIES

Hiking, wildlife watching, and snowshoeing are popular off-season activities at Rib Mountain. There are no cross-country ski trails in the park, but **Nine Mile County Forest**, located about seven miles south of the park, has 18 miles of groomed double-tracked ski trails. Downhill skiing is available at **Granite Peak Ski Area**, located at the base of Rib Mountain. www.skigranitepeak.com.

AREA ATTRACTIONS

The nearby city of Wausau has many visitor attractions, parks, and outstanding museums. (www.visitwausau.com) The **Woodson Art**

> **Ranger Note**
>
> Granite Peak offers chairlift rides on weekends in autumn for a small fee from late September through late October. This is a fun way to view the beautiful fall colors of surrounding forest areas. Access to the chairlift ride is available from at loading platform in the park's **Lower Picnic Area** and from the **Granite Peak Ski Area** below the hill.

Museum houses a priceless collection of sculptures, paintings, and other artwork that all feature birds as subjects (www.lywam.org). The **Dells of the Eau Claire River County Park and State Natural Area** features picturesque rapids and small waterfalls along a beautiful, rock-lined river gorge. The park has picnic areas, a campground, and hiking trails, including a section of the Ice Age Trail. The park is located about 26 miles northeast of Rib Mountain off County Highway Y west of the village of Aniwa. www.co.marathon.wi.us/parks

DIRECTIONS TO THE PARK
Rib Mountain is located in Marathon County seven miles south of the city of Wausau. From Interstate I-39/ Highway 51, exit onto County Highway N and then turn right (west) onto Park Road to the park entrance station on the summit of Rib Mountain.

Rib Mountain State Park
4200 Park Road
Wausau, WI 54401
(715) 842-2522

77. STRAIGHT LAKE STATE PARK

Wilderness Park—Paradise of Peace and Quiet

Straight Lake

PARK SNAPSHOT

Straight Lake State Park is located in northwestern Wisconsin about ten miles east of the St. Croix River in Polk County. This 2,066-acre, heavily forested park was intentionally designed as a quiet, semi-wilderness area with only limited development. Straight Lake has a small picnic area, ten walk-in campsites, and a few hiking trails, including one of the most scenic sections of the Ice Age National Scenic Trail. Other low-impact, silent activities here include fishing and non-motorized boating, canoeing and kayaking, plus outstanding wildlife viewing and nature study opportunities.

GEOLOGY AND HISTORY

The rolling hills and lakes of the park were formed during Wisconsin's last Ice Age when mile-high glaciers from the Canadian arctic pushed through this region, reshaping the earth under its massive weight. As the glaciers retreated back north

Ranger Note

A tunnel channel is a long valley formed by a large river that flowed beneath a glacier. The powerful currents of the river carve deep, wide channels beneath the ice and are gradually exposed as the ice above them finally melts.

about 12,000 years ago, a labyrinth of moraine hills, hummocks, rock-filled gorges, kettle lakes, swamps, and streams were left in it wake. The largest and most unique glacial feature of the park area is an immense seven-mile-long, U-shaped valley geologists refer to as a "tunnel channel."

The first Europeans to this region were French-Canadian fur traders who canoed down the nearby St. Croix River in the late 1600s. The French set up trading posts along rivers and streams throughout the area to acquire fur pelts from the Dakota Sioux and Ojibwa (Chippewa) hunters and trappers. As the fur trade began to wane in the early 19th century, timber interests and logging led to confrontations with Native people over land ownership. This conflict came to a head when tribal leaders were induced to sign what became known as the White Pine Treaty of 1837. As part of this agreement, both the Sioux and Chippewa gave up millions of acres of forest land to the U.S. government in return for annual cash payments and the right to hunt, fish, and gather plants on

Ranger Note

Timber crews logging in the Straight Lake area built an earthen dam across the outflow stream of the lake to raise the water level of the lake by several feet. In spring the dam was removed, allowing the floodwaters of the lake to float cut logs downstream to sawmills.

ceded lands. By the mid-1800s, hundreds of logging camps were set up through-out the Northwoods, including the Straight Lake area.

Ranger Trivia

The nearby village of Luck was once considered the "Yo-Yo Capital of the World." Duncan Toys Company built a factory here in 1946 to take advantage of the hard maple timber needed to manufacture their wooden toy yo-yos. By 1956, the factory was producing 3,600 yo-yos per hour to keep up with the world-wide demand for this iconic toy.

The first European settlers to the area arrived in the 1860s. Many were Danish immigrants who worked in local lumber camps and sawmills. The nearby village of Luck was established in 1869 by Daniel Smith, a native of New York. Before arriving here, Smith endured several business failures in the logging, mining, sawmill, and grocery business. When asked why he chose Luck as the name of his new town, he replied, "I propose to be in luck the balance of my life." Smith's fortunes did improve over time. He also established the town of Clam Falls 15 miles northeast of Luck and owned several sawmills.

The Straight Lake property changed ownership several times over the years. Past owners included DeWitt Wallace, founder of *Reader's Digest* magazine and the Boy Scouts of America, who operated a summer camp here. The property was acquired in 2005 by the State to establish Straight Lake State Park.

BEACHES, PICINIC AREAS, AND SHELTERS

A picnic area with tables, grills, a fire ring, and an open-air shelter facility is located along the shorelines of **Straight Lake** and **Rainbow Lake** near the park entrance. The park does not have a swimming beach, but one can be found at **Balsam Lake Beach Park**, located about five miles south of Straight Lake in the Village of Balsam Lake.

HIKING, BIKING, AND NATURE TRAILS

The park has 8.5 miles of hiking trails to explore, including a 3.65-mile-long segment of the **Ice Age National Scenic Trail**. **Straight Lake Trail** (0.9 miles) follows a forested ridge along the south shoreline of the lake connecting the picnic area to the walk-in campsite area. The **Rainbow Lake Loop Trail** (1.0 miles) begins near the picnic area and loops through an upland hardwood forest of oak and maple with scenic views of Rainbow Lake. **High Point Trail** (1.4 miles) turns south off Rainbow Lake Trail to the highest point in the park 135 feet above the level of the lake. **Glacial Trail** (.08 miles) exits the Rainbow Trail and connects to the Ice Age Trail at the eastern boundary of the park.

The nearby **Gandy Dancer State Trail** is a popular Rails-to-Trails crushed gravel bike route that extends for 47 miles connecting St. Croix Falls to Danbury. A 15-mile section of this trail leads from nearby village of Luck to St. Croix Falls.

Ranger Trail Pick

Many of the park's most outstanding features can be viewed by first hiking the Rainbow Lake Trail loop and then following Ice Age Trail for about two miles to the west. This section of the trail crosses the **Straight River** over a stepping-stone bridge and then follows a forested ridge with great views of Straight Lake. The trail crosses over a roadway (280th Avenue) and then heads north about 0.2 miles past outcrops of basalt rock formed by molten lava more than a billion years ago. From here, the trail leads through Boulder Valley, an amazing spring-fed gorge filled with hundreds of massive basalt and red granite boulders deposited by glaciers 12,000 years ago. The Ice Age Trail is a one-way route so hikers will need to retrace their steps to return to the picnic area.

WILDLIFE VIEWING

The forested hills, lakes, streams, and wetlands of this area provide ideal habitat for many wildlife species. White-tailed deer, squirrels, raccoon, coyotes, black bear, muskrat, otter, and occasionally fisher are spotted here. Straight Lake has been designated as an **Important Bird Area** by the **Wisconsin Bird Conservation Initiative**. Many species of woodland songbirds, waterfowl, shorebirds, owls, hawks, bald eagles, and game birds, such as ruffed grouse and wild turkey, are often seen here. Several uncommon birds have also been observed here including red-shouldered hawks, a nesting pair of trumpeter swans, and rarely seen songbirds like the Cerulean warbler.

The **Tunnel Channel Woods Natural Area** (457 acres) is located in the northeast section of the park. This site preserves the flora and fauna of this rare glacial valley, including 150-year-old hardwood trees.

INTERPRETIVE FACILITIES AND TRAILS

A kiosk located at the main entrance parking area has a property map and printed guides for hiking trails, hunting, and fishing. A brochure entitled the *Geology Hiking Guide for the Straight Lake Ice Age Trail* lists glacial formations and where to view them along the Ice Age Trail. The guide is available at the kiosk or can be downloaded online.

BOATING AND FISHING

Straight Lake (120 acres) is a wilderness lake that features an intact, undeveloped forested shoreline. Anglers can expect to catch northern pike, bass, and a variety of panfish in this lake. Adjacent to Straight Lake is **Rainbow Lake**. This spring-fed pothole lake also harbors northern pike, bass, and bluegills and is occasionally stocked with rainbow trout.

To maintain a quiet wilderness experience, motorboats are prohibited on Straight Lake and Rainbow Lake. Visitors with canoes, kayaks, and other non-motorized boats can access Straight Lake from carry-in boat launches near the entrance parking area and between campsites 8 and 9 on the west side of the lake. Rainbow Lake has carry-in launch near the parking area.

Big Butternut Lake (384 acres) and **Big Bone Lake** (1,667 acres) are located about five miles south of the park. Both have public boat launches and good populations of panfish, bass, northern pike, and walleye.

ACCESSIBLE FACILITIES AND TRAILS

The picnic area has accessible tables, vault-type restrooms, and a small open-air shelter. A nearby surfaced trail leads to a wheelchair-accessible fishing pier along

Rainbow Lake

Rainbow Lake. Due to the hilly topography and wilderness aspect of the park, some hiking trails may be challenging for mobility-impaired visitors.

CAMPING

The park has ten rustic, walk-in campsites located in a heavily forested area above the southeast shoreline of Straight Lake. Each site has a picnic table and a fire ring. Campers need to bring their own drinking water. A vault-type restroom and a self-registration station are located at the campground parking area off of 270th Avenue.

Campers who prefer more modern camping facilities will find them at **Interstate State Park**, located about 12 miles southwest of Straight Lake near St. Croix Falls. The park has 82 campsites (23 electric) with restrooms and showers. (See Chapter 70)

WINTER ACTIVITIES

Wildlife watching, hiking, snowshoeing, and cross-country skiing are popular off-season activities. None of the park's trails are groomed for skiing. Many anglers enjoy ice fishing on Straight Lake and Rainbow Lake. In compliance with the wilderness aspect of the park, ATV's and combustion-powered ice augers are not allowed on these lakes.

AREA ATTRACTIONS

Polk County has many parks, trails, museums, small towns, and fine resorts that offer excellent fishing and boating opportunities. The county has 400 lakes and 200 miles of rivers and trout streams. www.co.polk.wi.us/tourism

Wilke Glen and **Cascade Falls Park** feature a beautiful 25-foot-high waterfall perched within a forested gorge along Osceola Creek. The park has a staircase that descends to the bottom of the falls and a hiking trail that leads to a bluff overlook of the **St. Croix National Scenic Riverway**. The park is located 24 miles south of Straight Lake off of North Cascade Street in the village of Osceola. www.myosceolachamber.org.

DIRECTIONS TO THE PARK

Straight Lake State Park is located in northern Polk County about 12 miles northeast of the city of St. Croix Falls. From Highway 8, travel north on either Highway 35 or Highway 46 to the village of Luck. Take Highway 48 about 3.5 miles east of town and turn left (north) onto 120th Street about a mile to the park entrance.

Straight Lake State Park
2700 120th Street
Luck, WI 54853
(715) 483-3747 (Interstate State Park)

78. TURTLE-FLAMBEAU SCENIC WATERS AREA

A Wilderness Fishing and Boating Mecca

Turtle-Flambeau Flowage

PROPERTY SNAPSHOT

The Turtle-Flambeau Scenic Waters Area located in Iron County is one of the largest state properties of its kind in Wisconsin, encompassing 40,000 acres of near-wilderness forest land. At the heart of this scenic area is the 12,942-acre Turtle-Flambeau Flowage with its 330 miles of undeveloped shoreline and 377 islands.

The flowage is best known for its outstanding boating opportunities and excellent walleye fishing. Other popular activities here include wildlife watching, hiking, nature study, cross-country skiing, snowmobiling, and hunting. Boat-in campsites are available along the flowage shoreline and on several islands.

GEOLOGY AND HISTORY

The landscape of the Turtle-Flambeau is the result of both man-made and natural forces over time. Near the end of last Ice Age about 12,000 years ago, the massive glaciers that covered northern Wisconsin for more than 20,000 years began to retreat back to the Canadian Arctic. In their wake, a maze of moraine hills, kettle lakes, wetlands, and streams, including the Turtle and Flambeau rivers, were left behind. As grasslands, forests, and wildlife began to reclaim this barren landscape, several early Native American cultures migrated to the Northwoods to hunt big game, gather plants, and fish its many lakes and streams. By the time French-Canadian fur traders explored the Turtle and Flambeau rivers in the mid-17th century, the Ojibwe (Chippewa) people had established settlements throughout the area.

The discovery of iron ore in northern Wisconsin compelled the U.S. government to purchase mineral rights from the Chippewa in 1826. Iron mines provided jobs for many immigrants, especially those with experience in underground mining from Finland, Sweden, and England, who moved to Iron County in the mid-1800s. Underground mining continued in Iron County well into the 1960s until more efficient open-pit iron mines were developed in Minnesota.

The logging industry began about the same time as iron ore mining but unlike underground mines, cutting down entire pine forests and building sawmill dams across streams inevitably led to conflict with indigenous people. Logging interests eventually forced Chippewa leaders to sell over 24,000 square miles of northern forest lands to the U.S. government in 1842. Soon after, untold millions of pine logs were floated down the Flambeau and Turtle rivers to sawmills to the south.

The Turtle-Flambeau Flowage was created in 1926 when a dam was built across the Turtle River to provide a steady flow of water to the hydroelectric generating plants downstream. Many local people opposed the dam because the new flowage would submerge 16 natural lakes and thousands of acres of land, including several lakeside resorts. Opposition waned, however, when the completed flowage actually improved fishing and boating opportunities. New resorts built along the flowage also brought more tourism dollars into the local economy.

In 1990, the State of Wisconsin purchased 22,000 acres of the Turtle-Flambeau to establish a new scenic water area and provide public access this popular fishing and boating mecca.

BEACHES, PICNIC AREAS, AND SHELTERS

There are no developed picnic areas or swimming areas within the flowage. **Carow Park**, located about 15 miles north of the flowage on **Grand Portage Lake**, has a swimming beach and picnic area.

> **Ranger Trivia**
>
> The new high-end resorts built along the Turtle-Flambeau Flowage in the 1930s attracted fishermen from throughout the country, including a few notorious guests like gangster John Dillinger and mafia crime boss Al Capone.

HIKING, BIKING, AND NATURE TRAILS

There are 26 miles of hiking trails to explore within the flowage. The **Big Island Trail System** (8.8 miles) has several forested hiking paths, all of which can be accessed by boat or over a foot bridge between the Turtle River and Merkle Lake. **Deadhorse Trail** (4.6 miles) is located within the property's ruffed grouse demonstration area. Interpretive panels along this trail explain management techniques for providing habitat for ruffed grouse. Both trails are located off of Popko Circle Road West on the north side of the flowage.

Little Turtle Trail (5.2 miles) features the only native prairie area in Iron County and is also a great place to observe waterfowl, cranes, and shoreline birds. The trail is located off Joe's Shack Road southwest of Mercer. The **Wilson Hills Trail System** (5.5 miles) is a network of hiking trails that lead through forested moraine hills located between Wilson Lake and the Turtle-Flambeau Flowage.

Off-road biking is allowed on most flowage hiking trails. The nearby **Mercer Bike and Hike Trail** (7.8 miles) connects to the **Heart of Vilas Trail**, a 52-mile paved trail connecting the communities of Manitowish Waters, Boulder Junction, and Saint Germain with the **Northern Highland State Forest**. (See Chapter 74)

> **Ranger Trail Pick**
>
> The **Hidden Rivers Nature Trail** (2.0 miles) best exemplifies the beauty of the Turtle-Flambeau Flowage. This easy, walking path leads through a mature forest with interpretive panels that highlight the human history and development of the flowage. A side trail leads to **Turtle Point** at the end of a scenic peninsula with great views of the flowage. Hidden Rivers is located off Popkos Circle West Road near Fisherman's Landing on the north side of the flowage.

WILDLIFE VIEWING

The open waters, wetlands, lakes, and forests of the flowage provide ideal habitat for wildlife. White-tailed deer, squirrels, porcupine, and rabbits are commonly seen here. More secretive animals such as bobcats, bears, coyotes, gray wolves, and even an occasional moose, are also spotted from time to time.

> ### Ranger Trivia
>
> The nearby community of Mercer is known as the "Loon Capital of the World." Motorists entering the town along Highway 51 will see a 16-foot-tall roadside statue of a loon. Those who stop for a photo of this colossal bird will discover that it emits a recorded loon call.

More than 150 species of birds are known to nest or migrate though the flowage. Birders can observe many kinds woodland songbirds, waterfowl, owls, hawks, bald eagles, osprey, and rare trumpeter swans. The Turtle-Flambeau area has the highest concentration of nesting loons found anywhere within the continental United States.

INTERPRETIVE PROGAMS AND FACILITIES

Interpretive signs are located along nature trails and overlook areas throughout the flowage. A booklet entitled *Auto Tour of the Turtle-Flambeau Scenic Waters Area* can be picked up at the Mercer Ranger Station or is available online. This self-guided 24-mile auto tour features 15 roadside stops that highlight significant wetlands, lakes, bogs, and forests of the flowage. The tour begins at the **Little Turtle Waterfowl Management Area** southwest of Mercer and ends at the beautiful Turtle River waterfalls at **Lake of the Falls County Park**.

BOATING AND FISHING

The Turtle-Flambeau Flowage (12,942 acres) offers nearly unlimited boating and fishing opportunities with access from six boat launch areas. The **Springstead Boat Launch** on the south side of the flowage has paved ramps, boarding docks, vault-type restrooms, and the property's only drinking water source. The **Fisherman's Landing** on the north side of the flowage has a paved ramp with a boarding dock and restrooms.

> ### Ranger Note
>
> When the Turtle-Flambeau Flowage was created in 1926, it submerged several natural lakes, some as deep as 40 feet, but boaters still need to watch for underwater stumps, rocks, and logs in shallower areas.

Many anglers consider the Turtle-Flambeau Flowage to be the best lake in northern Wisconsin for walleye fishing, but it also has good populations of panfish, northern pike, crappie, and bass as well. Trophy muskellunge in excess of 50 pounds have also been caught in the flowage.

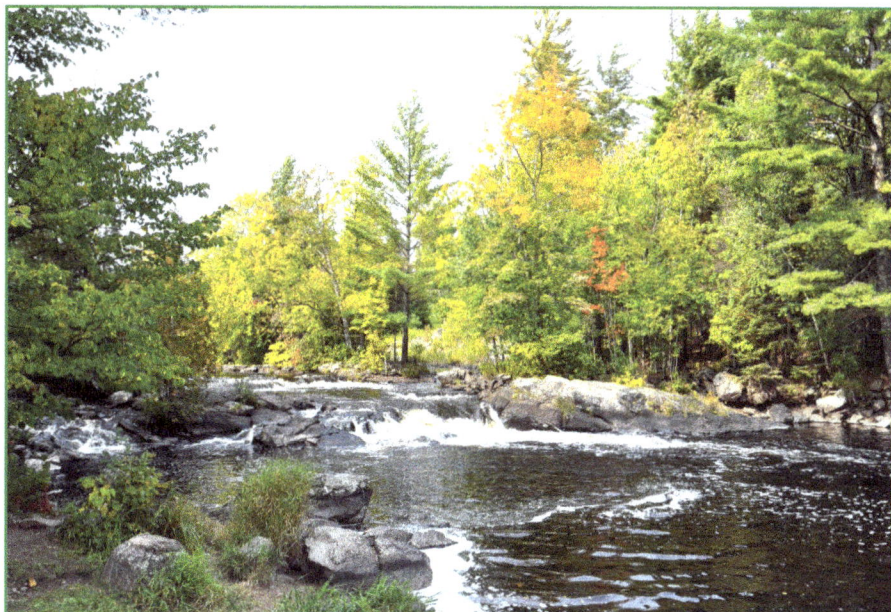

Turtle River waterfalls

ACCESSIBLE FACILITIES AND TRAILS

The **Hidden Rivers Nature Trail** has a fairly level gravel base along much of its route. A wheelchair-accessible, vault-type restroom is located at the nearby Fisherman's Landing. The Springstead Landing on the south side of the flowage also has accessible restrooms and a paved wheelchair walkway to a boat launch fishing pier.

Several boat-in group campsites with wheelchair access are located throughout the flowage. Certain group sites can be reserved in advance for mobility-impaired campers. The Turtle-Flambeau is a semi-wilderness area with rustic trails that may be difficult to access for visitors with walking disabilities.

CAMPING

The flowage has **58** boat-in-only family campsites and eight group sites located on both shoreline areas and on certain islands. All family campsites have a fire ring and open-air toilets. Group sites have fire rings, picnic tables, and vault-type restrooms. There is no fee to use family campsites and two of the group sites. All other group sites require a reservation and payment of fees in advance. Campers need to bring their own drinking water and pack out trash and recyclables.

Campers who prefer more modern camping facilities will find them at **Lake of the Falls County Park**, which offers **32** campsites (**22** electric). The park has flush toilets, showers, and a boat launch with access to the flowage. The park is located on the north side of the flowage along West County Park Road. www.ironcountyforest.org/lake-of-the-falls

WINTER ACTIVITIES

Off-season activities in include hiking, snowshoeing, wildlife watching, and cross-country skiing. The **Mercer Cross-Country Ski Club** grooms 17 miles of ski trails for both inline and skate skiers at the **Little Turtle Waterfowl Management Area**. A log cabin at the start of the trail serves as a warming house for skiers. Anglers enjoy ice fishing for panfish and walleye on the flowage. Several snowmobile trails lead through the property with connections to the 250-mile county trail system.

AREA ATTRACTIONS

Iron County has 74,000 acres of public forests and 300 inland lakes to explore. www.ironcountywi.com/recreation www.mercercc.com

North Lakeland Discovery Center is an interactive nature center with 12 miles of hiking, biking, and self-guided nature trails. The center is located along **Statehouse Lake** off Discovery Lane, north of Manitowish Waters. www.discoverycenter.net.

DIRECTIONS TO TURTLE-FLAMBEAU SCENIC WATERS AREA

The Turtle Flambeau is located south of the town of Mercer in Iron County. From Highway 51, exit southwest onto County Highway FF near Mercer or from Highway 13 take the northeast exit to County Highway F at Butternut.

Turtle-Flambeau Scenic Wates Recreation Area
Mercer Ranger Station location
5291 N. Statehouse Circle
Mercer, WI 54547
(715) 476-7846

79. WILLOW FLOWAGE SCENIC WATERS AREA

Wisconsin's "Near Canada" Wilderness Area

Willow Flowage

FLOWAGE SNAPSHOT

The Willow Flowage Scenic Waters Area is often compared to a remote Canadian wilderness lake region. In reality, this 30,000-acre property is located only a few miles southwest of Minocqua, one of the busiest tourist towns in Wisconsin. "The Willow," as locals call it, is a 6,400-acre flowage lake that features 73 miles of undeveloped shoreline and more than 100 islands. The Willow Flowage may be best known for its exceptional walleye fishing, but the property also offers camping, hiking, boating, nature study, hunting, and wildlife viewing. Paddlers flock here to test their skills on a 15-mile-stretch of the Tomahawk River, one of the best canoeing and kayaking rivers in the state.

GEOLOGY AND HISTORY

The Willow Flowage is located within an area of the Wisconsin geologists refer to as the **Wisconsin Dome** or the **Northern Highlands**. This massive region of the state is composed of elevated granite and basalt bedrock formed by volcanic activity about 1.1 billion years ago. Since then, several glaciers have pushed through this area over a period of 600 million years, leaving behind vast deposits of sand, gravel, moraine hills, and thousands of kettle lakes and ponds. Near the end of the last Ice Age about 12,000 years ago, melt water from disintegrating glaciers thundered down the Willow and Tomahawk rivers, carving new paths towards the Wisconsin River.

Native American people have hunted, fished, and lived in this part of the state for at least 8,000 years. The first French-Canadian fur traders explored this region in the 17th century. They set up trading posts along the Wisconsin River and its many tributaries, including the Willow and Tomahawk rivers. Fur trading for European goods with local Native American trappers and hunters, including the Ojibwe (Chippewa) people, lasted well into the early 19th century.

> **Ranger Note**
>
> Many Ojibwe (Chippewa) people prefer to be called the *Anishinaabe*, the "True People." The Anishinaabe consider northern Wisconsin and the Lake Superior region to be their ancestral home.

The Chippewa generally welcomed the European fur trade but conflicts over land rights arose in the 19th century when lumbermen began to clear cut pine forests and block river travel with sawmill dams. Under pressure by U.S. government agents, Chippewa leaders were induced to sign treaty settlements in 1837 and 1842, giving up their forest land in northern Wisconsin in return for annual cash payments and right to hunt, fish, and gather native plants on all ceded lands. Soon after, timber companies rushed to purchase forest land and set up dozens of logging camps in the local area. Throughout the mid- and late 1800s, untold thousands of pine logs were floated down the Willow and Tomahawk rivers to sawmills to the south. When the railroad line was extended to the Minocqua in 1887, the transportation of logs to mills became much more efficient and for the first time, hardwood trees such as maple and oak could be cut for market.

> **Ranger Note**
>
> Logs from softwood trees such as white and red pine could be floated downstream to sawmills. Logs from hardwood trees such as oak and maple have very dense wood and don't float well. As a result, they generally weren't cut until railroad cars could transport them.

The Willow Flowage was created in 1926 when a dam was built across the Tomahawk River by the Wisconsin Valley Improvement Company. The dam was intended to prevent flooding and ensure a steady flow of water downstream to power hydroelectric dams and paper mills. In 1998, the State of Wisconsin began to purchase forest land surrounding the Willow Flowage to ensure access to fishing and to provide other public recreation.

BEACHES, PICNIC AREAS, AND SHELTERS

The Willow Flowage is managed as a semi-wilderness area so there are no swimming beaches, shelters, flush toilets, drinking water fountains, or other day-use facilities here.

The **Willow Dam Rest Area**, located on the east side of the flowage, has a small picnic area with vault-type restrooms.

> ### Ranger Trail Pick
>
> The **Willow Nature Trail** (1.1 miles) leads to scenic overlook areas of the Willow Flowage. Interpretive signs along this route explain forest management techniques and highlight the wildlife of the area. The trail also serves as an access route to several walk-in campsites along the shoreline. The nature trail is located off Willow Dam Road about a quarter mile north of the dam.

HIKING, BIKING, AND NATURE TRAILS

Indian Shack Point Trail (1.0 mile) leads off the Willow Nature Trail along a peninsula jutting out into the Willow Flowage. The trail ends at a walk-in campsite at its terminus. The **Slaughterhouse Bay Hiking Trail** (2.0 miles) loops through an upland hardwood forest with great views of this scenic Willow Reservoir bay. The trail is located off Cedar Falls Road west of **Skunk Lake**.

Off-road bike riding is allowed on most property trails. The nearby **Bearskin State Trail** (21.5 miles) is a popular biking/hiking route that follows an historic logging railroad line. The trail leads through upland forests, a tamarack bog, and 16 wooden bridges including a 375-foot train trestle over **Lake Minocqua**.

WILDLIFE VIEWING

The forests, wetlands, and waters of the Willow Flowage attract a wide variety of wildlife. Hundreds of species of waterfowl, woodland songbirds, owls, hawks, herons, cranes, common loons, ruffed grouse, and woodcock can be observed here. Several pairs of bald eagles and osprey nest in trees along the flowage and can often be seen catching fish in its waters.

The Willow Flowage is also home to white-tailed deer, fox, porcupine, squirrels, black bear, coyote, and occasionally gray wolves. Backwater areas and lowland forests provide habitat for beaver, muskrat, mink, fisher, otter, and raccoon.

INTERPRETIVE PROGAMS AND FACILITIES

Interpretive panels are posted along the Willow Nature Trail and at the Willow Dam area on the east side of the flowage. The **Crystal Lake Nature Center** within the **Northern Highland State Forest** has several interpretive displays featuring the flora and fauna of the Northwoods. The center is located west of Sayner and is open from Memorial Day to the Labor Day weekend.

BOATING AND FISHING

The Willow Flowage (4,217 acres) is a designated **Outstanding Resource Water Area** (ORW) due to its high water quality and excellent fishery. Anglers can expect to catch walleye, panfish, perch, crappie, muskellunge, largemouth, and smallmouth bass in the flowage. There are six boat launch sites within the flowage. The **Sandy Landing** and **Willow Dam Landing** both have paved boat ramps and are located on the east side of the flowage. Four unpaved boat launches are situated on the west side of the flowage.

> **Ranger Note**
>
> The water level of the Willow Flowage and the Tomahawk River is controlled by the Wisconsin Valley Improvement Company and can change daily. Boaters need to watch for underwater hazards, such as stumps, sandbars, and rocks in shallower areas.

A 15-mile-stretch of the Tomahawk River within the flowage area is a favorite route for canoe and kayak enthusiasts. This scenic, undeveloped section of the river begins below the Willow Dam and flows south to **Lake Nokomis**. Most of the river has a gentle low-gradient flow except for two class II whitewater sections known as **Half-Breed Rapids** and **Prairie Rapids**. Paddlers are advised to portage around these two fast-moving, rocky stretches of the river.

ACCESSIBLE FACILITIES AND TRAILS

More than **95** percent of the Willow Flowage Scenic Waters Area is managed as a semi-wilderness area but there are two developed facilities for mobility-impaired visitors. The **Willow Nature Trail** (1.1 miles) is a near-level crushed gravel trail with interpretive panels and scenic overlook views of the flowage. This trail also provides access to campsite 2, a wheelchair-accessible site located along the shoreline of the Willow Flowage.

> **Ranger Trivia**
>
> The Tomahawk River is the largest tributary stream of the Wisconsin River and was once known as the "Little Wisconsin River." The river's name was derived from an Ojibwe word, *otamahuk*, which means "strike them." The term "tomahawk" also refers to a small, handheld ax or *tomahican*, as it was called by native people.

CAMPING

There are 30 campsites and seven group sites located on designated islands or shoreline areas of the flowage. All campsites are free and can be accessed by watercraft. Camping is limited to six people except for group sites, which can accommodate up to 15 people. Campers need to bring along their own water since there is no drinking water available within the flowage area.

Campsites 26 and 28, located along Indian Bay in the southwest section of the flowage, can be accessed by foot or 4-wheel-drive from a driveway off Iron Gate Road. The Willow Flowage also has four walk-in campsites that can be accessed along the Willow Nature Trail on the east side of the flowage. Each site has a picnic table, fire ring, and access to a primitive toilet. Campers who prefer more modern amenities will find them at **Cedar Falls Campground and Park**, which has 40 sites (29 electric), flush toilets, showers, and a boat launch with access to the Willow Flowage. The campground is located on the north side of the flowage off Cedar Falls Road. www.wvic.com/cedar-falls-campground

WINTER ACTIVITIES

Off-season activities include hiking, wildlife watching, snowshoeing, hunting, and cross-country skiing. None of the trails are groomed. Anglers enjoy ice fishing on the flowage for walleye, northern pike, and panfish.

Cedar Falls on the Tomahawk River

AREA ATTRACTIONS

Oneida County has more than 1,000 lakes and 800,000 acres of public forest land to explore. The nearby "island city" of Minocqua is one of the most popular tourist destination areas in Wisconsin.

www.oneidacountywi.com   www.minocqua.org

The **Rhinelander Logging Museum** and **Pioneer Park Historical Complex** is a free, open-air museum that preserves an original 1800s logging camp structure, equipment, and a Soo Line steam locomotive used to haul logs. The site also has a **Civilian Conservation Corps (CCC) Museum** and a display featuring the Hodag, the Northwoods' most iconic mythical creature. The museum is located 26 miles east of the flowage along Martin Lynch Drive in the city of Rhinelander. www.facebook.com/pphe.loggingmuseum.hodag

DIRECTIONS TO THE WILLOW FLOWAGE

The Willow Flowage is located in central Oneida County about 26 miles west of Rhinelander and 12 miles southwest of Minocqua. From Highway 51 south of Hazelhurst, take County Highway Y southwest to Willow Dam Road. A rest area with information about the flowage is located adjacent to the dam and boat launch.

Willow Flowage Scenic Water Recreation Area
8770 County Highway J
Woodruff, WI 54568
(715) 356-5211 Ext. 259

80. WILLOW RIVER STATE PARK
A Waterfall Wonderland

Willow River waterfalls

PARK SNAPSHOT

Willow River is one of the most popular state parks in Wisconsin and for good reason. This attractive 2,891-acre property features a beautiful, rock-lined river gorge with multi-tiered cascading waterfalls and a 170-acre lake, ideal for swimming, fishing, canoeing, and kayaking. The park also has shady forest campsites, picnic areas, and miles of scenic hiking trails to explore.

GEOLOGY AND HISTORY

Willow River is located in an area geologists refer to as the **Western Upland Region,** a mostly unglaciated (driftless) region of Wisconsin. The topography here is several hundred feet higher in elevation than central or southeastern Wisconsin as a result of continental collisions, which uplifted this part of the state a half billion years ago. This region was also once submerged beneath a warm, shallow inland sea, complete with coral reefs and primeval marine life, for at least 100 million years. Over time, thick deposits of sand, clay, silt, and decaying sea life sediments were transformed into sandstone and limestone. Exposed layers of this stone, along with embedded fossils, such as extinct trilobites, can be seen within the park's rock-lined river gorge

Near the end of the last Ice Age about 12,000 years ago, the Superior Lobe of the Wisconsin Glacier began to melt and retreat back to the Canadian Arctic. In its wake, a jumbled landscape of moraine hills, outwash plains of sand, wetlands, and kettle lakes were left behind. Glacial meltwater thundered down the Willow River, cutting through multiple layers of rock to create the park's 200-foot-deep river gorge and its waterfalls. Downstream, a few miles west of the park, the Willow joined the St. Croix River on its route to the Mississippi River.

Archeological excavations in the park area have revealed artifacts from several different Native American cultures that have lived in this area for here over the course of thousands of years. The abundance of game animals, water-fowl, fish, and plants, especially wild rice, attracted many indigenous people to this area, including the Ojibwe (Chippewa), the Dakota (Santee Sioux), and nine other Native American tribes. The first French-Canadian explorers arrived here in the late 1600s to set up fur trading posts along the Willow River and the nearby St. Croix River.

The Sioux and Chippewa were engaged in tribal warfare for decades over control of the St. Croix River region. A particularly bloody battle that occurred at the mouth of the Willow River was witnessed and recorded by a European explorer in 1795. The ongoing feud eventually led to the Sioux nation's decision to move westward into Minnesota and the Dakota Territory. The Chippewa's dominance of the Northwoods was short-lived, however. In 1837, under pressure from logging interests, the U.S. government forced Chippewa leaders to sell their land in return for annual cash payments and the right to hunt, fish, and gather plants on ceded lands. Soon after, logging camps began sending thousands of pine logs down the Willow River to sawmills along the St. Croix River.

> **Ranger Note**
>
> The St. Croix River was named after a French voyager by the name of Sainte-Croix who drowned and was buried beneath a wooden cross marker near the mouth of the river around 1689. The term St. Croix is French that literally means "Holy Cross."

In the early 1800s many European immigrants, especially from Germany, arrived in the local area to work in logging camps, sawmills, or to farm the rich prairie soils of the area. One of these early German immigrants was Christian Burkardt, who purchased 600 acres of land along the Willow River in 1868 to build a dam and a grist mill. After milling grain for local farmers for many years, Burkhardt built the first of four hydroelectric dams across the Willow River in 1891 to provide electrical service to the nearby village of Hudson.

80. Willow River State Park

The feasibility of generating electricity at small hydroelectric dams like those along the Willow River became obsolete in the 1960s. In 1967, the State purchased the 1,180-acre Willow River dam property from the Northern States Power Company to develop Willow River State Park, which opened in 1971.

BEACHES, PICNIC AREAS, AND SHELTERS
A wide, sandy swimming beach and a bathhouse facility are located along the southwest shoreline of **Little Falls Lake**. Adjacent to the beach is a large picnic area with restrooms, grills, and a playground. An attractive wood and stone shelter facility is located next to the Little Falls Dam on the far western edge of the lake.

HIKING, BIKING, AND NATURE TRAILS
The park has 17 miles of hiking trails to explore. The **Nelson Farm Trail** (3.7 miles) and the **Burkhart Trail** (4.0 miles) are located on the north side of the park. Both trails lead through restored prairies and upland forests with outstanding views of the river gorge below. A staircase off Burkhart Trail descends to the base of the Willow River waterfalls.

The **Pioneer Trail** (1.2 miles) leads past the gravesites of early pioneer settlers and to a scenic overlook above the waterfalls. The **Oak Ridge Trail** (1.1 miles) travels through a hilly, upland hardwood forest that showcases several glacial landforms created during Wisconsin's last Ice Age.

Little Falls Lake swimming beach

The **Hidden Ponds Nature Trail** (0.5 miles) is a paved walkway adjacent to the nature center with interpretive stops along its route. The **Trout Brook Trail** (1.4 miles) is a scenic, near-level walkway that begins west of the beach parking area and follows the shoreline of Willow River to the west.

Bike riding is allowed on the asphalt-paved section of **Little Falls Trail** (0.7 miles), which connects the swimming beach to campground areas. The nearby **Lakeland to Afton Biking/Hiking Trail** (1.6 miles) is a paved trail that begins in the community of Hudson and crosses the I-94 bridge over the St. Croix River to the cities of Lakeland and Afton in Minnesota.

Ranger Trail Pick

The **Willow Falls Hill Trail** (0.4 miles) is a short, paved route that descends from the upper parking area to the base of the Willow River waterfalls. The waterfalls can also be accessed via the **Little Falls Trail** (0.7 miles) near the beach area and connects to **Willow Falls Trail** (1.0 miles) within the **300 Campground**. This route follows the shoreline of Little Falls Lake and the Willow River to a bridge overlook area near the base of the waterfalls.

WILDLIFE VIEWING

The wetlands, prairies, and upland forests of the park provide ideal habitat for a variety of wildlife. More than 220 species of birds have been recorded in the park, including many types of woodland songbirds, waterfowl, shorebirds, woodpeckers, game birds, and turkey vultures. Avian raptors such as red-shouldered hawks, osprey, and bald eagles are often seen here as well.

White-tailed deer, squirrels, raccoon, and chipmunks are commonly seen in the park. More secretive animals such as badgers, fox, coyotes, and occasionally black bears are spotted here from time to time.

INTEPRETIVE PROGRAMS AND FACILITIES

The park's nature center located within the picnic area is open daily from Memorial Day through Labor Day. The center has displays featuring the flora, fauna and local history of the area plus live animal exhibits. Interpretive programs and hikes are presented throughout the summer season.

BOATING AND FISHING

Little Falls Lake (170-acres) is a backwater reservoir of the Willow River. A boat launch for non-motorized watercraft is located east of the swimming beach area. Canoe and kayak rentals are available through the park office. **Lake Mallalieu** (289 acres) is located downstream from the park. Motorboats and waterskiing are allowed on this lake. Anglers will find good populations of bluegill, crappie, perch, northern pike, and largemouth bass in both lakes.

The Willow River (61 miles) is stocked annually with brown trout. Shore fishing is popular both below and above the waterfall area. Sections of the Willow River are accessible for kayaking, but paddlers need to be aware of underwater rocks, fallen trees, rapids, and waterfalls.

ACCESSIBLE FACILITIES AND TRAILS

The park office, bathhouse, and most restrooms at Willow River are accessible to mobility-impaired visitors. The park's nature center and the **Hidden Ponds Nature Trail** are both wheelchair-accessible. Campsites **325** and **120** are wheelchair-accessible and have electric service. Both are located adjacent to shower/restroom facilities.

CAMPING

Willow River has **155** campsites (**55** electric) located within three separate campgrounds.

The **100 Campground** has a central restroom/shower facility and open, sunny campsites. The **200 Campground** has flush restrooms but no shower facilities. The **300 Campground** is the most popular camping area in the park due to its close proximity to the swimming beach and well-shaded sites. The campground has a restroom/shower facility and an RV water fill and sanitary station. The park has four walk-in group campsites that can accommodate between 30 to 40 people. The group campground has vault-type restrooms and a shared pavilion-style shelter.

WINTER ACTIVITIES

Park staff groom 12 miles of cross-country ski trails for both traditional and skate skiing. The nature center serves as a warming house for skiers on weekends. A two-mile, hard-packed trail on the north side of the river is available for hiking, snowshoeing, and dog sledding. Other off season activities include hiking, sledding, ice fishing, wildlife watching, and walk-in winter camping.

AREA ATTRACTIONS

St. Croix County is known for its outstanding scenery and small historic river towns. The community of Hudson offers a variety of trendy restaurants, coffee houses and boutique shops for visitors.

www.sccwi.gov   www.discoverhudsonwi.com

The **Hudson Walking Dike** (0.4 miles) was originally built as a tollway road in 1903 for U.S. Highway 12. The road provided vehicle access across the

St. Croix River connecting Wisconsin to the state of Minnesota. The tollway was closed to vehicle traffic in 1951 upon the completion of the Interstate I-94 Bridge south of Hudson. The dike and roadway have since been repurposed into a scenic river walkway. The Hudson Walking Dike is located along the St. Croix River off First Street in downtown Hudson.

Ranger Trivia

The nearby community of Hudson was originally called "Willow River" and later renamed "Buena Vista." In 1852, Alfred D. Gray, the city's first mayor, petitioned to have the town renamed Hudson because the high bluffs along the St. Croix River reminded him of the Hudson River Valley in his home state of New York.

DIRECTIONS TO THE PARK

Willow River State Park is located in St. Croix County about 26 miles east of Minneapolis/St. Paul near the city of Hudson. From Interstate I-94, take the U.S. Highway 12 exit north for 1.6 miles. Turn onto County Highway U for a short distance and turn north onto County Highway A about 1.5 miles to the park entrance.

Willow River State Park
1034 County Highway A
Hudson, WI 54016
(715) 386-5931

INDEX

Wisconsin State Parks, Forests, and Recreation Areas are shown below alphabetically in **Bold** type. Alphabetical listings of the State Parks, Forests, and Recreation Areas by Region can be found in the Table of Contents. City, county, and private facilities are listed separately under the index heading "camping areas."

A

about using this book, 2–3
admission fees, 6, 45
American Legion State Forest, 5, 433–38, 466
Amnicon Falls State Park, 331–36
Apostle Islands National Lakeshore, 337–42
area maps, iv, 8, 155, 329
attractions. See interpretive programs and attractions
Aztalan State Park, 156–60

B

Badger Army Ammunitions Plant, 300–303
Bass Lake Peatlands State Natural Area, 381
bats/bat colonies
    Governor Dodge State Park, 220
    Kohler-Andrae State Park, 72
    Menominee River State Recreation Area, 430
    Peninsula State Park, 116
    Yellowstone Lake State Park, 325–26
Baxter's Hollow State Natural Area, 215–16
Belmont Mound State Park, 161–66
Belmont Mound Woods State Natural Area, 164
Belmont Prairie State Natural Area, 164
Berg Prairie State Natural Area, 205
Big Bay Sand Spit and Bog State Natural Area, 340

Big Bay State Park, 337–42
Big Foot Beach State Park, 9–14
Billy Goat Ridge State Natural Area, 205
birding/bird-watching
    Bald Eagle Days, 273
    Cross Plains State Park, 210
    Great Mississippi River Migration Route, 418
    Great Wisconsin Birding and Nature Trails, 17, 188
    International Crane Foundation, 263, 270
    Kettle Moraine Forest, 60, 66, 84, 94
    Lake Michigan migration route, 28, 33, 72, 77, 109–10, 128
    Madison Metropolitan Sewerage District, 199
    National Migratory Bird Stopover, 116
    Sandhill State Wildlife Area, 178
    Turtle-Flambeau Scenic Waters Area, 460
    Wyalusing State Park, 320–21
Black River State Forest, 173–78
Blackhawk Lake Recreation Area, 167–72
Blue Mound State Park, 179–84
Brady's Bluff Dry Prairie State Natural Area, 284
Browntown Oak Forest Natural Area, 187–88
Browntown-Cadiz-Springs Recreation Area, 185–89
Brule River State Forest, 343–48
Brunet Island State Park, 349–54
Buckhorn State Park, 190–95

475

Index

ACKNOWLEDGMENTS

Many thanks go out to my wife Kathy for her tireless editing of this book. Also thanks to our children Jim, Krissy, and Joey for their encouragement and support in completing this book. They, along with their spouses and our grandchildren, spent many memorable outings with us exploring and camping in state parks and forests throughout the state.

And a big shout-out to Kira Henschel for her expertise in editing, book design, and seeing this project to completion.

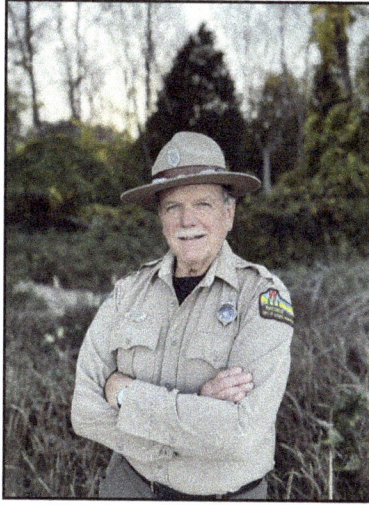

Jim Buchholz served as a Park Ranger and Park Supervisor at several state parks and forests throughout Wisconsin during his 40-year career with the Wisconsin Department of Natural Resources. In addition to his law enforcement and park management duties, Jim enjoyed helping park visitors learn more about the flora, fauna, history, geography and recreational amenities available at each property. His previous book *Wild Wisconsin Notebook*, published by Prairie Oak-Black Earth Press, features 144 illustrated nature essays he wrote over the course of 15 years for a daily newspaper.

Jim is an avid outdoorsman and enjoys hiking, biking, camping, nature study and photography. He and his wife Kathy live in rural Plymouth, Wisconsin. They are the proud parents of three grown children and six grandchildren, all of whom are state park and forest campers, hikers, and nature lovers.

Facebook: www.facebook.com/Jim.Buchholz.Ranger
Website: www.ParkRangerWrites.com